Spintronic 2D Materials

Spintronic 2D Materials

Fundamentals and Applications

Edited by

Wenqing Liu

Royal Holloway University of London, Egham, United Kingdom

Yongbing Xu

University of York, Heslington, United Kingdom

Elsevier
Radarweg 29, PO Box 211, 1000 AE Amsterdam, Netherlands
The Boulevard, Langford Lane, Kidlington, Oxford OX5 1GB, United Kingdom
50 Hampshire Street, 5th Floor, Cambridge, MA 02139, United States

British Library Cataloguing-in-Publication Data
A catalogue record for this book is available from the British Library

Library of Congress Cataloging-in-Publication Data
A catalog record for this book is available from the Library of Congress

ISBN: 978-0-08-102154-5

For Information on all Elsevier publications
visit our website at https://www.elsevier.com/books-and-journals

Publisher: Matthew Deans
Acquisition Editor: Kayla Dos Santos
Editorial Project Manager: Joshua Mearns
Production Project Manager: Swapna Srinivasan
Cover Designer: Christian J. Bilbow

Typeset by MPS Limited, Chennai, India

Contents

List of contributors

Matthew T. Bryan Department of Electronic Engineering, Royal Holloway, University of London, Egham, United Kingdom

Jie Chen Physics Department, National Laboratory of Solid State Microstructures, Jiangsu Provincial Key Laboratory for Nanotechnology, Nanjing University, Nanjing, P.R. China

Rebekah Chua Department of Physics, National University of Singapore, 2 Science Drive 3, Singapore, Singapore; NUS Graduate School for Integrative Sciences and Engineering, Centre for Life Sciences, National University of Singapore, 28 Medical Drive, Singapore, Singapore

Youwei Du Physics Department, National Laboratory of Solid State Microstructures, Jiangsu Provincial Key Laboratory for Nanotechnology, Nanjing University, Nanjing, P.R. China

Chun-Gang Duan State Key Laboratory of Precision Spectroscopy and Key Laboratory of Polar Materials and Devices, Ministry of Education, Department of Optoelectronics, East China Normal University, Shanghai, P.R. China; Collaborative Innovation Center of Extreme Optics, Shanxi University, Taiyuan, P.R. China

Liang He York-Nanjing Joint Center (YNJC) for Spintronics and Nanoengineering, School of Electronics Science and Engineering, Nanjing University, Nanjing, P.R. China

He Hu State Key Laboratory of Precision Spectroscopy and Key Laboratory of Polar Materials and Devices, Ministry of Education, Department of Optoelectronics, East China Normal University, Shanghai, P.R. China

Muhammad Zahir Iqbal Nanotechnology Research Laboratory, Faculty of Engineering Sciences, GIK Institute of Engineering Sciences and Technology, Topi, Pakistan

Wenqing Liu York-Nanjing Joint Center (YNJC) for Spintronics and Nanoengineering, School of Electronics Science and Engineering, Nanjing University, Nanjing, P.R. China; Department of Electronic Engineering, Royal Holloway, University of London, Egham, United Kingdom

Aurelien Manchon Material Science and Engineering, Physical Science and Engineering Division, King Abdullah University of Science and Technology, Thuwal, Saudi Arabia

Hongzhe Pan Physics Department, National Laboratory of Solid State Microstructures, Jiangsu Provincial Key Laboratory for Nanotechnology, Nanjing University, Nanjing, P.R. China; School of Physics and Electronic Engineering, Linyi University, Linyi, P.R. China

Minghu Pan School of Physics, Huazhong University of Science and Technology, Wuhan, P.R. China

Yong Pu Nanjing University of Posts and Telecommunications, Nanjing, Jiangsu, China

Xin-Wei Shen State Key Laboratory of Precision Spectroscopy and Key Laboratory of Polar Materials and Devices, Ministry of Education, Department of Optoelectronics, East China Normal University, Shanghai, P.R. China

Yuanyuan Sun Physics Department, National Laboratory of Solid State Microstructures, Jiangsu Provincial Key Laboratory for Nanotechnology, Nanjing University, Nanjing, P.R. China; School of Physics and Electronic Engineering, Linyi University, Linyi, P.R. China

Nujiang Tang Physics Department, National Laboratory of Solid State Microstructures, Jiangsu Provincial Key Laboratory for Nanotechnology, Nanjing University, Nanjing, P.R. China

Tao Tang Physics Department, National Laboratory of Solid State Microstructures, Jiangsu Provincial Key Laboratory for Nanotechnology, Nanjing University, Nanjing, P.R. China; College of Science, Guilin University of Technology, Guilin, P.R. China

Xuefeng Wang School of Electronic Science and Engineering, and Collaborative Innovation Center of Advanced Microstructures, Nanjing University, Nanjing, P.R. China

Yiren Wang School of Materials Science and Engineering, Central South University, Changsha, P.R. China

Andrew Thye Shen Wee Department of Physics, National University of Singapore, 2 Science Drive 3, Singapore, Singapore; Centre for Advanced 2D Materials (CA2DM) and Graphene Research Centre (GRC), National University of Singapore, 6 Science Drive 2, Singapore, Singapore

Ping Kwan Johnny Wong Centre for Advanced 2D Materials (CA2DM) and Graphene Research Centre (GRC), National University of Singapore, 6 Science Drive 2, Singapore, Singapore

Yongbing Xu York-Nanjing Joint Center (YNJC) for Spintronics and Nanoengineering, School of Electronics Science and Engineering, Nanjing University, Nanjing, P.R. China; Department of Electronic Engineering, The University of York, York, United Kingdom; Spintronics and Nanodevice Laboratory, Department of Electronic Engineering, University of York, York, United Kingdom; Nanjing-York Joint Center in Spintronics, Nanjing University, Nanjing, China

Jiabao Yi Global Innovative Centre for Advanced Nanomaterials, School of Engineering, The University of Newcastle, Callaghan, NSW, Australia

Hui Yuan School of Physics, Huazhong University of Science and Technology, Wuhan, P.R. China

Minhao Zhang School of Electronic Science and Engineering, and Collaborative Innovation Center of Advanced Microstructures, Nanjing University, Nanjing, P.R. China

Wen Zhang Department of Physics, National University of Singapore, 2 Science Drive 3, Singapore, Singapore

Yafei Zhao York-Nanjing Joint Center (YNJC) for Spintronics and Nanoengineering, School of Electronics Science and Engineering, Nanjing University, Nanjing, P.R. China

Preface

"Spin" is an intrinsic form of angular momentum universally carried by elementary particles, composite particles, and atomic nuclei. It is a solely quantum phenomenon and has no counterpart in classical mechanics. Many fundamental questions of the electrons' spin remain open issues up to this date: the spin–orbital coupling, spin–photon interaction, and spin-wave transmission to name a few. Significantly, the discovery of giant magnetoresistance effect, celebrated by the 2007 Nobel Prize, has generated a revolutionary impact on the data storage technologies. This triggered the rise of spintronics, an interdisciplinary subject dedicated for the study of spin-based other than or in addition to charge only–based physical phenomena of electronic systems.

Two-dimensional (2D) materials with strong intrinsic spin fluctuations have introduced new physical paradigms and enabled the development of novel spintronic devices. These cleavable materials provide an ideal platform for exploring spin ordering in the 2D limit, where new phenomena are expected, and represent a substantial shift in our ability to control and investigate nanoscale phases. With steady improvement in growth techniques, spintronic 2D materials can be now assembled in a chosen sequence with one-atomic-plane precision. Combining them in vertical stacks further offers enormous amount of possibilities to broaden the versatility of 2D materials, allowing for achieving unusual material properties that cannot be obtained otherwise.

With contributions from the world-leading scientists, this book aims to provide an overview of the research in spintronic 2D materials. This starts with 2D magnetism fundamental and covers the development of the most representative spintronic 2D materials up to date. It provides background, introduction, the latest research results, and an extensive list of references in each chapter. It is hoped that the collection of these materials in one book will enable the challenges and research progress in 2D spintronics to be seen in context, while the individual chapters are designed to be self-contained. The editors wish to thank all the authors for writing their chapters and the dedicated Elsevier publication team for publishing the book.

Wenqing Liu and Yongbing Xu

CHAPTER 1

Introduction to spintronics and 2D materials

Wenqing Liu[1], Matthew T. Bryan[1] and Yongbing Xu[2,3]

[1]Department of Electronic Engineering, Royal Holloway, University of London, Egham, United Kingdom, [2]Spintronics and Nanodevice Laboratory, Department of Electronic Engineering, University of York, York, United Kingdom, [3]Nanjing-York Joint Center in Spintronics, Nanjing University, Nanjing, China

Chapter Outline

Spin-based phenomena are rich in fundamental physics and important to information technology. Many aspects of electron spin dynamics remain open questions, including spin–orbital coupling, spin–photon interaction, spin ordering at low dimensions, and spin-wave transmission to name a few. Since the discovery of the giant magnetoresistance effect (GMR), celebrated by the 2007 Nobel Prize in Physics, spin-based electronics—or "spintronics" as it quickly became known—has developed into an interdisciplinary field dedicated to the study of spin-based effects rather than purely charge-based physical phenomena previously associated with electronic systems (Fig. 1.1). This eventually led to a revolutionary impact on device concepts, particularly in data storage technologies where the introduction of spintronic technologies enabled a rapid increase in the capacity of hard drives.

In parallel with the development of spintronics research, new two-dimensional (2D) materials have become available. Beginning with the discovery of graphene, acknowledged by the 2010 Nobel Prize in Physics, a number of 2D materials have been produced with unique properties not seen elsewhere, even in bulk (three-dimensional) analogs of the same materials, due to confinement of their electronic band structure. Combining these properties with spintronics could produce the next revolution in electronics technology, with potential

Spintronic 2D Materials.
DOI: https://doi.org/10.1016/B978-0-08-102154-5.00001-1

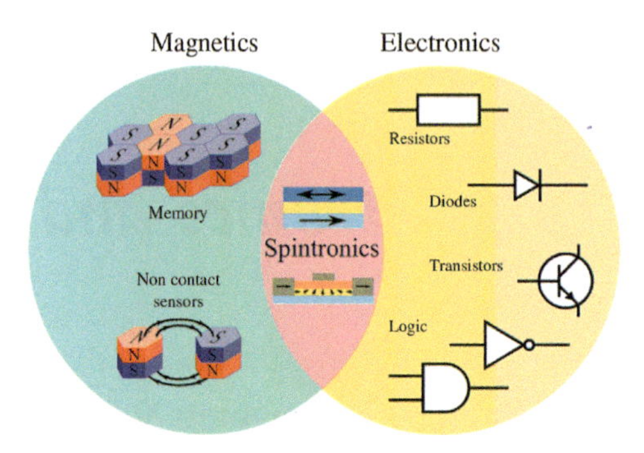

Figure 1.1
Spintronics combines the nonvolatile and remote sensing properties of magnetic materials with the processing functionality of electronics.

for minimal power dissipation or extreme sensitivity to magnetic fields. This book covers recent advances in spintronic 2D materials, showing how magnetism and low-dimensional electronics can be coupled to produce new device concepts and novel applications.

1.1 Spin and spin ordering

Spin is an intrinsic form of angular momentum universally carried by elementary particles, composite particles and atomic nuclei. It is a solely quantum phenomenon and has no counterpart in classical mechanics. The earliest sign of "spin" can be traced back to the 1880s, when Albert A. Michelson observed closely-spaced, but discrete, lines in the emission spectra of sodium gas. When atomic spectra were first discovered, the sodium spectrum was thought to be dominated by a bright line known as the sodium D-line at wavelength $\lambda_D = 589.3$ nm. However Michelson was able to resolve the spectrum in finer detail and found that the D-line was in fact split into two lines, namely $\lambda_{D1} = 589.6$ nm and $\lambda_{D2} = 589.0$ nm, called the fine structure. Today, we know that the sodium D-line arises from the transition from the 3p to the 3s levels, with the fine structure caused by slight differences in energy levels of opposite spins due to the spin–orbit interaction.

While Michelson did not recognize spin as the origin of the fine structure, his observations mark the beginning of the study of spin-based phenomena. Following Joseph J. Thomson's discovery of the electron as a particle in 1897, next experimental hint of spin came in 1912, when Friedrich Paschen and Ernst E. A. Back observed that in the presence of a strong magnetic field the sodium D1- and D2-lines further split into four and six lines, respectively. This field-induced splitting of fine structure spectra due to spin–orbit effects is sometimes referred to as the anomalous Zeeman effect (or the Paschen—Back effect at high fields). In the following year, Niels Bohr published his atom theory, which included quantized energy shells and electron orbital momentum. It provided a framework to understand many of the new quantum phenomena being discovered at the time, but still lacked a spin angular momentum term. Quantization of angular momentum was demonstrated by Otto Stern and Walther Gerlach in 1922, by measuring the deflection of a collimated beam of gaseous, electrically neutral silver atoms passing through a nonhomogeneous magnetic field into two distinct bands (rather than the single, broad band expected from a classical distribution of angular momentum). However it was not until 1925, when Samuel Goudsmit and George Uhlenbeck suggested that the electron had an intrinsic quantized angular momentum, that the concept of spin was grasped and used to explain the fine structure, and anomalous Zeeman effect. By 1929, Paul A.M. Dirac had developed his theory of relativistic quantum mechanics, demonstrating that unlike orbital angular momentum, electronic spin was restricted to just two quantized values: $S = \pm 1/2$.

In explaining the anomalous Zeeman effect, electronic spin is directly linked to the magnetic moment of each atom. Indeed, spin is responsible for magnetic ordering within a crystal via a quantum mechanical interaction, called exchange. Fundamentally, the exchange interaction manifests as an electrostatic interaction between neighboring spins: by the Pauli exclusion principle, particles in identical quantum states cannot occupy the same position, so electrons in the same band are repelled if they have aligned spins. Therefore the spatial distribution of charge within the crystal is dependent on the alignment of neighboring spins. Naturally, the interaction is reciprocal; the charge distribution determined by the crystal lattice influences spin direction giving rise to magneto-crystalline anisotropy (having a preferred magnetization "easy" axis or axes) and mechanical distortions in the lattice may cause, or be caused by, changes in the spin direction (an effect called magnetostriction). Strong interaction between an electron's spin and the magnetic field it experiences due to its orbit around a charged nucleus (the "spin–orbit interaction") can create an additional exchange term, called the Dzyaloshinskii–Moriya interaction (DMI), which acts to align spins perpendicular to each other. Since it competes with normal exchange, DMI tends to introduce a topological chirality to a magnet, favoring a particular sense of spin rotation whenever magnetization becomes nonuniform.

Each interaction contributes to the overall energy of the system, with the lowest energy state determining the magnetization profile. At a fundamental level, the exchange Hamiltonian, $\mathscr{H}$, may be described by the Heisenberg model:

$$\mathscr{H} = -J\sum_{i,j}\vec{s}_i \cdot \vec{s}_j \tag{1.1}$$

where J is the exchange constant and $s_{i,j}$ are nearest-neighbor spins, and the summation is over all nearest-neighbor pairs. In a system exhibiting a DMI, this exchange Hamiltonian is augmented with a DMI term:

$$\mathscr{H} = -J\sum_{i,j}\vec{s}_i \cdot \vec{s}_j + D_{ij} \cdot \left(\vec{s}_i \times \vec{s}_j\right) \tag{1.2}$$

where D_{ij} is a constant describing the strength of the DMI between the neighboring spins. As an alternative to the Hamiltonian form, the exchange energy density (neglecting DMI), E_{ex}, may be expressed in terms of the material magnetization, M:

$$E_{ex} = A\left(\nabla\vec{M}\right)^2 \tag{1.3}$$

where A is the exchange stiffness (proportional to J). The exchange stiffness is a measurable property of a material and fundamentally affects the magnetic ordering. When $A > 0$, magnetization divergence increases the energy, so exchange energy is minimized when neighboring spins align parallel with each other to give ferromagnetic ordering. Therefore ferromagnetic materials characteristically have a spontaneous net magnetic moment.

While exchange aligns spins, anisotropy determines the spin directionality. There are a number of different (and sometimes competing) mechanisms that induce anisotropy, and therefore determine spin or magnetization direction. Magneto-crystalline anisotropy dominates in many materials, such that the alignment direction depends on the crystal structure. For uniaxial materials, magneto-crystalline energy, E_K, is given by:

$$E_K(uniaxial) = K_1 sin^2\theta \tag{1.4}$$

whereas cubic materials have

$$E_K(cubic) = K_1\left(sin^2\theta cos^2\theta + sin^4\theta sin^2\phi cos^2\phi\right) + K_2 sin^4\theta cos^2\theta sin^2\phi cos^2\phi \tag{1.5}$$

where $K_{1,2}$ are anisotropy constants that define the strength of the alignment, and θ and ϕ are the spherical polar coordinate directions (with polar direction aligned with an easy axis). Internal demagnetizing fields caused by free surface poles encourage divergence in magnetization, resulting in the formation of domain structure (in which magnetization within a domain is aligned, but neighboring domains are unaligned and separated by boundaries called domain walls) and also an effect called "shape anisotropy" that energetically favors magnetization aligned parallel to structural surfaces and edges. Anisotropy can also arise from external stimuli, such as stress (magnetostriction), providing a mechanism to tune magnetic behavior in some materials.

In contrast to ferromagnetic order, materials with $A < 0$ have antiferromagnetic ordering in which neighboring spins are aligned antiparallel and there is no net magnetization. Note that not all neighbors need to be antiparallel; an antiferromagnet may enter a number of different antiferromagnetic phases depending on stoichiometry and crystal structure (Fig. 1.2A). In some cases, the antiferromagnetic state becomes "canted" and the spins misalign from antiparallel, developing a slight tilt in one direction to give a small net magnetization (Fig. 1.2B). From a theoretical viewpoint, antiferromagnets may be considered as two sublattices that have opposite magnetizations (Fig. 1.2A). This

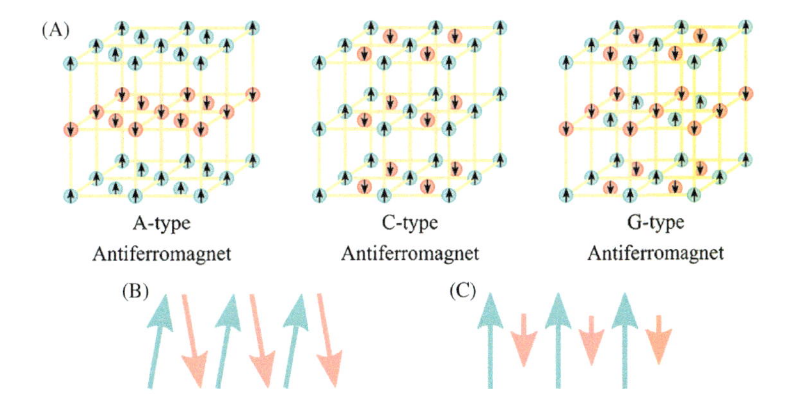

Figure 1.2
Examples of (A) common antiferromagnetic phases, (B) canted antiferromagnet alignment, and (C) ferrimagnetic configuration. Coloring indicates the different sublattices.

description is also useful to describe a third magnetic order, ferrimagnetism, where the sublattices are aligned antiparallel, but have unequal magnetizations (Fig. 1.2C).

Magnetic ordering can also be caused by interactions that occur indirectly, via intermediaries. Superexchange may occur between two nonneighboring magnetic atoms when they have a common nonmagnetic neighbor (Fig. 1.3A). In essence, the response of the nonmagnetic atom to the one magnetic atom affects its interaction with the other magnetic atom, leading to an indirect exchange coupling between the magnetic atoms. Double exchange occurs in compounds with chemical bonding structure that shares electrons between atoms in their ground state (Fig. 1.3B). Since these "itinerant" electrons are not localized to a particular atom and since spin is preserved as the electron hops from one atom to the next, the electrons mediate exchange between magnetic ions. Ruderman–Kittel–Kasuya–Yosida (RKKY) interactions involve exchange between core electrons localized to an atom and conduction electrons [1], which are not localized to a particular atom (Fig. 1.3C). By interacting with several atoms, the electrons in the conduction band may couple atoms that are several unit cells apart. Importantly, progressively increasing the separation between atoms not only diminishes the strength of the interaction, but also causes oscillations in the favored alignment of the atoms. That is depending on the separation, the RKKY interaction can promote either ferromagnetic or antiferromagnetic spin alignment (Fig. 1.3D). Since RKKY interactions are mediated by conduction electrons, they only occur in conductive materials. Insulators may experience long-range indirect magnetic ordering via van Vleck paramagnetism. Whereas the previous indirect exchange mechanisms discussed involved materials in their electronic ground states, the van Vleck mechanism occurs due to excited states of valence electrons (Fig. 1.3E). In the absence of other interactions, no correlation between spins of neighboring ions in the ground state would be expected. However perturbing the ground

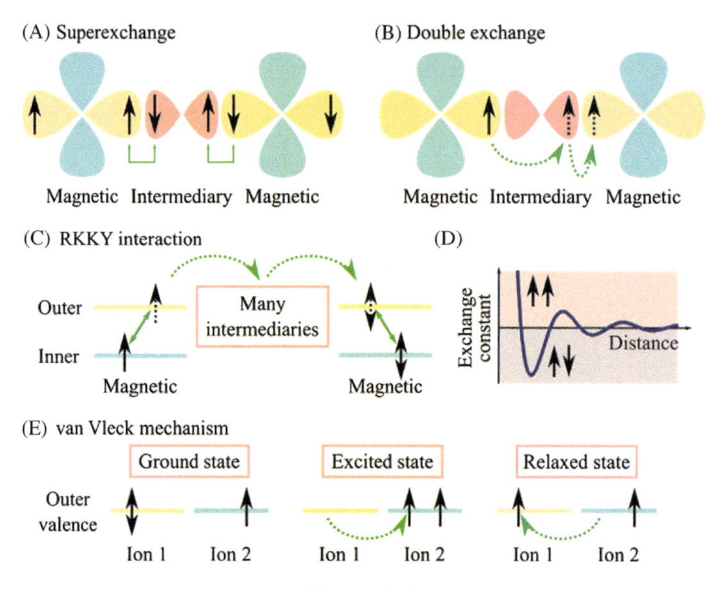

Figure 1.3

Schematic diagrams of (A) superexchange between electrons in outer shells of magnetic ions via a nonmagnetic intermediary, (B) double exchange through hopping of itinerant electrons in outer shells between magnetic ions via a nonmagnetic intermediary, (C) RKKY interactions between inner core electrons of magnetic ions, via exchange with delocalized outer shell electrons, (D) the variation of the RKKY exchange constant with distance between interacting ions (arrows show the favored spin alignment between ions), and (E) the van Vleck mechanism of exchange between two ions in a relaxed state following excitation from a ground state with randomly aligned spins. In all figures, solid arrows represent localized spins or direct exchange interactions, dotted arrows represent spins of itinerant electrons or the hopping that mediates the interaction. *RKKY*, Ruderman–Kittel–Kasuya–Yosida.

state (e.g., by thermal excitation) may result in the transfer of an electron between neighboring atoms, forming either a triplet ($S = 1$) or singlet ($S = 0$) state within the unfilled shall of the recipient ion. Crucially, in the van Vleck mechanism triplet states are energetically favored, promoting spin alignment [2]. Upon relaxation to the ground state, the aligned spin is retained as the electron moves back to its original host ion, so that there is an effective ferromagnetic coupling between the ions. Although weak in many materials, the van Vleck mechanism is very important for understanding ferromagnetism in insulators exhibiting inversion of the conduction and valence bands.

Another factor that plays a critical role in determining magnetic order is temperature. Many materials may adopt different magnetic ordering at different temperatures. Crucially, magnetic ordering is lost in all materials at temperatures exceeding a characteristic ordering temperature, called the Curie temperature in ferromagnetic materials and the Néel temperature in antiferromagnets. Above the ordering temperature, materials become paramagnetic; individual atoms retain magnetic moments, but there is no systematic alignment between neighbors and therefore no spontaneous magnetization. While individual moments will align with an applied magnetic field, inducing a magnetization, removal of the field destroys the alignment and the material will return to zero magnetization at remanence.

Materials may also display paramagnetic-like characteristics below the ordering temperature if their dimensions are small (well below the size needed to ensure single-domain magnetization), an effect called superparamagnetism. Superparamagnetism is the mechanism of spontaneous magnetic reversal caused by thermal perturbations overcoming the energy barrier between magnetic states and is important when the thermal energy (k_BT, where k_B is the Boltzmann constant, and T is the temperature) is greater than around 2.5%–4% of the anisotropy energy (KV, where K is the anisotropy constant, and V is the activation volume, which if reversed will lead to the whole sample magnetization switching). Since thermal fluctuations are random, superparamagnetic magnetization reversal may be expected to occur within a time-period, τ, described by the Arrhenius–Néel equation:

$$\tau = \frac{1}{f_0} exp\left(\frac{KV}{k_BT}\right) \tag{1.6}$$

where f_0 is a characteristic time-scale called the attempt frequency (on the order of 1–100 GHz). Whether or not superparamagnetism is important depends on the time-scale over which stability is required. In the recording industry, data must be preserved for several years, so the anisotropy energy must be at least 40–45 k_BT [3]. Practically, stability times required for research is much shorter so samples with anisotropy energy of 30 k_BT may be expected to retain their magnetic state during measurements (lasting minutes to hours).

1.2 The discovery of giant magnetoresistance and tunnelling magnetoresistance

For over half a century after its discovery, spin was largely neglected from electronics applications. That changed in 1988, when groups led by Albert Fert [4] and Peter Grünberg [5] independently discovered GMR and triggered the birth of the field of spintronics, for which they were later awarded the 2007 Nobel Prize in Physics. GMR is the dependence of resistance on the relative orientation of two magnetic layers connected by a conducting spacer. The term "giant" refers to the fact that the first GMR measurements demonstrated relative changes in resistance that was 10–100 times larger than the anisotropic magnetoresistance (variation of resistance with alignment between current and magnetization direction) found in the constituent materials. Although the earliest GMR devices were demonstrated using the current in plane geometry, GMR may also be seen in the current perpendicular to plane (CPP) geometry. Indeed, CPP devices display longer spin diffusion length and an even larger GMR effect [6].

To provide a simple theoretical understanding of GMR, let us neglect resistance contributions of the thin spacer layer and consider identical magnetic layers such that the only difference in resistance is due to the alignment of the layer magnetizations. Furthermore, let us restrict the magnetization of each magnetic layer to point either "up" or "down" and split the electrical conductivity into two independent channels corresponding to spins pointing up and down, with negligible mixing between the conduction channels. According to Mott's s-d scattering theory [7], resistivity is low for spins aligned with the magnetization of the host layer (ρ_p) and high for spin opposing the host magnetization (ρ_a). Fig. 1.4 shows the circuit diagrams of this "two-current" model for the parallel and antiparallel alignment of the magnetic layers. The equivalent resistance for the parallel ($R_{\uparrow\uparrow}$) and antiparallel ($R_{\uparrow\downarrow}$) arrangements are given by:

$$R_{\uparrow\uparrow} \propto \frac{2\rho_p\rho_a}{\rho_p + \rho_a}, R_{\uparrow\downarrow} \propto \frac{\rho_p + \rho_a}{2} \tag{1.7}$$

Since $\rho_p < \rho_a$, Eq. 1.7 shows that $R_{\uparrow\uparrow} < R_{\uparrow\downarrow}$: greater scattering occurs when the magnetic layers are antiparallel. To compare different devices, GMR is often expressed as a ratio:

$$GMR = \frac{R_{\uparrow\downarrow} - R_{\uparrow\uparrow}}{R_{\uparrow\uparrow}} \tag{1.8}$$

In the GMR stacks created by Fert and Grünberg, the coupling between the magnetic layers was antiferromagnetic in the absence of a magnetic field. High fields were needed to overcome the antiparallel coupling and align the layer magnetizations to

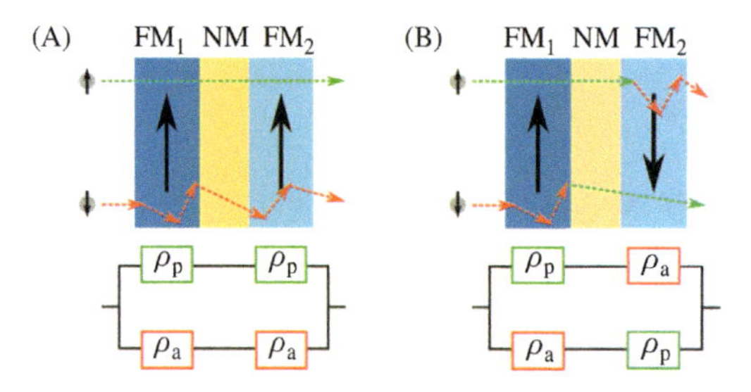

Figure 1.4
Schematic illustrations and corresponding circuit diagrams of two-current model in ferromagnet (FM_1)/(NM)/ ferromagnet (FM_2) trilayers when the ferromagnetic layers are (A) parallel, and (B) antiparallel. Low scattering paths are marked in green; high scattering paths are marked in red. ρ_p is the resistivity of spins aligned parallel with host magnetization, and ρ_a is the resistivity of spins aligned antiparallel with host magnetization. *NM*, nonmagnetic spacer.

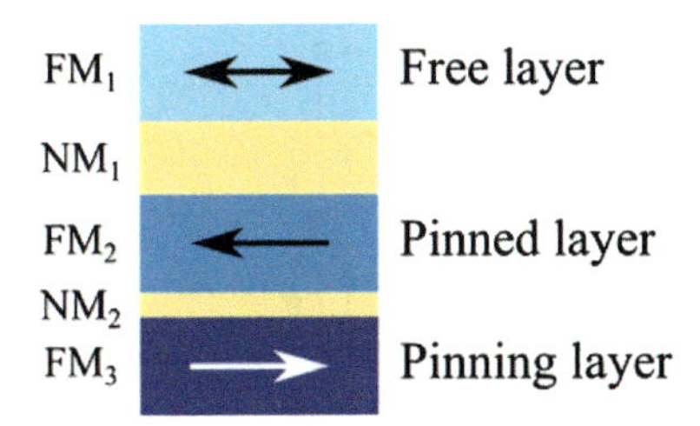

Figure 1.5
Schematic diagram of a spin-valve structure. Alternative configurations feature an antiferromagnetic pinning layer. Modern spin-valves also utilize additional sublayers to the free and pinned layers to enhance performance [11].

achieve the low resistance state. Shortly after the discovery of GMR, Stuart S. P. Parkin developed an architecture, called the spin-valve [8–10], which enabled arbitrary alignment between the layers at zero-field and low-field switching between resistance states, paving the way for GMR to be used in practical applications, notably in hard disk drive read-heads. In spin-valves, a third magnetic "pinning" layer is coupled to the basic GMR structure via a thin conducting nonmagnetic spacer (Fig. 1.5). Due to the thickness of the nonmagnetic spacer (labeled NM_2 in Fig. 1.5), the pinning layer and the bottom ("fixed" or "pinned") layer of the basic GMR structure are strongly antiferromagnetically coupled. This coupling causes the pinned layer to be magnetically hard, requiring a large field to induce magnetization reversal. Furthermore, the thickness of the pinning layer may be tailored to compensate for the magnetization in the other layers to give a negligible magnetic moment for the structure as a whole when in the low resistance state. Zero-field alignment between the pinned and the top ("free") layers is determined by RKKY interactions, enabling either parallel, antiparallel or decoupled remanent states to be chosen during fabrication, via control of the spacer layer thickness (NM_1 in Fig. 1.5). Under low fields, the magnetization of the magnetically soft free layer can be reversed independently of the pinned layer to switch between the high and low resistance states. Figuratively, this means the spin of the free layer is used like a valve to turn the current through the device "on" or "off".

Tlthough GMR is credited with beginning the spintronics revolution, it was not the first spin-engineered phenomena. In 1975 Michel Julliere released a brief paper describing a 14% difference in the resistance of a trilayer consisting of two ferromagnets connected by an *insulator* (Fe/Ge/Co at 4.2K) [12], which was almost completely overlooked prior to the discovery of GMR (receiving just three citations). Now known as tunneling magnetoresistance (TMR), the magnetoresistance discovered by Julliere requires quantum tunneling in order for electrons to pass through the insulating layer. Therefore TMR operates through a completely different mechanism to GMR, which works via classical diffusive scattering.

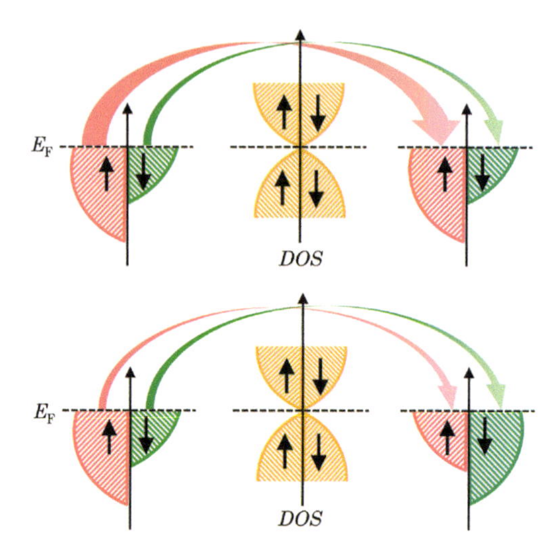

Figure 1.6
Conceptual illustration of the origin of differential spin scattering in a magnetic tunnel junction, associated with the relative asymmetry in the DOS near the E_F of the two FM layers. *FM*, ferromagnet; *DOS*, density of states.

Tunneling can only occur when there are available states to tunnel into, so the tunneling probability is dependent on the density of states (DOS) near the Fermi energy, E_F, in each of the ferromagnetic layers. The DOS is divided between the two spin-states, leading to an asymmetry in the DOS of magnetic materials. Fig. 1.6 presents a conceptual illustration of spin tunneling through the insulating layer of a TMR trilayer (the structure, analogous to the architecture of the GMR spin-valve, is known as a magnetic tunnel junction, MTJ). Due to the asymmetry in the DOS, each ferromagnetic layer has more states available in one spin state than the other, corresponding to its magnetization direction. When a bias voltage is placed across the MTJ, the probability of tunneling through the insulator depends on the availability of free states for each spin direction. If the two magnetic layers are parallel (Fig. 1.6, upper row), the DOS for each spin-state is similar in both layers, so there will be many available states (for both spin-channels) to tunnel into, resulting in a large tunneling current and low overall resistance. On the other hand, if the layers are antiparallel (Fig. 1.6, lower row), there is a mismatch between the source and sink DOS of each spin channel: the majority spin-channel in the source layer has fewer states available in the sink layer, whereas the minority spin-channel has excess of states available in the sink layer, but too few source electrons to fill them. This bottleneck causes the antiparallel magnetization configuration to have a high resistance.

Despite the differences in physical mechanism, the architecture underlying GMR and TMR is conceptually similar, which may be why interest in TMR grew after the discovery of GMR. Indeed, the two-current model described in Eqs. (1.7) and (1.8) may readily be applied to TMR. Comparing the magnetoresistance ratios, TMR is much stronger (with room-temperature magnetoresistance ratios of up to 600% [13]) than GMR (up to 65% at room temperature [14]). This is a result of the extreme sensitivity of the quantum mechanical mechanism. On the other hand, the same sensitivity presents a challenge for manufacturing MTJs at industrial scale [11], since minor variation in the thickness of the insulating layer can dramatically alter device resistance.

Both GMR and TMR technology have been commercially exploited into hard disk drive read-heads, and this remains their predominant realization [15]. When used in field sensing, the free layer is made to lie orthogonal to the pinned layer at remanence, using either shape anisotropy, a localized biasing field or weak pinning [11]. The orthogonal configuration enables the sensors to operate within a linear regime, since the magnetoresistance is proportional to the projection of the free layer magnetization along the direction of the pinned layer magnetization.

Beyond field sensing and data read-out applications, the bi-stable nature of spin valves and MTJs has led to GMR and TMR being implemented for memory technologies. Since the magnetic state of GMR and TMR cells is nonvolatile (the memory state is retained even when the power is turned off), is electrically addressable and can be accessed on nanosecond timescales, there has been great research effort in the development of devices capable of performing as magnetic random access memory (MRAM). If the potential of MRAM could be realized, it would produce a new type of memory that combined features of magnetic hard disk drives (nonvolatile, low power) with those of semiconducting random access memory (fast read/write times, no mechanical failure). However commercial realizations of MRAM have lagged behind semiconductor memory, with the leading producer, Everspin Technologies, only announcing pilot production of 1 Gb MRAM in June 2019. The main challenges to increasing MRAM data density are switching individual cells as their footprint decreases and the space taken up by architecture supporting the MRAM cells, since transistors are needed to control current paths through the cells. Currently, MRAM designs make use of spin-transfer torque (STT), a current-based mechanism that makes use of the transfer of angular momentum when spin-polarized electrons moving from the pinned layer. Further progress in MRAM technology will require improvements in the switching current or efficiency of spin transfer or discovery of alternative mechanisms that can retain the ability to switch smaller cells.

1.3 Semiconductor spintronics: dilute magnetic semiconductor and spin field-effect transistor

The success of GMR and TMR devices (often classified as the first generation of spintronic devices) has inspired a huge research effort into spin-based phenomena, reaching into fields previously considered unrelated to magnetism. One of the most fascinating areas that have developed has focussed on inducing and controlling spin polarization in nonmagnetic or paramagnetic semiconductors. Semiconductors display enhanced spin-properties compared to metallic spintronic materials. For example, semiconductor electron spin relaxation times are several orders of magnitude longer than the electron momentum and energy relaxation times [16], enabling electrons to propagate up to 100 μm without losing their spin coherence [17]. This demonstrates that semiconductors have the potential to efficiently transport spin information over long channel lengths. Furthermore, carrier profiles in semiconductors can be modified with dopants, indicating that device properties could be tailored toward specific spintronic applications. Doping also opens up opportunities for realizing novel physical phenomena, such as the generation of a dissipationless spin current in the absence of a net charge current (the spin Hall effect) [18–20]. These promising properties, combined with the already established dominance of nonspintronic semiconductors in signal processing and the computing hierarchy, indicate that development of spintronic semiconducting technology could well lead the next generation of spintronic devices.

Broadly, there are three strategies employed to induce spin polarization in semiconductors. Firstly, hybrid structures, consisting of a ferromagnet in contact with the semiconductor, provide a mechanism of direct spin injection since the spin polarization from the ferromagnet is retained when the electrons enter into the semiconductor [21–23]. Secondly, optical pumping by irradiation with circularly polarized light takes advantage of magneto-optical effects to preferentially excite spin polarization within the semiconductor [16,24]. Whereas both of these strategies manipulate nonequilibrium electrons, in the final strategy generates spin polarization by doping with magnetic ions to create a new class of semiconductor that has a magnetic moment, the DMS [25].

Much work on hybrid spintronics has been stimulated by the idea of the spin field-effect transistor (FET), first proposed by Supriyo Datta and Biswajit Das in 1990 [26]. Similarly to a conventional FET (such as the metal-oxide-semiconductor FET, MOSFET), the spin FET injects electrons from a source electrode through a two-dimensional electron gas (2DEG) channel that can be electrically gated and into a sink electrode (Fig. 1.7A). However spin FETs function though a remarkably different mechanism because both the source and sink electrodes are ferromagnetic, providing additional control based on the electronic spin. Transport of the electron spins is confined in the high mobility 2DEG channel, so is dependent on the applied gate voltage [27]. Without bias from the gate, the relative magnetization directions in the source and drain dominate the conductivity in the device (Fig. 1.7B). When a gate voltage is applied across the channel, the spin-polarized electrons experience an effective magnetic field due to the Rashba spin–orbit interaction, leading to a precession of spin and consequently a change in the spin polarization of the current (Fig. 1.7C). Therefore the gate voltage modulates the amount of spin-scattering at

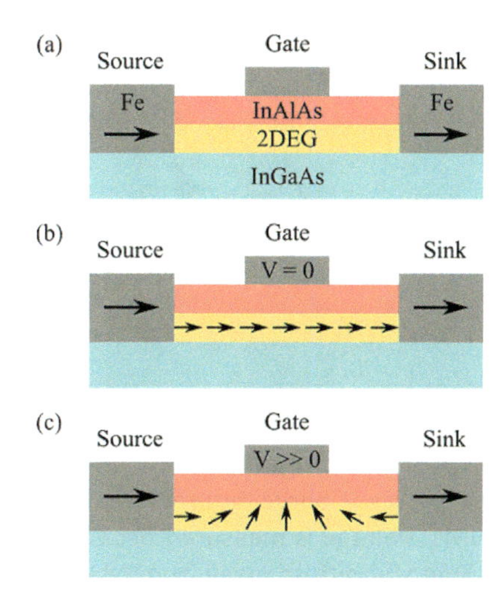

Figure 1.7
Not-to-scale schematic diagrams of (A) the spin FET proposed by Datta and Das [26], together with the spin polarization in the 2DEG channel with (B) no gate voltage applied, and (C) when a gate voltage is used to induce precession of the spin polarization in the 2DEG. *FET*, field-effect transistor; *2DEG*, two-dimensional electron gas.

the sink electrode and therefore the current passing though the transistor. Since only a small amount of energy and a short time periods are needed to switch the current through the device (compared to that required in a MOSFET where the channel needs to be under inversion), spin FETs are expected to combine high computing speed with low power consumption [27–30].

One of the major advantages of the spin FET design is that it has an all-electrical device design (all control of spin exists within electronic circuits), so is intrinsically compatible with existing circuit architecture. An alternative mechanism of incorporating all-electrical control into a spintronic device is to utilize "nonlocal" device geometries. Fig. 1.8 shows an example of a nonlocal device with down-spin-polarized current injected at magnetic electrode M1 and flowing through a semiconductor between M1, and nonmagnetic electrode NM1 [31,32]. Since the semiconductor is nonmagnetic, the accumulation of the down-spin polarization around the magnetic injection contact causes the up-spin polarization to diffuse uniformly away from the contact (without charge flow), both towards NM1 and towards the second nonmagnetic electrode, NM2. Crucially, the flow of up-spin polarization (due to the diffusion) towards NM2 creates a potential difference, even though no charge current is flowing. Measurement of the voltage across the second magnetic electrode, M2, and NM2 is "nonlocal" because it is outside of the charge current circuit (between M1 and NM1). Since spin polarization diminishes with distance from the injection point, and is negligible around NM2, the lack of magnetization in NM2 does not

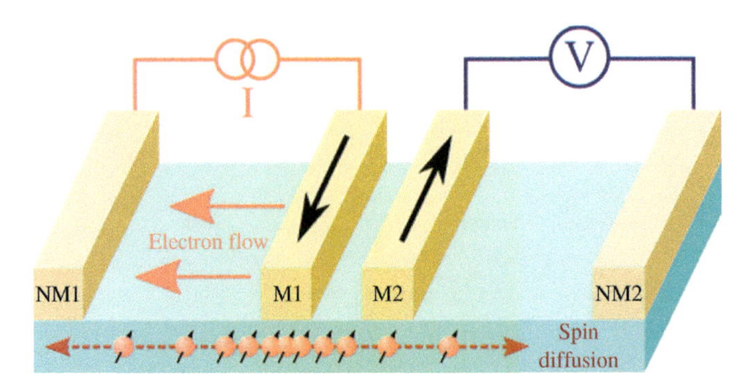

Figure 1.8

Schematic diagram of a nonlocal device. Charge current is passed from magnetic electrode M1 to nonmagnetic electrode NM1, voltage is measured between magnetic electrode M2 to nonmagnetic electrode NM2. Spin will diffuse away from M1 uniformly, such that spin currents flow both with and in the opposite direction to the charge current. The voltage measured is dependent on the relative magnetization orientation of M1 and M2, leading to the electrical spin detection. Source: *Image adapted from the literature X.H. Lou, et al., Electrical detection of spin transport in lateral ferromagnet-semiconductor devices, Nat. Phys. 3 (2007) 197–202.*

affect the nonlocal voltage. However M1 and M2 form a lateral spin-valve structure, in that their alignment determines how M2 filters the up-spin polarization diffusion and therefore the sign of the nonlocal voltage. Therefore nonlocal voltage measurements provide a mechanism for sensing spin polarization without generating additional charge currents. In addition to reducing power consumption, this also minimizes effects that can interfere with detection of spin polarization, such as electrode magnetoresistance and Hall effects.

While great progress has been made with hybrid semiconductor spintronics, there remain major obstacles to spin injection that currently limit room temperature efficiency to below 10%. Intermixing between the ferromagnetic and semiconductor layers can interfere with the electronic properties of the semiconductor. Formation of "magnetic dead layers", in which magnetization is lost close to a nonmagnetic interface, can disrupt the injected spin polarization. Finally, the mismatch in conductivity between magnetic and semiconductor layers causes a large amount of scattering and therefore spin depolarization [33]. While spin depolarization due to conductivity mismatch may be negated through use of tunnel barriers [34], this solution comes at the price of dramatically increasing the resistance of the device, which may itself limit operation. Faced with these challenges, hybrid spintronics research has changed focus from interfacing magnets with classical semiconductors, such as GaAs [22,35,36], InAs [30,37], and GaN [38,39], towards using novel materials that have only recently been discovered, such as topological insulators, graphene and other 2D materials [40,41].

An alternative to directly injecting spin from magnetic electrodes is to first generate highly polarized carriers in one semiconductor through optically pumping with circularly polarized light, before injecting the spin to another semiconductor [16,24]. At a fundamental level, generation of a spin polarization through optical pumping is a manifestation of magnetic circular dichroism, where an electron preferentially absorbs light with angular momentum matched to its spin direction, but the polarization achieved is determined by a competition between several factors, including the rate of nonequilibrium spin creation, rates of carrier recombination and spin relaxation time. Recombination in this case occurs between the photo-excited spin-polarized electrons and the unpolarized holes and is accompanied by a partially polarized luminescence (Fig. 1.9) [43,44]. Since the inter-band transition probabilities for polarized electrons follow the optical selection rule, quantities such as spin relaxation time, recombination time and spin orientation can be obtained by analyzing such electroluminescence. Over the years, this scheme has been routinely used as a detection methodology for measuring spin injection efficiency, as it is less ambiguous than those based on resistance measurements [45,46].

Many demonstrations of optically pumped spin polarization were made possible by doping a semiconductor layer with a magnetic ion to form a DMS, since the spin-imbalance in the ground-state enhances differences in spin-dependant photo-absorption via the Zeeman effect. Indeed, compared to spin injection efficiencies of around 2% when injecting electrons from a ferromagnetic electrode [22], injection from a DMS layer is striking, achieving spin injection efficiencies of up to 70% using a ZnMnSe/AlGaAs structure [42]

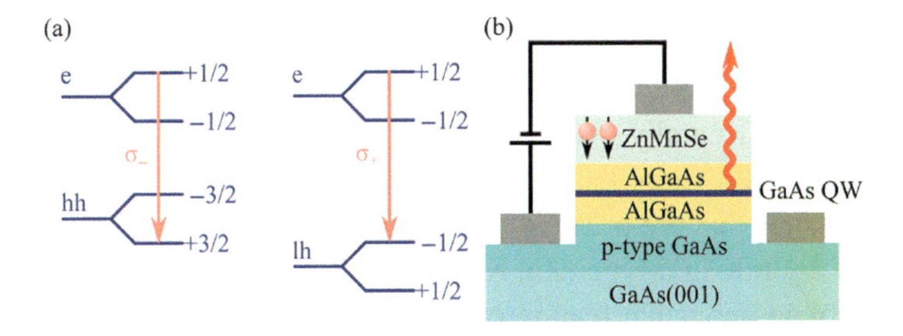

Figure 1.9

(A) Schematic diagram of electron (e), heavy hole (hh), and light hole (lh) spin-dependent energy levels, together with the radiative transitions for $m_j = +1/2$ electrons allowed by the sum rule $\Delta m_j = \pm 1$, where m_j is the total angular momentum. By conservation of angular momentum, the emitted light has polarization $\sigma = \pm 1$ opposite to Δm_j. Transitions to heavy hole states are more probable, leading to circular polarization of the emitted light. (B) Schematic diagram of a reverse-biased spin LED using the DMS ZnMnSe as the spin injecting electrode and AlGaAs/GaAs as the quantum well. *DMS*, dilute magnetic semiconductor. Source: *Images adapted from the literature H.J. Zhu, et al., Room-temperature spin injection from Fe into GaAs., Phys. Rev. Lett. 87 (2001) 016601 [22], and B.T. Jonker, et al., Robust electrical spin injection into a semiconductor heterostructure, Phys. Rev. B62 (2000) 8180–8183 [42].*

(see Fig. 1.9B). Such high efficiencies are possible because unlike hybrid ferromagnetic metal/semiconductor structures, there is little conductivity mismatch between DMS and normal semiconductors.

Dilute magnetic semiconductors (DMSs) have the potential to enable spintronic technology to link seamlessly with current electronic devices, which are based on nonmagnetically doped Si and GaAs. They also display novel effects, such as the ability to manipulate magnetic reversal by using an electric field to modulate magneto-crystalline anisotropy [47], which may enhance functionality. However there are several challenges that must be met before this potential can be realised. For many years, technologically useful semiconductor materials were thought incompatible with magnetic materials, since there are large differences in the crystal structure of traditional magnetic and semiconductor materials. Even though DMS materials have now been demonstrated, this fundamental mismatch still presents difficulties for ongoing development and the introduction of magnetic dopants can easily disrupt the optoelectronic properties that make semiconductors useful. Similarly, while magnetic properties are generally enhanced at higher doping concentrations, the limited solubility of magnetic impurities in semiconductors restricts the amount of doping that can occur before inducing a phase change. To an extent, some of these problems can be overcome through improvements in growth techniques, such as the use of low temperature molecular beam epitaxy (MBE), which was employed in 1996 to produce the first ferromagnetic DMS, (Ga, Mn)As [25].

Possibly the single biggest challenge facing the DMS community is that room-temperature ferromagnetism has remained elusive. Indeed, the highest Curie temperature achieved in a technologically compatible DMS, 178K in (Ga, Mn)As [48], is well below room temperature. Furthermore, the Curie temperature varies hugely depending on the electronic properties of the DMS. Generally, magnetically doped III-V semiconductors, such as (Ga, Mn)As, produce the largest Curie temperatures. By comparison, II-VI-based compounds have Curie temperatures that are orders of magnitude smaller, just 1.45K in p-type (Zn, Mn)Te:N and only 160mK in n-type (Zn, Mn)O:Al [49]. Nevertheless, low Curie temperatures are not an inevitable feature of magnetic doping, as shown by a number of magnetically doped oxide materials that have been reported displaying ferromagnetism at room temperature [50,51]. Therefore applying the principles of magnetic doping to novel 2D materials may yield breakthroughs that have been lacking in bulk semiconductors.

1.4 2D materials and magnetism in 2D materials

For many years, experimental research was restricted to large and bulk samples. Bulk crystals have a repeating symmetry that defines their properties. However the symmetry is broken at the crystal surface, which can lead to the formation of edge states displaying properties that are different from the bulk. As microfabrication techniques developed, it became possible to study the emergent effects of dimensional constraints on material properties. For example, creating thin films of the low-anisotropy ferromagnet Permalloy ($Ni_{80}Fe_{20}$) constrains its magnetization to lie within the plane of the film due to magnetostatic considerations (shape anisotropy). Additionally reducing lateral dimensionality to form wires with widths comparable to the size of a domain wall causes the structure to become single domain, with magnetic reversal occurring through the propagation of a domain wall along the length of the wire. In such structures, the precise wire shape determines the magnetic switching properties, enabling a variety of functions to be performed using domain wall propagation, including diodic operations [52,53], processing logic [54] and memory storage [55,56]. Therefore dimensionality can be considered as a controllable material parameter that strongly affects magnetic behavior.

In electronics, control of dimensionality was taken to the extreme in 2004 when Novoselov and Geim discovered the method of atomically extracting thin graphite (carbon) sheets called graphene [57] (for which they were later awarded the 2010 Nobel Prize in Physics). Graphene was the first experimental realization of a free-standing 2D material; a material typically comprising a layered structure formed via van der Waals bonding in which electronic charge, spin and heat transport are confined to a plane and displaying properties that are distinct from those in the bulk substance.

Like the discovery of GMR, the isolation of graphene ignited interest in related materials, with massive advances made in just a few years. The reason for this excitement was that graphene exists entirely in the surface state, without a bulk component, and therefore exhibits properties very different from bulk graphite. Therefore studies of graphene and other 2D materials not only enrich the world of low-dimensional physics, but also provide a platform for transformative technical innovations.

Electronically, single-layer graphene exhibits a linear dispersion relation around a Dirac point, where the conduction and valence bands meet at a single place in reciprocal-space (Fig. 1.10A), which causes charge carriers to behave as *massless* Dirac fermions. This is important for potential applications, as it not only imbues the charge carriers with ultra-high mobility, and enables to propagate submicrometre distances without scattering, but also makes quantum phenomena, such as the half-integer quantum Hall effect (integer steps at half-integers of the conductance quantum e^2/h, Fig. 1.10B), accessible at room temperature [60]. However if the graphene consists of two atomic layers, the dispersion relation changes to a parabola (Fig. 1.10E) [61], so that charge carriers become *massive* Dirac fermions and different electrical properties emerge, such as an alternative form of quantum Hall effect (integer steps of the conductance quantum, but skipping the zero-conductance plateaux, Fig. 1.10D) [58]. Adding additional atomic layers further modifies the electronic properties, particularly resulting in a significant semimetallic overlap between conduction and valence

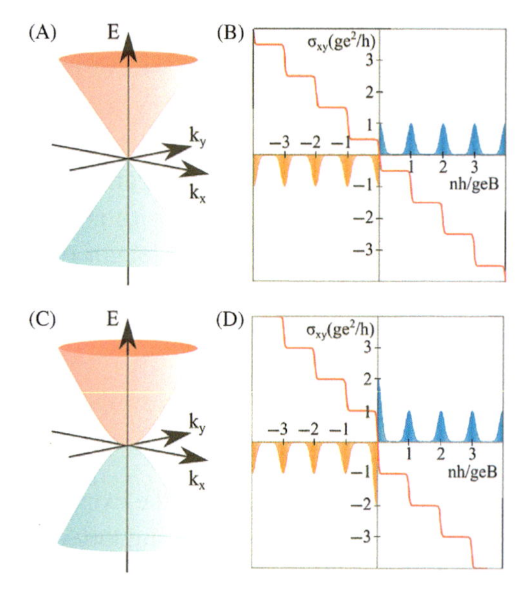

Figure 1.10

Schematic diagrams of (A) the dispersion relation of the conduction band (*rose*) and valence band (*blue*), and (B) the change in the Hall conductivity, σ_{xy}, with magnetic field, B, for single-layered graphene. (B) also shows the respective Landau levels for electrons (*blue*) and holes (*orange*). (C) and (D) show the corresponding relationships for bi-layered graphene. E is the energy, $k_{x,y}$ is the x-, y-component of wavevector, g is the system degeneracy, e is the electron charge, is Planck's constant, and n is the number of carriers. Source: *Adapted from K.S. Novoselov, et al., Unconventional quantum Hall effect and Berry's phase of 2 pi in bilayer graphene. Nat. Phys. 2 (2006) 177–180 [58]; A.K. Geim, Graphene: status and prospects. Science 324 (2009) 1530–1534 [59].*

bands as bulk properties appear and begin to dominate, until by 11 atomic layers the properties only slightly differ from bulk values.

Graphene was first isolated from graphite using micromechanical cleavage. This takes advantage of graphite's crystal structure, which is essentially several layers of graphene sheets layered linked with weak bonds between layers, by using adhesive tape (the method is also known as the scotch-tape technique) to iteratively strip off layers from bulk graphite until a single atomically layer is left. This remains the "gold-standard" of graphene production in terms of quality, although it is relatively slow. Alternative methods include chemical exfoliation, in which molecules infuse into the graphite to break bonds between graphene layers, and epitaxial growth on top of a bulk material, which is then etched at a later stage to release the graphene.

Inspired by the feasibility of producing novel, technologically relevant phenomena from relatively simple fabrication techniques, research into graphene and related analogs soared. During this intense period of exploration, a multitude of materials were found that could be chemically or mechanically exfoliated to single- or few- atomic layer thicknesses in a similar manner to graphene [62]. Collectively known as van der Waals materials, the list of members (and families) is still growing as improved fabrication protocols enable new members to be added. Indeed, progress is so fast that many van der Waals materials have only recently been discovered and their properties and stability not yet fully established. Broadly, van der Waals materials may be divided into four families (Fig. 1.11): graphene-based materials, 2D chalcogenides, 2D halides and 2D oxides. As well as graphene itself, the graphene family includes a number of materials derived by reacting graphene with particular molecules, such as fluorinated graphene (fluorographene) and oxidized graphene (graphene oxide), and materials which share graphene's 2D hexagonal crystal structure (Fig. 1.11), such as hexagonal boron nitride and boron carbon nitride. The 2D chalcogenides refer generally to any 2D material containing atoms from the chalcogen group of the periodic table (S, Se, Te), but the most commonly studied structure is the transition metal dichalcogenide (TMD) family, which

Graphene family Top Side 	**Graphene** Graphene, Irradiated graphene Stressed graphene	**Derivatives** Fluorographene, Graphene oxide	**Analogues** hBN, BCN
2D chalcogenides Top Side 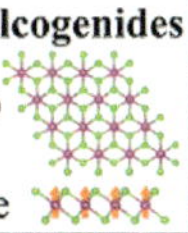	**Non-magnetic** ZrS_2, $ZrSe_2$, WS_2, WSe_2	**Ferromagnetic** MoS_2, $MoSe_2$,VSe_2, $MnSe_2$, Fe_3GeTe_2, $Cr_2Ge_2Te_6$, $Cr_2Si_2Te_6$	**Antiferromagnetic** $FePS_3$, $FePSe_3$, $MnPS_3$, $MnPSe_3$, $NiPS_3$,$NiPSe_3$, $AgVP_2S_6$, $AgVP_2Se_6$, $CrSe_2$, $CrTe_3$
2D halides Top Side 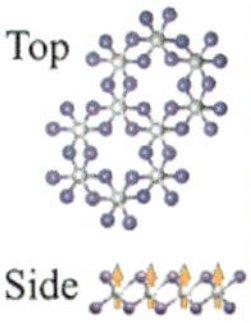	**Nonmagnetic** $(C_6H_9C_2H_4NH_3)_2PbI_4$	**Ferromagnetic** CrI_3 (single layer), $CrBr_3$, GdI_2	**Antiferromagnetic** CrI_3 (bi-layer), $FeCl_2$, $CoCl_2$, $NiCl_2$, VCl_2, $CrCl_3$, $FeCl_3$, $FeBr_2$, $MnBr_2$, $CoBr_2$, VBr_2, $FeBr_3$, FeI_2, VI_2, CrOCl, CrOBr, CrSBr
2D oxides Top Side 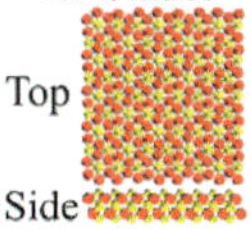	**Non-magnetic** MoO_3, WO_3, BSCCO	**Ferromagnetic** ZnO, hematene (2D α-Fe_2O_3), K_2CuF_4,	**Antiferromagnetic** $Ni(OH)_2$

Figure 1.11

Examples of van der Waals materials grouped into families [62–68]. Source: *Representative crystal structures shown for each family are adapted from literature. Graphene (graphene family): A.K. Geim, K.S. Novoselov, The rise of graphene. Nat. Mater. 6 (2007) 183–191; MnSe2 (2D chalcogenides): C. Gong, X. Zhang, Two-dimensional magnetic crystals and emergent heterostructure devices, Science 363 (2019) 706; CrI3 (2D halides): C. Gong, X. Zhang, Two-dimensional magnetic crystals and emergent heterostructure devices, Science 363 (2019) 706; hematene (2D oxides): A.P. Balan, et al., Exfoliation of a non-van der Waals material from iron ore hematite, Nat. Nanotech. 13 (2018) 602.*

is based around the formula MX_2, where M is a transition metal (such as Mo or W) and X is a chalcogen. The 2D TMDs have a hexagonal lattice (like graphene), but the arrangement of the unit cell means that they could be considered as a trilayer with the transition metal between two chalcogen layers (Fig. 1.11). Unlike graphene, 2D TMDs have a direct band gap and exhibit strong spin−orbit coupling, making them particularly interesting for optics and spintronics research. 2D halides contain atoms from the halide group of the periodic table (Cl, Br, I), and generally follow a similar formula to 2D chalcogenides, MX_2 or MX_3 (where M is a transition metal and X is a halogen), complete with a similar crystal structure (Fig. 1.11). The 2D oxides come in a variety of forms (Fig. 1.11), from relatively simple metallic oxides (e.g., MoO_3, WO_3) to 2D layers of the high-temperature superconductor bismuth strontium calcium copper oxide. Although 2D materials have mostly been studied as individuals, there has recently been advances in fabrication that enable different 2D materials to be layered into a heterostructure [62]. If this technique is perfected, it could enable electronic and structural properties to be tailored, so as to produce a new composite material not seen in nature: a 3D structure composed of 2D materials.

As understanding of 2D materials has developed, it has become clear that spintronic effects may be introduced to enhance functionality and induce novel phenomena. Key to achieving magnetic order in 2D materials is the ability to generate large anisotropy. 2D chalcogenides and halides offer a variety of systems with sufficient anisotropy to generate magnetic order, from ferromagnetic semiconductors (including VSe_2 and $MnSe_2$, crystal structure shown in Fig. 1.11) to insulating antiferromagnets (of the form MPX_3, where M is a transition metal, P is phosphorus, and X is either S or Se) [64]. 2D CrI_3 (crystal structure shown in Fig. 1.11) is particularly interesting, since single- and tri-layered CrI_3 is ferromagnetically ordered, but bi-layered CrI_3 is an antiferromagnet [69]. Furthermore, not only can bi-layered CrI_3 be switched to a ferromagnetic state at a critical field, but the critical field can be tuned by the application of an electric field [70]. The 2D oxides may yield surprising properties, since the recently isolated hematene (2D α-Fe_2O_3, crystal structure shown in Fig. 1.11) was demonstrated to be ferromagnetic, even though hematite (bulk α-Fe_2O_3) is antiferromagnetic [68]. Spintronic effects can also be induced in graphene, even in the absence of doping with magnetic ions. Irradiation with hydrogen results causes alterations in the electronic band structure, resulting in net spin polarization at room temperature [71]. In addition, application of strain to graphene can induce pseudo-magnetic fields, in which electron dynamics behave as if a magnetic field greater than 300 T is present [72].

Stability is a major challenge in the search for new 2D materials, not least because the entirety of the 2D structure is exposed to potential reactants that could modify bonding and so interfere with the material properties. Oxygenation can easily lead to degradation in graphene- or chalcogenide-based structures, while 2D oxides are susceptible to moisture. 2D materials can also be sensitive to temperature, since melting temperature decreases as atomic thicknesses are approached. Therefore while several 2D materials have been discovered, considerations of long term stability may limit the number of materials suitable for applications.

A more stable alternative to free-standing 2D materials may lie in the surface region within the first few nanometers of bulk materials. Due to the broken symmetry that exists when the material terminates, the localized electronic properties of the surface can be dramatically different from those of the bulk material. Therefore in a sense, a surface state can be considered as a substrate-dependent 2D material. Topological insulators highlight the stark contrast that can exist, since they exhibit an insulating bulk state, but a conductive surface state (Fig. 1.12A) (2D topological insulators also exist, in which the surface state is insulating and the 1D edge state is conductive). Furthermore, strong spin–orbit coupling within surface state acts to cause an inversion of the conduction and valance band that, together with considerations of time-reversal symmetry, is topologically protected (Fig. 1.12B). The topological nature of the surface state induces electronic properties that are particularly interesting for the development of spintronic technologies. Firstly, like graphene, topological insulators have a linear dispersion relation, which results in highly

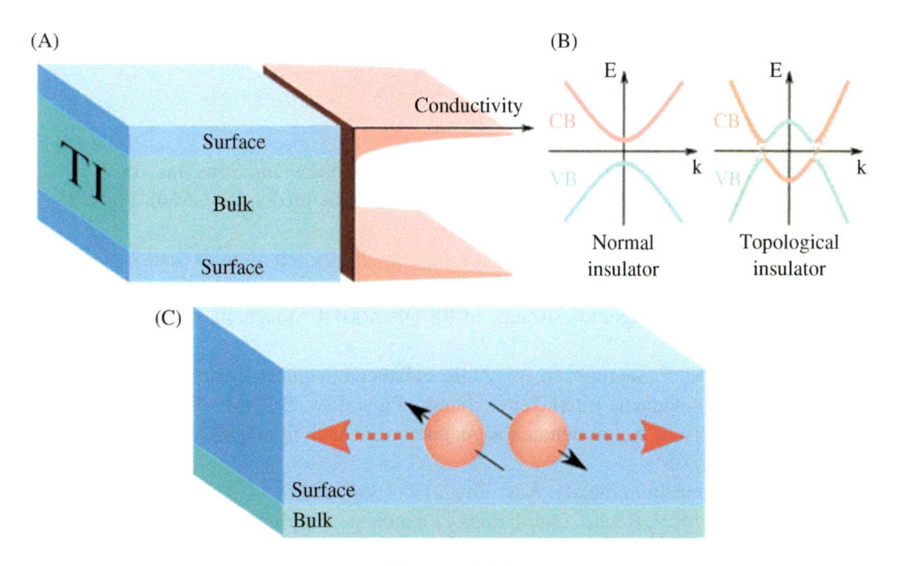

Figure 1.12

(A) Schematic diagram of surface and bulk regions of a TI and the depth-variation of the conductivity. (B) Schematic diagram of CB and VB in a normal insulator compared to the band inversion (CB falls below VB) in a TI, in which a band gap forms in the bulk (*solid line*), while conductive surface states (*dotted lines*) form at the surface. (C) Spin-locking fixes the spin orientation (*solid arrow*) perpendicular to the electron momentum (*dotted arrow*), and the surface normal, such that opposite spins propagate in opposite directions. Conversely, current direction determines spin-polarization, so back-scattering is suppressed without concurrent spin-reversal. *TI*, topological insulator; *VB*, valence band; *CB*, conduction band.

mobile carriers. Secondly, spin–orbit interactions result in an electron's spin being "locked" at an angle perpendicular to its momentum and the surface normal, which supresses scattering and enables efficient conversion between spin and charge currents (Fig. 1.12C). Finally, breaking of the time-reversal symmetry, by doping the topological insulator with magnetic ions or placing the surface state in proximity to a magnetic material, introduces an energy gap in the density of states and can result in exotic phenomena such as the quantum anomalous Hall effect.

1.5 Overview of this book

This book starts with the theoretical model of 2D magnetism, followed by the introduction of a selection of emerging magnetic 2D materials and systems. These include magnetic graphene, magnetic 2D transition metal chalcogenides, magnetic topological insulators, and topological semimetals. The book does in no way try to give a complete overview of the emerging 2D systems, as many of them are still developing rapidly every day. The purpose is, instead, to present an up-to-date understanding of 2D magnetism, and to relate it with some of the most important findings of the contemporary material research.

References

[1] C. Kittel, Introduction to Solid State Physics, fifth ed., John Wiley and Sons, Inc, 1976.
[2] J.H. Van Vleck, Models of exchange coupling in ferromagnetic media, Rev. Mod. Phys. 25 (1953) 220–227.
[3] A. Baker, et al., A study of hard:soft layer ratios and angular switching in exchange coupled media, J. Appl. Phys. 106 (2009) 053902.
[4] M.N. Baibich, et al., Giant magnetoresistance of (001)Fe/(001)Cr magnetic superlattices, Phys. Rev. Lett. 61 (1988) 2472–2475.
[5] G. Binasch, P. Grunberg, F. Saurenbach, W. Zinn, Enhanced magnetoresistance in layered magnetic structures with antiferromagnetic interlayer exchange, Phys. Rev. B 39 (1989) 4828–4830.
[6] T. Valet, A. Fert, Theory of the perpendicular magnetoresistance in magnetic multilayers, Phys. Rev. B 48 (1993) 7099–7113.
[7] N. Mott, Electrons in transition metals, Adv. Phys. 13 (1964) 325.
[8] S.S.P. Parkin, N. More, K.P. Roche, Oscillations in exchange coupling and magnetoresistance in metallic superlattice structures: Co/Ru, Co/Cr, and Fe/Cr, Phys. Rev. Lett. 64 (1990) 2304–2307.
[9] S.S.P. Parkin, D. Mauri, Spin engineering: direct determination of the Ruderman-Kittel-Kasuya-Yosida far-field range function in ruthenium, Phys. Rev. B 44 (1991) 7131–7134.
[10] B. Dieny, et al., Giant magnetoresistance in soft ferromagnetic multilayers, Phys. Rev. B: Condens. Matter Mater. Phys. 43 (1991) 1297–1300.
[11] P.P. Freitas, R. Ferreira, S. Cardoso, Spintronic Sensors, Proc. IEEE 104 (2016) 1894–1918.
[12] M. Julliere, Tunneling between ferromagnetic films, Phys. Lett. A 54 (1975) 225–226.
[13] S. Ikeda, et al., Tunnel magnetoresistance of 604% at 300 K by suppression of Ta diffusion in CoFeB/MgO/CoFeB pseudo-spin-valves annealed at high temperature, Appl. Phys. Lett. 93 (2008) 082508.
[14] S.S.P. Parkin, Z.G. Li, D.J. Smith, Giant magnetoresistance in antiferromagnetic Co/Cu multilayers, Appl. Phys. Lett. 58 (1991) 2710–2712.

[15] A. Hirohata, K. Takanashi, Future perspectives for spintronic devices, J. Phys. D: Appl. Phys 47 (2014) 193001.
[16] J.M. Kikkawa, D.D. Awschalom, Resonant spin amplification in n-type GaAs, Phys. Rev. Lett. 80 (1998) 4313–4316.
[17] J.M. Kikkawa, D.D. Awschalom, Lateral drag of spin coherence in gallium arsenide, Nature 397 (1999) 139–141.
[18] J.E. Hirsch, Spin Hall effect, Phys. Rev. Lett. 83 (1999) 1834–1837.
[19] Y.K. Kato, R.C. Myers, A.C. Gossard, D.D. Awschalom, Observation of the spin hall effect in semiconductors, Science 306 (2004) 1910–1913.
[20] J. Wunderlich, B. Kaestner, J. Sinova, T. Jungwirth, Experimental observation of the spin-Hall effect in a two-dimensional spin-orbit coupled semiconductor system, Phys. Rev. Lett. 94 (2005).
[21] W. Clark, G. Feher, Nuclear polarization in InSb by a DC current, Phys. Rev. Lett. 10 (1963) 134–138.
[22] H.J. Zhu, et al., Room-temperature spin injection from Fe into GaAs, Phys. Rev. Lett. 87 (2001) 016601.
[23] I. Žutić, J. Fabian, S. Das Sarma, Spintronics: fundamentals and applications, Rev. Mod. Phys. 76 (2004) 323–410.
[24] F. Meier, B.P. Zakharchenya, Optical Orientation In Solid State Physics, North Holland, 1984.
[25] H. Ohno, et al., (Ga, Mn) As: a new diluted magnetic semiconductor based on GaAs, Appl. Phys. Lett. 69 (1996) 363–365.
[26] S. Datta, B. Das, Electronic analog of the electro-optic modulator, Appl. Phys. Lett. 56 (1990) 665–667.
[27] Y.A. Bychkov, E.I. Rashba, Properties of a 2D electron gas with lifted spectral degeneracy, Sov. J. Exp. Theor. Phys. Lett. 39 (1984) 78–81.
[28] S. Sugahara, M. Tanaka, A spin metal-oxide-semiconductor field-effect transistor using half-metallic-ferromagnet contacts for the source and drain, Appl. Phys. Lett. 84 (2004) 2307–2309.
[29] K.C. Hall, M.E. Flatte, Performance of a spin-based insulated gate field effect transistor, Appl. Phys. Lett. 88 (2006) 162503.
[30] H.C. Koo, et al., Control of spin precession in a spin-injected field effect transistor, Science 325 (2009) 1515–1518.
[31] X.H. Lou, et al., Electrical detection of spin transport in lateral ferromagnet-semiconductor devices, Nat. Phys. 3 (2007) 197–202.
[32] X. Lou, et al., Electrical detection of spin accumulation at a ferromagnet-semiconductor interface, Phys. Rev. Lett. 96 (2006) 176603.
[33] G. Schmidt, D. Ferrand, L.W. Molenkamp, A.T. Filip, B.J. van Wees, Fundamental obstacle for electrical spin injection from a ferromagnetic metal into a diffusive semiconductor, Phys. Rev. B 62 (2000) R4790–R4793.
[34] E.I. Rashba, Theory of electrical spin injection: tunnel contacts as a solution of the conductivity mismatch problem, Phys. Rev. B 62 (2000) R16267–R16270.
[35] M. Ramsteiner, et al., Electrical spin injection from ferromagnetic MnAs metal layers into GaAs, Phys. Rev. B 66 (2002) 081304.
[36] R. Mattana, et al., Electrical detection of spin accumulation in a p-type GaAs quantum well, Phys. Rev. Lett. 90 (2003) 166601.
[37] M. Ferhat, K. Yoh, High quality Fe3-delta O4/InAs hybrid structure for electrical spin injection, Appl. Phys. Lett. 90 (2007) 112501.
[38] B. Beschoten, et al., Spin coherence and dephasing in GaN, Phys. Rev. B 63 (2001) 121202.
[39] S. Krishnamurthy, M. van Schilfgaarde, N. Newman, Spin lifetimes of electrons injected into GaAs and GaN, Appl. Phys. Lett. 83 (2003) 1761–1763.
[40] D. Culcer, E.H. Hwang, T.D. Stanescu, S. Das Sarma, Two-dimensional surface charge transport in topological insulators, Phys. Rev. B 82 (2010) 155457.
[41] D. Pesin, A.H. MacDonald, Spintronics and pseudospintronics in graphene and topological insulators, Nat. Mater. 11 (2012) 409–416.
[42] B.T. Jonker, et al., Robust electrical spin injection into a semiconductor heterostructure, Phys. Rev. B 62 (2000) 8180–8183.

[43] I. Zutic, J. Fabian, S. Das Sarma, Spin injection through the depletion layer: A theory of spin-polarized p-n junctions and solar cells, Phys. Rev. B 64 (2001) 121201.
[44] Y. Ohno, et al., Electrical spin injection in a ferromagnetic semiconductor heterostructure, Nature 402 (1999) 790–792.
[45] M. Oestreich, et al., Spin injection into semiconductors, Appl. Phys. Lett. 74 (1999) 1251–1253.
[46] R. Fiederling, et al., Injection and detection of a spin-polarized current in a light-emitting diode, Nature 402 (1999) 787–790.
[47] D. Chiba, et al., Magnetization vector manipulation by electric fields, Nature 455 (2008) 515–518.
[48] M. Wang, et al., Determining Curie temperatures in dilute ferromagnetic semiconductors: high Curie temperature (Ga, Mn)As, Appl. Phys. Lett. 104 (2014) 132406.
[49] T. Dietl, A ten-year perspective on dilute magnetic semiconductors and oxides, Nat. Mater. 9 (2010) 965–974.
[50] Y. Matsumoto, et al., Room-temperature ferromagnetism in transparent transition metal-doped titanium dioxide, Science 291 (2001) 854–856.
[51] C. Martinez-Boubeta, et al., Ferromagnetism in transparent thin films of MgO, Phys. Rev. B 82 (2010) 024405.
[52] D.A. Allwood, G. Xiong, R.P. Cowburn, Domain wall diodes in ferromagnetic planar nanowires, Appl. Phys. Lett. 85 (2004) 2848–2850.
[53] M.T. Bryan, T. Schrefl, D.A. Allwood, Symmetric and asymmetric domain wall diodes in magnetic nanowires, Appl. Phys. Lett. 91 (2007) 0142502.
[54] D.A. Allwood, et al., Magnetic domain-wall logic, Science 309 (2005) 1688–1692.
[55] S.S.P. Parkin, M. Hayashi, L. Thomas, Magnetic domain-wall racetrack memory, Science 320 (2008) 190–194.
[56] J. Dean, M.T. Bryan, T. Schrefl, D.A. Allwood, Stress-based control of magnetic nanowire domain walls in artificial multiferroic systems, J. Appl. Phys. 109 (2011) 023915.
[57] K.S. Novoselov, et al., Electric field effect in atomically thin carbon films, Science 306 (2004) 666–669.
[58] K.S. Novoselov, et al., Unconventional quantum Hall effect and Berry's phase of 2 pi in bilayer graphene, Nat. Phys 2 (2006) 177–180.
[59] A.K. Geim, Graphene: status and prospects, Science 324 (2009) 1530–1534.
[60] K.S. Novoselov, et al., Room-temperature quantum hall effect in graphene, Science 315 (2007). 1379.
[61] B. Partoens, F.M. Peeters, From graphene to graphite: Electronic structure around the K point, Phys. Rev. B 74 (2006) 075404.
[62] A.K. Geim, I.V. Grigorieva, Van der Waals heterostructures, Nature 499 (2013) 419–425.
[63] C. Gong, X. Zhang, Two-dimensional magnetic crystals and emergent heterostructure devices, Science 363 (2019) 706.
[64] K.S. Burch, D. Mandrus, J.G. Park, Magnetism in two-dimensional van der Waals materials, Nature 563 (2018) 47–52.
[65] N. Sethulakshmi, et al., Magnetism in two-dimensional materials beyond graphene, Mater. Today 27 (2019) 107–122. Available from: https://doi.org/10.1016/j.mattod.2019.03.015.
[66] W. Niu, A. Eiden, G.V. Prakash, J.J. Baumberg, Exfoliation of self-assembled 2D organic-inorganic perovskite semiconductors, Appl. Phys. Lett. 104 (2014) 171111.
[67] A.K. Geim, K.S. Novoselov, The rise of graphene, Nat. Mater. 6 (2007) 183–191.
[68] A.P. Balan, et al., Exfoliation of a non-van der Waals material from iron ore hematite, Nat. Nanotech. 13 (2018) 602.
[69] B. Huang, et al., Layer-dependent ferromagnetism in a van der Waals crystal down to the monolayer limit, Nature 546 (2017) 270.
[70] B. Huang, et al., Electrical control of 2D magnetism in bilayer CrI3, Nat. Nanotech. 13 (2018) 544.
[71] H. Ohldag, et al., The role of hydrogen in room-temperature ferromagnetism at graphite surfaces, New J. Phys. 12 (2010).
[72] N. Levy, et al., Strain-Induced Pseudo-Magnetic Fields Greater Than 300 Tesla in Graphene Nanobubbles, Science 329 (2010) 544–547.

CHAPTER 2

Rashba spin–orbit coupling in two-dimensional systems

Aurelien Manchon

Material Science and Engineering, Physical Science and Engineering Division, King Abdullah University of Science and Technology, Thuwal, Saudi Arabia

Chapter Outline

Spintronic 2D Materials.
DOI: https://doi.org/10.1016/B978-0-08-102154-5.00002-3

2.1 The origin of Rashba spin–orbit coupling

2.1.1 From Dirac to Pauli...

Spin–orbit coupling is a direct consequence of the relativistic motion of a particle in a potential gradient [1]. In special relativity, the energy functional $\hat{H}$ is defined as:

$$\hat{H}^2 = c^2\hat{\mathbf{p}}^2 + m^2c^4, \tag{2.1}$$

where the first and second terms in the right-hand side of Eq. (2.1) are the kinetic and potential energies, respectively. The simplest solution is found in four-dimensional (4D) Hilbert space, where the Hamiltonian reduces to

$$\hat{H} = c\hat{\alpha}\cdot\hat{\mathbf{p}} + \hat{\beta}mc^2, \tag{2.2}$$

$$\hat{\alpha} = \hat{\tau}_x \otimes \hat{\sigma}, \hat{\beta} = \hat{\tau}_z \otimes \hat{1}, \tag{2.3}$$

$\hat{\sigma}$, $\hat{\tau}$ are two vectors of 2×2 Pauli matrices referring to the spin-half and particle/antiparticle subspaces, respectively, and $\otimes$ is the direct product. Dirac showed that the solutions Ψ of this equation are of the form $\Psi = (\psi^+, \psi^-, \phi^+, \phi^-)$, where $\psi^\pm$ and $\phi^\pm$ are the particle and antiparticle wave functions. Installing the electromagnetic field ($\mathbf{A}$, Φ) through the electromagnetic gauge field transformation that is $\hat{\mathbf{p}} \to \hat{\mathbf{p}} + e\mathbf{A}$ and $\hat{H} \to \hat{H} - e\Phi$ (here, we set $e > 0$) and solving Eq. (2.2) for the electron wave function $\psi_e = (\psi^+, \psi^-)$ yields, to the lowest order in electromagnetic field ($\mathbf{E}$, $\mathbf{B}$)

$$\hat{H}_e = \underbrace{\frac{1}{2m}(\hat{\mathbf{p}}+e\mathbf{A})^2 - e\Phi}_{\text{Schrodinger equation}} + \underbrace{\frac{e\hbar}{2m}\hat{\sigma}\cdot\mathbf{B}}_{\text{Zeeman term}} - \underbrace{i\frac{e\hbar}{4m^2c^2}\hat{\mathbf{E}}\cdot\hat{\mathbf{p}}}_{\text{Darwin term}} + \underbrace{\frac{e\hbar}{4m^2c^2}\hat{\sigma}\cdot(\mathbf{E}\times\hat{\mathbf{p}})}_{\text{Spin–orbit coupling}} \tag{2.4}$$

where we define

$$\mathbf{B} = \nabla\times\mathbf{A}, \quad \mathbf{E} = -\nabla\Phi - \partial_t\mathbf{A} \tag{2.5}$$

The first two terms in the right-hand side of Eq. (2.4) constitute the low-energy Schrödinger's equation, the third term is Zeeman's energy, the fourth term is called Darwin's term, and the last term is the spin–orbit coupling. To the lowest order in electromagnetic field, the spin–orbit coupling term is therefore similar to a Zeeman term in the presence of a momentum-dependent magnetic field, also called the spin–orbit field, $\mathbf{B}_{so} = \mathbf{E}\times\hat{\mathbf{p}}/2mc^2$. The factor 2 in the denominator comes from Thomas precession. In the presence of a central potential (e.g., the hydrogen potential), $\mathbf{E} = -\partial_r\Phi\hat{\mathbf{r}}/r\mathbf{E}$ and the spin–orbit field becomes $\mathbf{B}_{so} \approx -\langle\partial_r\Phi/r\rangle\hat{\mathbf{r}}\times\hat{\mathbf{p}}/2mc^2$, which results in the well-known Russell–Saunders form of the spin–orbit coupling, $\hat{H}_{so} = \xi\hat{\mathbf{L}}\cdot\hat{\sigma}$, where $\hat{\mathbf{L}} = \hat{\mathbf{r}}\times\hat{\mathbf{p}}$ is the operator of orbital angular momentum, and $\xi = -e\hbar\langle\partial_r\Phi/r\rangle/4m^2c^2$.

2.1.2 ... and back again

In solid state, the potential Φ arises from the ionic environment of the electron that is crystal structure, defects, impurities, etc. Therefore through spin–orbit interaction the information of this environment gets imprinted in the spin-dependent part of the wave function. An instructive manner to apprehend this effect involves expanding the Hamiltonian close to high symmetry points in the Brillouin zone via $\mathbf{k}\cdot\mathbf{p}$ theory. If the crystal structure possesses a center of inversion, such as in diamond crystals (Si, Ge, C), the spin–orbit coupling only contributes to terms that are *even* in momentum $\mathbf{k}$ [2]. However if the structure has broken inversion symmetry, such as in ZnSe or wurtzite crystals, the spin–orbit coupling gives rise to terms that are *odd* in momentum. In a ground-breaking article, Dresselhaus predicted that ZnSe crystals possess a spin–orbit term of the form [3]

$$\hat{H}_{\mathrm{D}_3} = \beta_3((k_x^2 - k_y^2)k_z\hat{J}_z + (k_z^2 - k_x^2)k_y\hat{J}_y + (k_y^2 - k_z^2)k_x\hat{J}_x) \tag{2.6}$$

This Hamiltonian is linear in angular momentum operator $\hat{\mathbf{J}}$ and cubic in linear momentum $\mathbf{k}$. If strain is applied along the (001) axis for instance, the z-axis is no more equivalent to the x- and y-axes, and $\langle k_z^2\rangle > \langle k_x^2\rangle = \langle k_y^2\rangle$. Then [4],

$$\hat{H}_{\mathrm{D}_1}^{001} \approx -\beta_1(k_x\hat{J}_x - k_y\hat{J}_y), \tag{2.7}$$

where $\beta_1 = \beta_3\langle k_z^2\rangle$. These terms are now *linear* in momentum $\mathbf{k}$ and drive a wealth of nonequilibrium phenomena in semiconductors [5]. In wurtzite crystals such as GaN or ZnO, the elongated hexagonal structure results in a bulk $\mathbf{k}$-linear spin–orbit coupling, as predicted by Rashba [6]. About 20 years later, Ohkawa and Uemura [7] and Vas'ko [8] used the $\mathbf{k}\cdot\mathbf{p}$ theory to demonstrate that two-dimensional electron gases (2DEG) also experience a k-linear splitting in the presence of inversion symmetry breaking. Assuming a cylindrical symmetry around the z-axis, normal to the plane of the 2DEG, the potential drop creates an effective field $\nabla\Phi \approx -Ez$ on the itinerant electrons, which consequently feel an interaction of the form

$$\hat{H}_{\mathrm{R}} = \frac{\alpha_{\mathrm{R}}}{\hbar}\hat{\boldsymbol{\sigma}}\cdot(\hat{\mathbf{p}}\times\mathbf{z}), \tag{2.8}$$

where $\alpha_{\mathrm{R}} \approx \hbar^2\partial_z\Phi/4m^2c^2$. Eq. (2.8) is now widely referred to as Rashba or Rashba–Vas'ko spin–orbit coupling, and strikingly resembles the (massless) Dirac Hamiltonian given in Eq. (2.2). In other words, under certain conditions Bloch electrons behave like massless Dirac fermions, as illustrated in the context of topological materials (see below). The typical energy dispersion of a massive 2DEG with Rashba spin–orbit coupling is displayed in Fig. 2.1A together with its corresponding spin texture in momentum space. These aspects are further discussed in Section 2.2.1.

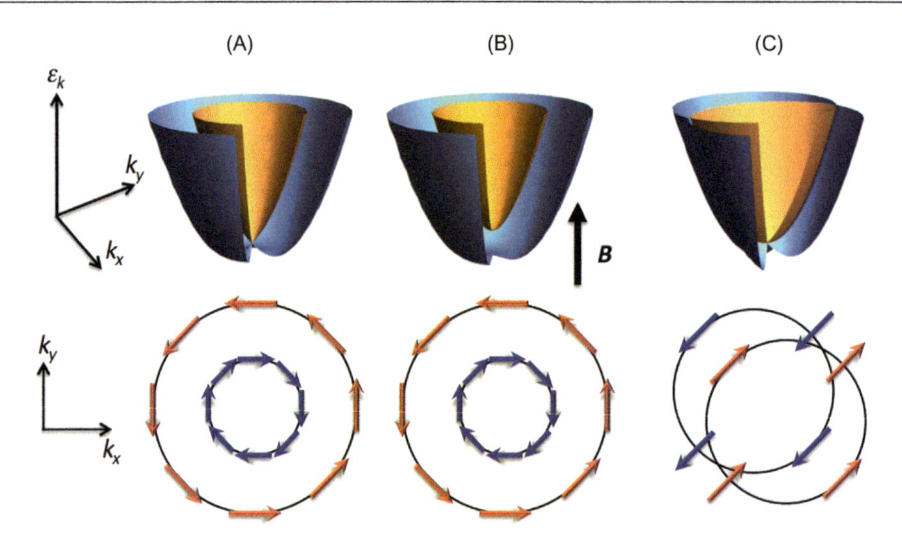

Figure 2.1
Dispersion and corresponding spin texture in three illustrative cases: a Rashba 2DEG (A) in the absence and (B) in the presence of a perpendicular magnetic field, and (C) a Rashba gas with k-linear Dresselhaus spin–orbit coupling in the spin helix state ($\alpha_R = \beta_1$).

In this picture, the Rashba parameter α_R is governed by the potential drop at the interface. If one considers the extreme case of a metallic surface characterized by a work function Φ_W ($\sim$5 eV in metals) and a screening distance of the order of the Fermi wavelength λ_F ($\sim$5 Å), the resulting spin-splitting becomes $\alpha_R k_F \approx 4\mu$ eV. This value of the spin–orbit splitting is actually far below the values commonly observed in semiconducting 2DEG or at metallic surfaces ($\approx$meV), which indicates that the magnitude of the Rashba splitting is not simply proportional to the interfacial potential drop, as discussed below.

2.1.3 Spin-splitting in semiconducting quantum wells

The Rashba parameter, α_R, can be calculated using the $\mathbf{k} \cdot \mathbf{p}$ theory [9] in low-doped semiconducting quantum wells accounting for only a few bands around Γ-point [7,10,11]. In these calculations, the potential gradient $\sim \partial_z \Phi$ distorts the envelope function of the Bloch states close to the interface. Then, perturbation theory is applied on the atomic spin–orbit coupling $\sim \hat{\mathbf{L}} \cdot \hat{\boldsymbol{\sigma}}$ which, once projected on this distorted wave function, produces a k-linear term identified to Rashba spin–orbit coupling.

The Rashba spin-splitting has been investigated using various techniques such as electron dipole spin resonance [12], Shubnikov–de Haas oscillations [13–15], and weak antilocalization [16,17]. These phenomena are reviewed in the next section. Of major interest from an applied perspective, the strength of Rashba spin–orbit coupling in semiconducting

Table 2.1: Cubic Dresselhaus β_3, linear Dresselhaus β_1, and Rashba spin–orbit coupling α_R for various semiconductor quantum wells. Nakamura et al. [20] found a cubic Rashba spin–orbit coupling of 1–2 eV Å^3 at $SrTiO_3$(001) surface.

Structure Reference	InAlAs/InGaAs [18]	GaAs/GaAlAs [16]	GaAs/InGaAs [21]	$SrTiO_3/LaAlO_3$ [19]
β_3 (eV Å^3)	–	~30	–	–
β_1 (eV.Å)	–	0.004	−0.0014	–
α_R (eV.Å)	0.07	0.005	0.0015	0.01–0.05

quantum wells can be tuned by applying a gate voltage that modulates the interfacial potential drop, and the associated wave function [16,18]. Interestingly, the existence of Rashba spin–orbit coupling has been recently confirmed at hetero-oxides interfaces [19,20]. Standard values of spin–orbit parameters for semiconducting 2DEG are given in Table 2.1.

We conclude this rapid tour d'horizon by mentioning recent progress on hybrid halide perovskytes, very promising candidates for photovoltaic applications [22]. These materials have the typically composition $CH_3NH_3(Pb,Sn)(I,Br)_3$ and display ferroelectricity thanks for their inversion symmetry breaking. A large spin-splitting of 3.7 eV Å for both conduction and valence bands in $CH_3NH_3PbI_3$ has been reported [23]. This spin-splitting enables the coupling between light and spin transport [23,24] and could be effective in suppressing the carrier recombination during photovoltaic operation [25].

2.1.4 Metallic surfaces

While Shubnikov–de Haas oscillations or weak antilocalization methods are difficult to apply to metals, Rashba spin-splitting has been successfully revealed using angle-resolved photoemission spectroscopy (ARPES) experiments, suitable for clean metallic surfaces. In 1996 Lashell et al. reported the spin-splitting of the surface states of Au [26,27], soon extended to Ag [28], Bi compounds [29,30], Gd [31], or metal-based quantum wells [32]. A few characteristic values are reported in Table 2.2. Metallic surfaces usually display much stronger Rashba spin-splitting than semiconductor quantum wells due to their larger atomic spin–orbit coupling and sharper potential drop. The physics underlying the emergence of the Rashba spin–orbit coupling in metals has been investigated theoretically [34–36]. In a very insightful work, Bihlmayer et al. [34] quantified the distortion of the wave function by computing the admixture between orbitals of different characters (say d-p or p-d) induced by the inversion symmetry breaking. The largest the admixture, the stronger the spatial gradient of the wave function and the larger the Rashba spin-splitting. In systems dominated by d-orbitals, such as transition metal interfaces, inversion symmetry breaking creates an admixture between different d-orbitals sitting on different atoms (say, d_{xy} and d_{yz}). Via the atomic spin–orbit coupling, this admixture leads to an effective Rashba term [37].

Table 2.2: Rashba spin–orbit coupling α_R for various metallic surfaces.

Structure Reference	Bi [29]	Ag [27]	Au [26]	Bi/Ag [30]	Bi/Cu [33]	Pb/Ag [28]	Gd/GdO [31]
α_R (eV Å)	0.56	0.03	0.3	3.05	1	1.52	0.25

2.1.5 Induced Rashba spin–orbit coupling

Inducing Rashba spin–orbit coupling in hexagonal monolayers (graphene and its siblings, transition metal dichalcogenides [38], etc.,) has been an important research direction lately. For instance, a spin-splitting of about 100 meV has been reported in graphene in contact with Au (111), [39], although other substrates such as Ni(111) or Bi(111) have proven unsuccessful. Similarly, a large spin-splitting was predicted at MoS_2/Bi(111) interface [40]. Most interestingly, Cheng et al. [41] proposed that "Janus" monolayers, that is transition metal dichalcogenides with dissimilar chalcogen elements such as MoSSe or WSSe, could promote the emergence of Rashba spin–orbit coupling by breaking the spatial inversion symmetry. This idea was achieved experimentally by Lu et al. who synthetized MoSSe monolayers [42].

2.1.6 Topological insulators

Topological materials constitute a novel class of systems displaying exotic transport properties directly related to the topology of their bulk ground states. This vast field of research is in continuous expansion and we encourage the reader to refer to some of the excellent reviews available on this topic [43–46]. What makes these materials particularly appealing in the context of the present chapter is their ability to display strong spin-momentum locking at their edges, surfaces, or hinges. Among the wide zoology of materials discovered to date, we selected two main classes of systems of highest interest: the topological insulators and the topological semimetals.

Topological insulators are characterized by insulating bulk states and conductive helical edge or surface states protected by symmetry. In Z_2 topological insulators for instance, time-reversal symmetry is preserved such that these helical states appear in odd numbers in the Brillouin zone and form Dirac cones (see Fig. 2.2A), displaying a linear dispersion and adopting the same spin-momentum locking scheme as in the Rashba case (see Eq. 2.8 and Fig. 2.2C). Although predicted originally in graphene [47,48] and HgTe quantum wells [49,50], where the conducting edges carry two counter propagating one-dimensional (1D) spin-polarized helical states, the concept of topological insulators was soon extended to three dimensions [51–54]. Due to time-reversal symmetry, these Dirac cones possess opposite spin chirality on opposite surfaces, and they become gapped in the presence of a magnetic field (or magnetic exchange) applied perpendicular to the surface (see Fig. 2.2B).

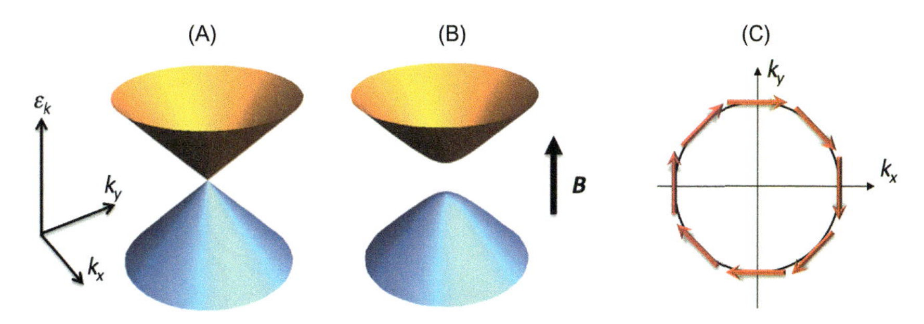

Figure 2.2
Dispersion of the surface state of a Z_2 topological insulator (A) in the absence and (B) in the presence of a perpendicular magnetic field, and (C) its corresponding spin texture.

The Rashba parameter governing the surface states range from 4.1 eV Å for Bi_2Se_3 [55] to 5.7 eVÅ for Kawazulite [i.e., $Bi_2(Te,Se)_2(Se,S)$] [56].

Topological crystalline insulators present an alternative paradigm: instead of being protected by time-reversal symmetry, the surface states are protected by selected crystal symmetries [45,57,58]. Since time-reversal symmetry is not required anymore, the Dirac cones appear in even number in the Brillouin zone. Such a behavior has been predicted [59] and observed in $Sn_xPb_{1-x}Te$-based compounds [60–62], as well as in KHg(As,Sb,Bi) [63], pyrochlore oxides [64], $Tl(Bi,Sb)S_2$ compounds [65], $A_3(Pb,Sn)O$ antiperovkites [66], FeS [67], and full Heusler alloys [68]. Because crystal symmetries determine the nature of the surface states, the Fermi contour and the related spin-momentum locking scheme significantly depend on the surface considered. Fig. 2.3A and Fig. 2.3B display the Fermi contour and spin texture at the (111) and (001) surfaces of SnTe [59,69,70]. In this material, the symmetry that protects the surface states is the reflection with respect to the (110) mirror plane. Any surface state that conserves such a symmetry is topologically protected and therefore can be gapped by applying proper strain [71].

Topological Kondo insulators are yet another class of topological insulators. In Kondo insulators, the bulk (narrow) gap emerges from the hybridization between conduction d electrons and localized f electrons [72]. The rational put forward by Dzero et al. [73] is that f states are odd parity while the d-like conduction bands are even parity. Therefore band crossing between the two is accompanied by a change of Z_2 index, opening the path to topological insulating regime. A hypothetical candidate is SmB_6, a known Kondo insulator in which spin-momentum locked surface states have been experimentally observed through ARPES [74,75]. The typical Fermi contour and corresponding spin-momentum locking at the (001) surface of SmB_6 is illustrated on Fig. 2.3C.

While the prospects opened by topological Kondo insulators are certainly inspiring, whether SmB_6 is one of them remains of matter of debates among experts [76–79].

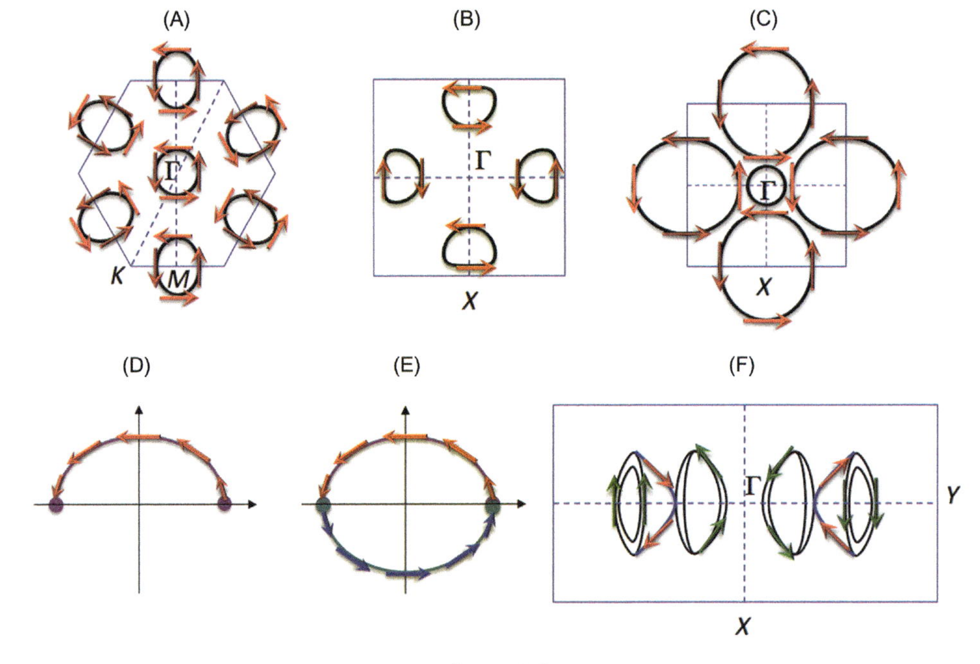

Figure 2.3
(Top panel) Schematics of the spin texture in momentum space of (A) the (111) surface and (B) the (001) surface of a topological crystalline insulator, SnTe, (C) the (001) surface of a topological Kondo insulator, SmB_6. (Bottom panel) Schematics of the spin texture in momentum space of (D) a type-I Weyl semimetal, (E) a Dirac semimetal, and (F) a type-II Weyl semimetal, WTe_2. In (D and E), the dots represent the projection of the Weyl nodes on the surface connected by the Fermi arcs. In (F), the black closed contours represent the projection of the bulk electron and hole pockets on the surface with their respective spin texture (*green arrows*), while the topologically protected surface states are represented in blue with their spin texture (*red arrows*).

2.1.7 Topological semimetals

Besides topological insulators, topological semimetals also display intriguing surface states coexisting with bulk conducting bands. Weyl semimetals are three-dimensional (3D) materials for which the conduction and valence bands touch at an even number of selected points of the Brillouin zone, called the Weyl nodes [80–82]. At these points, the energy dispersion is linear in the three momentum directions, $\hat{H} = v_{ij}\hat{\sigma}_i p_j$, involving the three Pauli spin matrices, and providing the condensed matter realization of the Weyl fermions [83]. Each node acts like a source or sink of Berry curvature (i.e., it is analogous to a magnetic monopole), such that two paired nodes possess opposite chirality. The emergence of degenerate Weyl nodes can only occur when the system is not symmetric under the product of parity and time reversal. In other words, either spatial inversion or time-reversal symmetry must be broken. Weyl semimetals have been predicted in pyrochlore iridates [80] (such as $Y_2Ir_2O_7$), in ferromagnetic $HgCr_2Se_4$ [84], multilayers of topological insulators [85–87], $TlBiS_2$ class of compounds [88], and nonmagnetic transition metal monophosphides [89,90] (TaAs, etc.).

An outstanding feature of Weyl semimetals is the emergence of very unusual bound states at their surfaces [80,91]. Instead of forming a closed Fermi surface, as expected in conventional metals, the surface states form so-called Fermi arcs that connect bulk Weyl nodes of opposite chirality (see Fig. 2.3D). Recent theoretical investigations have proposed that such Fermi arcs are only the equal-energy contours of helicoidal dispersion [92,93]. These Fermi arcs have been the hallmark of Weyl semimetals and observed experimentally using ARPES in TaAs [94–97], NbAs [98], and TaP [99] for instance. One characteristic of the Weyl semimetals described above is that they respect Lorentz invariance, which establishes a direct correspondence with high-energy physics [100,101]. However as emphasized by Soluyanov et al. [102], Lorentz symmetry does not need to be fulfilled in solid state. Following this idea, the authors demonstrated the existence of a second class of systems, tagged the type-II Weyl semimetals. In these materials, the bulk Weyl cones are inclined in such a way that the bulk Fermi surface does not reduce to a *point*, but rather to electron and hole *pockets*. As a result, the surface states are composed of electron and hole pockets connected via a Fermi arc (see Fig. 2.3F). The candidate materials are transition metal dichalcogenides, such as $W_xMo_{1-x}Te_2$ [102–104] as well as transition metal diphosphides, $(W,Mo)P_2$ [105]. Until now, surface Fermi arcs of type-II Weyl semimetals have been reported in WTe_2 [106–108] and $MoTe_2$ [109–113]. These states are topologically protected that is they cannot be removed without destroying the bulk Weyl nodes.

If both time-reversal symmetry and inversion symmetry are preserved, Weyl nodes carrying opposite chirality can be stabilized at the same crystal momentum, resulting in a fourfold singular degeneracy. Because the net Chern number is zero, such a degeneracy is generally not topologically protected, unless space group symmetries prevent mixing between the nodes. In the latter case, the material forms a Dirac semimetal whose surface states can be seen as two copies of the Fermi arcs that is "double Fermi arcs" [114] as illustrated on Fig. 2.3E. Dirac semimetals have been identified in Na_3Bi [115] and Cd_3As_2 [116], and possibly CuMnAs [117].

Notice that other classes of topological materials exhibiting strong spin-momentum locking have been identified recently, with the particularly intriguing case of so-called higher order topological insulators, displaying topological hinge states in 3D [118,119]. A thorough discussion of all these fascinating situations goes beyond the scope of the present chapter and we encourage the reader to refer to specialized reviews [43–46].

2.2 Fundamental signatures

2.2.1 Energy dispersion and spin texture

To understand how Vas'ko–Rashba spin–orbit coupling impacts spin transport in a 2DEG, let us consider the following model Hamiltonian:

$$\hat{H} = \frac{\hat{\mathbf{p}}^2}{2m} + \frac{\alpha_R}{\hbar}\hat{\boldsymbol{\sigma}}\cdot(\hat{\mathbf{p}}\times\mathbf{z}) + \frac{\Delta}{2}\hat{\sigma}_z \tag{2.9}$$

where the first term is the kinetic energy, the second term is Rashba spin–orbit coupling, and the last term accounts for the Zeeman coupling between the itinerant spin $\hat{\boldsymbol{\sigma}}$ and a perpendicular magnetic field $\sim\mathbf{z}$. The eigenstates of this Hamiltonian are

$$|+\rangle = \begin{pmatrix} ie^{-i\varphi_{\mathbf{k}}}\cos\frac{\chi_{\mathbf{k}}}{2} \\ \sin\frac{\chi_{\mathbf{k}}}{2} \end{pmatrix}, |-\rangle = \begin{pmatrix} -ie^{-i\varphi_{\mathbf{k}}}\sin\frac{\chi_{\mathbf{k}}}{2} \\ \cos\frac{\chi_{\mathbf{k}}}{2} \end{pmatrix}, \tag{2.10}$$

$$\epsilon_{\mathbf{k}}^s = \frac{\hbar^2 k^2}{2m} + s\sqrt{\frac{\Delta^2}{4} + \alpha_R^2 k^2} \tag{2.11}$$

where $\cos\chi_{\mathbf{k}} = \Delta/\sqrt{\Delta^2 + 4\alpha_R^2 k^2}$ and $\mathbf{k} = k(\cos\varphi_{\mathbf{k}}, \sin\varphi_{\mathbf{k}}, 0)$. In the absence of an external magnetic field ($\Delta = 0$), the energy dispersion reduces to $\epsilon_{\mathbf{k}}^s = \frac{\hbar^2 k^2}{2m} + s\alpha_R k$, as represented in Fig. 2.1A. The band structure is composed of two states of opposite spin chirality, degenerate at $k = 0$ and the spin density for state s ($= \pm 1$) reads

$$\mathbf{s}_{\mathbf{k}}^s = -s\mathbf{z}\times\hat{\mathbf{k}}, \tag{2.12}$$

where $\hat{\mathbf{k}} = \mathbf{k}/k$. The spin momentum is systematically in-plane, lying perpendicular to the momentum $\mathbf{k}$. When applying an external magnetic field perpendicularly to the 2DEG plane a gap opens at $k = 0$ (see Fig. 2.1B), and the spin density for state s reads

$$\mathbf{s}_{\mathbf{k}}^s = -s\frac{2\alpha_R}{\sqrt{\Delta^2 + 4\alpha_R^2 k^2}}\mathbf{z}\times\mathbf{k} + s\frac{\Delta}{\sqrt{\Delta^2 + 4\alpha_R^2 k^2}}\mathbf{z}. \tag{2.13}$$

The spin density can be parsed into a contribution that is *odd*-in-k and remains in-plane, normal to $\mathbf{k}$, and a contribution that is *even*-in-k and lies out-of-the-plane. The former produces the nonequilibrium spin density, sometimes called the Rashba or Rashba–Edelstein field (see the discussion on the inverse spin galvanic effect), while the latter is the equilibrium spin density, aligned on the external magnetic field.

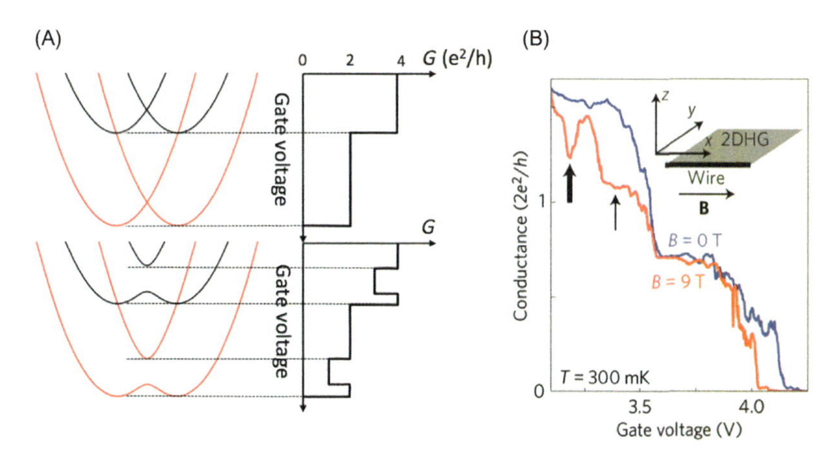

Figure 2.4
(A) One-dimensional band structure of a Rashba nanowire in the absence (top) and presence of a longitudinal magnetic field (bottom), and their corresponding conductance trace. (B) Conductance of a GaAs/AlGaAs nanowire as a function of the gate voltage [122]. The arrows indicate the reentrant behavior (i.e., drop of conductance) due to the spin-orbit gap.

As an illustration, let us contemplate the impact of Rashba spin-splitting on the physics of 1D conductors. Such systems are currently the object of intensive research as they have the potential to host Majorana fermions [120,121]. The spin–orbit gap discussed above can be detected via quantized conductance measurements. In the absence of magnetic field, the conductance of the wire is quantized and exhibits $2e^2/h$ steps as a function of the gate voltage (see Fig. 2.4A). When a magnetic field is applied along the nanowire (i.e., transverse to the Rashba field), it opens a gap between the two chiral spin states, resulting in conductance drops by e^2/h when the Fermi energy lies in the gap. This "reentrant" behavior characterizes spin–orbit gaps and has been observed in GaAs/AlGaAs [122] and InAs nanowires [123], as displayed in Fig. 2.4B.

2.2.2 Electron dipole spin resonance

In his 1960 influential work, Rashba suggested that the spin–orbit coupling arising from inversion symmetry breaking promotes electron spin resonance via ac *electric fields* [124,125]. In fact, the electron charge is several orders of magnitude more sensitive to an external magnetic field than its spin angular momentum: Upon cyclotron resonance, an electron is excited at the cyclotron frequency $\omega_c = eB/m^*$ (in SI units), and describes orbits of radius $r_B = \sqrt{\hbar/eB}$ corresponding to an electric dipole $p_c \sim er_B$. In turns, paramagnetic resonance occurs at Larmor frequencies $\omega_s = g\mu_B B/\hbar$, which corresponds to a magnetic dipole $\mu_s \sim e\lambda_c$, $\lambda_c = e\hbar/mc$ being the Compton length. As a result, Rashba estimates that the ratio between the electron cyclotron and paramagnetic resonance intensities is of the order of $I_c/I_s \approx (r_B/\lambda_c) \sim 10^{10}$. Therefore "even small [spin–orbit] coupling interaction leading to the coupling of orbital and spin motions will cause intensive electric excitation of [spin resonance]" [125]. This effect is coined electron dipole–induced spin resonance (EDSR and also called "combined resonances").

More precisely, the velocity operator deduced from Hamiltonian (2.9), $\hbar\mathbf{v} = \hbar\mathbf{k}/m + \alpha_R \mathbf{z} \times \hat{\boldsymbol{\sigma}}$, possesses off-diagonal elements in spin space that stimulates spin-flip transitions. Hence, exciting an electron using an ac electric field $\mathbf{E}(t)$ results in spin excitation, somewhat similarly to paramagnetic resonance. In the clean limit ($\tau\omega \gg 1$, where τ is the scattering time and ω is the ac field frequency), the effective Bloch equation describing the dynamics of the average spin density $\bar{\mathbf{s}}$ reads [126]

$$\partial_t \bar{\mathbf{s}} \approx \frac{2}{\hbar}\left(\mu_B \mathbf{B}_0 - \frac{\alpha_R}{\hbar\omega}\mathbf{z} \times e\mathbf{E}(t)\right) \times \bar{\mathbf{s}}, \tag{2.14}$$

where $\mathbf{B}_0$ is a static magnetic field, while $\mathbf{E}(t)$ is the oscillating electric field. When applying the electric field *along* the in-plane magnetic field $\mathbf{B}_0$ and tuning its frequency, one arrives at the resonance condition $\omega = \omega_s$ (see Fig. 2.5A), where the spin precesses with a Rabi frequency $\omega_R = \alpha_R eE/\hbar\omega_s$. The resonance frequency ω_s depends on whether the electron is in a conduction or bound state, yielding different resonance frequencies as displayed in Fig. 2.5B. The EDSR signal is maximum when the ac electric field is applied parallel to the external dc magnetic field (Fig. 2.5C), a feature that allows for distinguishing the EDSR signals from cyclotron and paramagnetic resonances. This theory

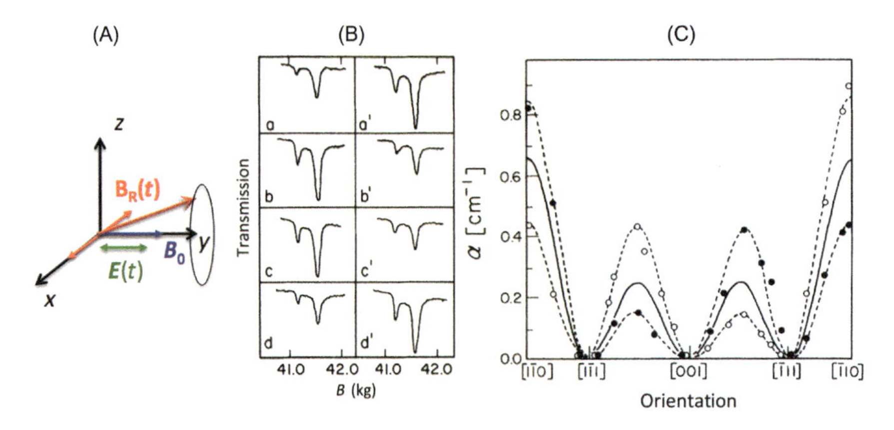

Figure 2.5
(A) Schematics of the EDSR setup. An ac electric field $\mathbf{E}(t)$ applied parallel to an external dc field $\mathbf{B}_0$ generates an ac Rashba field $\mathbf{B}_R(t)$ that excites the electron spin, resulting in EDSR. (B) Absorption spectrum in *n*-InSb showing two EDSR peaks corresponding to bound and conduction electrons, and (C) angular dependence as reported in [127]. *EDSR*, Electron dipole–induced spin resonance.

was first confirmed by experimenting in InSb [127,128], and then extended to asymmetrically grown structures such as GaAs/$Al_{0.3}Ga_{0.7}As$ 2DEG [12], as well as AlAs [129], and Si/Ge quantum wells [130]. In such systems, an important question that arises is the role of carrier scattering [126]. In a nutshell, in the clean limit EDSR is independent on the mobility and decreases with frequency ($\sim 1/\omega$) (see Eq. 2.14), while in the disordered regime, the EDSR is independent on the frequency, and proportional to the mobility ($\sim\tau$).

The idea behind EDSR has been inspirational for the electric manipulation of the spin degree of freedom in quantum wells and quantum dots [131]. For instance, Meier et al. [21] have used optically induced spin dynamics to quantify the Rashba and Dresselhaus spin–orbit coupling in GaAs/InGaAs quantum wells, while Frolov et al. [132] have used ballistic scattering in a confined nanowire to generate an effective ac Rashba field and obtain so-called ballistic spin resonance. These observations build on the original ideas developed by Rashba [124,125]. Concepts related to EDSR have been extended to the electric manipulation of spin in quantum wells through the control of the g-factor [133], as well as oscillating gate field inducing EDSR [134].

2.2.3 Quantum oscillations

The most popular experimental method to identify and quantify the Rashba spin-splitting is probably the analysis of Shubnikov–de Haas oscillations. In a 2DEG submitted to a perpendicular magnetic field, the electrons form quantized orbitals, known as Landau levels, $\epsilon_n^s = \hbar\omega_c(n+1/2)$, where $\hbar\omega_c = eB/m^*$ is the cyclotron frequency, $n \in N^*$ and s is the spin index. Increasing the magnetic field enhances the energy splitting between two consecutive levels, progressively pushing them above the Fermi level (see Fig. 2.6A). The degeneracy of a Landau level being $N_L = g_v\, g_s eB/h$ (where $g_{s,v}$ accounts for the spin and valley degeneracy), the number of filled degenerate Landau levels is given by the filling factor, $\nu = n_s/N_L$, n_s being the carrier density of the 2DEG. When the Landau levels are filled, that is, the Fermi level lies at a minimum of the density of states and the filling factor reaches an integer value, the conductance reaches a minimum. On the contrary, when the Landau level is half-filled the Fermi level lies in a maximum of the density of states and the conductivity is maximum. The magneto conductance is then [135]

$$\sigma_{xx} = \frac{e^2}{\pi^2\hbar}\sum_{n,s}\left(n + s\frac{1}{2}\right)\exp\left(-\frac{(\epsilon_F - \epsilon_n^s)^2}{\Gamma^2}\right). \tag{2.15}$$

where Γ is the impurity broadening, $\epsilon_0 = \hbar\omega_c/2$, and ϵ_n^s is given above. The resistivity is computed in the large Hall effect regime, $\rho_{xx} \approx \sigma_{xx}/\sigma_{xy}^2 \sim \sigma_{xx}(B/en_s)^2$, and yields the

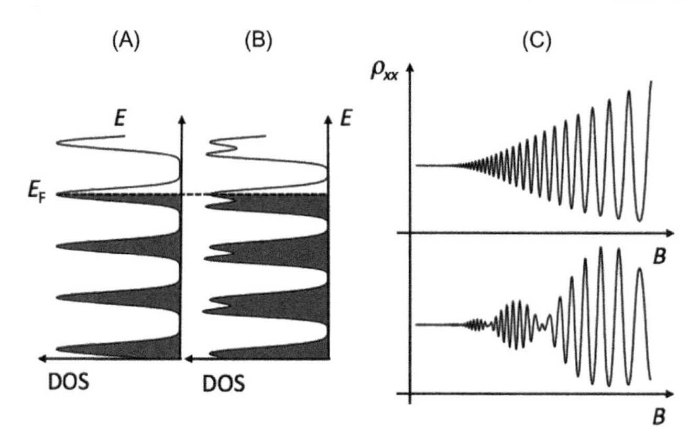

Figure 2.6
Density of states of a 2DEG under a perpendicular magnetic field (A) in the absence and (B) presence of Rashba spin-splitting. (C) Corresponding Shubnikov–de Haas oscillations in the absence (top) and presence (bottom) of Rashba spin-splitting.

characteristic Shubnikov–de Haas oscillations displayed on Fig. 2.6C (top). The carrier density n_s can be extracted from the difference between two consecutive oscillation minima

$$\Delta\left(\frac{1}{B_{\min}}\right) = g_v g_s \frac{e}{h n_s}. \tag{2.16}$$

In the presence of both Zeeman and Rashba spin-splitting, the Landau levels are not spin-degenerate anymore (see Fig. 2.6B) and their dispersion becomes [135,136]

$$\epsilon_n = \hbar\omega_c\left[(n + 1/2) \pm \sqrt{\delta^2 + n\gamma}\right], \tag{2.17}$$

where $\delta = (1 - gm^*/2m_0)/2$ and $\gamma = 2m^*\alpha_R^2/\hbar^3\omega_c$ account for the Zeeman and Rashba spin-splitting respectively. Here m^* is the effective mass, while m_0 is the free electron mass. Hence, the oscillatory behavior of the longitudinal conductivity exhibits a *beating pattern* that reflects the presence of two nondegenerate populations of carriers, as displayed on Fig. 2.6C (bottom).

Original signatures of such a beating pattern were investigated in bulk inversion asymmetric systems such as HgSe [137] or GaSb [138]. A decade later, analyzing the behavior of the conductance maxima at low and high fields in GaAs/AlGaAs heterostructures, Stormer et al. [13] and Eisenstein et al. [139] first identified the presence of two such populations of carriers, associated with a lift in spin degeneracy at $k \neq 0$. Shubnikov–de Haas oscillations were subsequently investigated by Luo et al. [14,135] and Das et al. [15] in InAs/GaSb and $In_xGa_{1-x}Al/In_{0.52}Al_{0.48}As$ quantum wells using different approaches. In their 1988 work, Luo et al. [14] identified the two carrier densities, $n_{\pm}$, by performing a Fast Fourier Transform of the beating pattern. Then, the Rashba parameter α_R can be expressed [10,14]

$$\alpha_R = \Delta n \frac{\hbar^2}{m^*} \sqrt{\frac{\pi}{2(n - \Delta n)}}, \tag{2.18}$$

where $n = n_- + n_+$ and $\Delta n = n_- - n_+$. This method is only valid in the limit $B \to 0$. Another approach consists in analyzing the zeros of the beating pattern [11,15]. Indeed, in the presence of spin-splitting the Shubnikov–de Haas oscillation is modulated by $\sim \cos(\pi \delta_s / \hbar \omega_c)$, where δ_s is the spin-splitting energy [140], resulting in nodes in the oscillation pattern (see Fig. 2.6C, bottom). The ith node is attained when [11]

$$i = \frac{2m^* \alpha_R k_F}{\hbar e} \frac{1}{B} + \left(\frac{g m^*}{2m_0} + \frac{1}{2} \right). \tag{2.19}$$

Hence, identifying three nodes is necessary and sufficient to determine consistently m^*, α_R, and k_F.

Analyzing the Shubnikov–de Haas oscillation pattern in 2DEG has proven particularly useful to demonstrate the control of the Rashba spin-splitting using a gate voltage. This control has been achieved first by Nitta et al. [18], Engels et al. [10], and Schäpers et al. [11] in $In_{0.53}Ga_{0.47}As/In_{0.52}Al_{0.48}As$ and $In_xGa_{1-x}As/InP$ heterostructures. For instance, a 50% modulation of α_R has been obtained by varying the gate voltage between ± 1.5 V [18]. This gate control is extremely important for applications as it offers an additional degree of freedom and enables the design of Rashba devices [141].

2.2.4 Weak antilocalization

When electrons diffuse on random scatterers they form a variety of trajectories in real space, some of which consist in closed loops. Such loops support self-interference processes between counter propagating wave packets, as displayed in Fig. 2.7A. When time-reversal symmetry is preserved, the two wave packets accumulate the exact same phase, resulting in constructive interferences. Therefore if the momentum scattering time τ is shorter than the electron dephasing time τ_φ, the quantum correction to the conductivity is [142]

$$\delta\sigma \approx -\frac{e^2}{2\pi^2 \hbar} \ln\left(\frac{\tau_\varphi}{\tau}\right) \tag{2.20}$$

However when a magnetic field is applied perpendicularly to the plane of the 2DEG, the wave function acquires an Aharonov–Bohm phase $\Psi \to \Psi e^{\pm i\pi eB/h}$, where $\pm$ corresponds to the (anti)clockwise propagation. Therefore one can define a magnetic time, $\tau_B = h/eB\mathcal{D}$, such that if $\tau < \tau_B < \tau_\varphi$, the quantum correction becomes

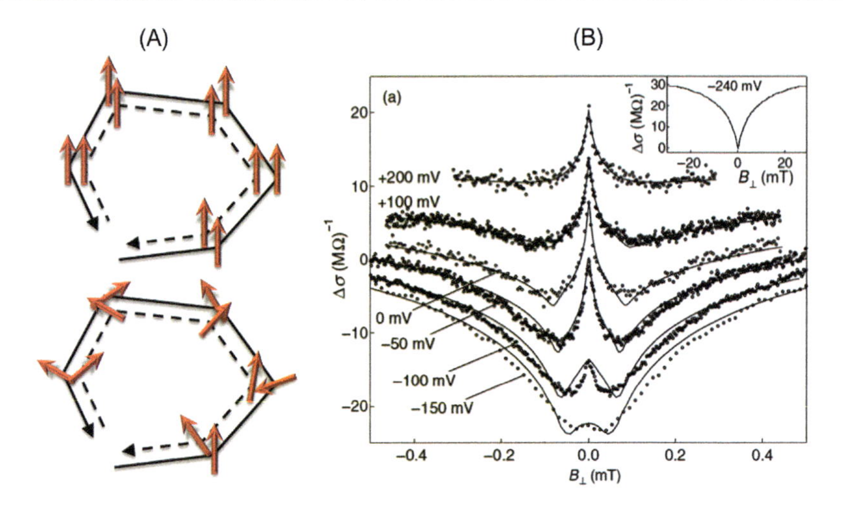

Figure 2.7
(A) Schematics of weak localization process in the absence (top) and presence of spin–orbit coupling (bottom) and (B) measurement of the weak localization signature in GaAs/AlGaAs quantum well for various gate voltages [16].

$$\delta\sigma(B) \approx -\frac{e^2}{2\pi^2\hbar}\left[\psi\left(\frac{1}{2}+\frac{\hbar}{4e\mathscr{D}B\tau}\right)-\psi\left(\frac{1}{2}+\frac{\hbar}{4e\mathscr{D}B\tau_\varphi}\right)\right]. \tag{2.21}$$

Here, $\psi(x)=\Gamma'(x)/\Gamma(x)$ is the digamma function, and $\mathscr{D}=v_F^2\tau/2$ is the diffusion coefficient. Increasing the magnetic field reduces the localization and enhances the conductivity [142].

When spin–orbit coupling is present in the 2DEG, momentum scattering is accompanied by spin relaxation, characterized by a spin-flip relaxation time τ_{sf}. After traveling around the loop, the spin states of the two counter propagating wave packets may not be the same, so that some of the loops lead to destructive interferences that is *antilocalization* (see Fig. 2.7A). As a result, two regimes are observed experimentally. As long as $\tau_B > \tau_{sf}$, the spin momenta of the two wave packets progressively align on the magnetic field, enhancing the localization and decreasing the conductivity. Then, for $\tau_{sf} > \tau_B$, spin relaxation is quenched and the magnetic field adds an Aharonov–Bohm phase to the wave function, enhancing the antilocalization and increasing the conductivity, as shown in Fig. 2.7B. In Rashba 2DEG the quantum correction becomes [143]

$$\delta\sigma(B) \approx -\frac{e^2}{4\pi^2\hbar}\left[\psi\left(\frac{1}{2}+\frac{\hbar}{4e\mathscr{D}B\tau_\varphi}\frac{1}{1+\xi^2}\right)+\ln\left(\frac{4e\mathscr{D}B\tau_\varphi}{\hbar}(1+\xi^2)\right)\right] \tag{2.22}$$

where $\xi=\alpha/\hbar v_F$. Weak antilocalization has been successfully used to distinguish the various types of spin–orbit coupling (Rashba, *k*-linear and *k*-cubic Dresselhaus, etc.) in semiconducting 2DEG [16,144] as well as in oxide heterostructures [19,20] and topological insulators [145].

2.3 Spin–orbit transport

2.3.1 Spin-charge conversion effects

The locking between the linear and spin angular momenta, given by Eq. (2.12), results in the so-called inverse spin galvanic effect. This relation shows that a charge current $\mathbf{j}_c \sim \langle \mathbf{k} \rangle$ is always accompanied by a nonequilibrium spin density, $\mathbf{S} = (\alpha_R m/e\hbar)\mathbf{z} \times \mathbf{j}_c$. This effect has been predicted by Ivchenko and Pikus [146], Vas'ko [8], and later Edelstein [147]. It was subsequently observed in tellurium [148], and more recently in quantum wells [149] and strained bulk semiconductors [150] using optical Faraday rotation, as well as at the surface of topological insulators [151]. The reciprocal effect is the spin galvanic effect, where a nonequilibrium spin density induces a charge current, $\mathbf{j}_c = -e(\alpha_R/\hbar)\mathbf{z} \times \mathbf{S}$. This effect was also predicted by Ivchenko and Pikus [152] and observed in various systems. In the original setup, the nonequilibrium spin density is generated by circularly polarized light in a n-GaAs/AlGaAs 2DEG, resulting in an (nonequilibrium) occupation imbalance between the two Rashba subbands, producing a nonequilibrium charge current [153]. Spin galvanic effect has also been realized using a magnetic drive. In this case, a ferromagnetic layer excited at the ferromagnetic resonance pumps a spin current into an adjacent 2DEG, for example, the metallic surface of Bi/Ag [154] and α-Sn [155], which is itself converted into a charge current via the spin-momentum locked states of the metallic surface.

Rashba spin–orbit coupling also enables the conversion of a unpolarized charge current $\mathbf{j}_c$ into a transverse pure spin current $\mathcal{J}_s^z \propto (\hbar/2e)\mathbf{z} \times \mathbf{j}_c$ polarized along the normal $\mathbf{z}$, a phenomenon called spin Hall effect [156]. This process has been the subject of intense debates lately because it was demonstrated that Rashba-driven spin Hall effect vanishes upon disorder [157,158]. This cancellation is now considered specific to the unrealistic quadratic band dispersion of the Rashba model [156]. An important distinction between spin Hall effect and inverse spin galvanic effect resides in the different spin accumulation profiles they generate. Spin Hall effect induces an antisymmetric spin accumulation at opposite *edges*, while inverse spin galvanic effect generates a constant spin accumulation in the *center* of the 2DEG. The spin transport in Rashba 2DEG has been investigated theoretically using various techniques [159,160]. In the limit of strong disorder ($\epsilon_F \gg \Gamma \gg \alpha_R k_F$), the spin-charge coupled drift-diffusion equations in the Rashba 2DEG read

$$\frac{\partial n}{\partial t} = \mathcal{D}\nabla^2 n + 4\mathcal{K}_{sc}(\mathbf{z} \times \nabla)\cdot\mathbf{S}, \tag{2.23}$$

$$\frac{\partial \mathbf{S}}{\partial t} = \mathcal{D}\nabla^2\mathbf{S} - \frac{S_x\mathbf{x} + S_y\mathbf{y}}{2\tau_z} - \frac{S_z\mathbf{z}}{\tau_z} + \mathcal{K}_{sc}(\mathbf{z} \times \nabla)n - 2\mathcal{K}_p(\mathbf{z} \times \nabla) \times \mathbf{S} \tag{2.24}$$

The terms $\sim \mathcal{K}_{sc}$ account for spin galvanic and inverse spin galvanic effects, while the term $\sim \mathcal{K}_p$ accounts for the precession of the diffusing spin accumulation around the Rashba field. Notice that the spin relaxation is anisotropic that is the out-of-plane spin component S_z relaxes twice faster than the in-plane components $S_{x,y}$. This is D'yakonov–Perel's relaxation [161], inherent to noncentrosymmetric materials. In a Rashba 2DEG, $\tau_z = \hbar^2/(4\alpha_R^2 k_F^2 \tau)$.

Spin-charge conversion processes have also been studied at the surface of topological insulators [162] and in Weyl semimetals [163]. In the former case, inverse spin galvanic effect is quite similar to the Rashba case discussed above, while in the latter case, in spite of the coexistence of conductive bulk states the current-driven spin density is predicted to be one order of magnitude larger than at the surface of topological insulators. Finally, Weyl and Dirac semimetals also feature so-called chiral, or Adler–Bell–Jackiw anomaly [101]. This effect comes from the fact that when a magnetic field B is applied parallel to the driving electric field, the chirality current is not conserved [164–166]. A direct consequence is the emergence of a very large negative magnetoresistance, quadratic in B, and observed in both Weyl [167,168] and Dirac semimetal candidates [169–171].

2.3.2 Persistent spin helix

Suppressing spin relaxation is a major challenge in semiconductor spintronics, and an effective way to achieve this goal is to utilize the so-called persistent spin helix condition [172–174]. This condition is obtained by tuning Rashba and k-linear Dresselhaus spin–orbit coupling, Eqs. (2.7) and (2.8), such that $\alpha_R = \beta_1$. Then, the eigenstates and their associated spin density read

$$|s\rangle = (se^{-i\pi/4}|\uparrow\rangle + |\downarrow\rangle)/\sqrt{2} \Rightarrow \mathbf{s}_{\mathbf{k}}^{s} = s(\mathbf{x}+\mathbf{y})/\sqrt{2} \tag{2.25}$$

They are both *independent* on the wave vector k, in contrast with the Rashba states given in Eq. (2.10) (see Fig. 2.1C). A similar condition is obtained for $\alpha_R = -\beta_1$. Because the spin state is independent on the wave vector, it is immune to D'yakonov–Perel's spin relaxation. In fact, in real materials cubic Dresselhaus and impurity-driven spin–orbit coupling are usually present and limit the spin lifetime. The persistent spin helix has been confirmed in 2DEG by both optical [175] and magneto conductance measurements [176], showing substantial enhancement of the spin relaxation length, as well as reduced weak antilocalization [177]. Since the spin helix state can be controlled by a gate voltage, it can serve to design nonballistic field-effect spin transistor [172].

2.3.3 Spin and charge pumping

Adiabatic quantum pumping is a mechanism by which a current is pumped out of a system upon slow periodic variation its parameters [178]. A concrete example is the spin pumping using precessing ferromagnets [179], now widely used in the field of spintronics to generate pure chargeless spin currents [180]. Equivalently, it is possible to pump an ac spin current in a Rashba 2DEG by modulating the gate voltage [181]. Other scenarios propose to pump a dc spin current out of quantum dots using synchronized ac gate voltages [182], or in 1D wires by tuning both the potential and Rashba spin-splitting [183]. Charge pumping using precessing magnetization in noncentrosymmetric ferromagnets has also been proposed [184] and observed in (Ga,Mn)As [185]. Such an effect can also be obtained at the interface between a ferromagnet and a topological insulator [186,187], providing a charge current of the form $\mathbf{j}_c \propto m_z \partial_t \mathbf{m} + \mathbf{z} \times \partial_t \mathbf{m}$ [188]. A direct consequence of this pumped current is the induction of an anisotropic magnetic damping on the magnetic layer [189].

2.3.4 Zitterbewegung effect

In systems possessing spin-momentum locking, propagating wave packets experience a *trembling* motion, called *Zitterbewegung*. This effect, initially investigated in the context of relativistic electron theory [190], stems from the coupled dynamics of the spin and position operators. As discussed in Section 2.2.1, the spin and velocity operators are interdependent, which results in a coupled, oscillatory motion. For the case of a Rashba 2DEG, the time-dependent position operator reads [191]

$$\begin{aligned}\hat{\mathbf{r}}(t) &= \hat{\mathbf{r}}(0) + \frac{\hbar \mathbf{k}}{m} t - \frac{1}{k^2}(\mathbf{k} \times \mathbf{z})\hat{\sigma}_z(1 - \cos\omega_R t) \\ &\quad - \frac{1}{2k^3}\mathbf{k}[\hat{\sigma} \cdot (\mathbf{z} \times \mathbf{k})](\omega_R t - \sin\omega_R t) + \frac{1}{2k}\mathbf{z} \times \hat{\sigma} \sin \omega_R t\end{aligned} \quad (2.26)$$

where $\omega_R = 2\alpha_R k/\hbar$. Zitterbewegung of nonequilibrium wave packets is an ubiquitous effect, present in all systems displaying some form of (pseudo)spin-momentum locking and is therefore present in Dirac systems, including honeycomb lattices such as graphene [192]. In semiconducting nanowires, Schliemann et al. [191] showed that the oscillation *transverse* to the wire direction reaches a resonance when the spin precession length matches the quantum well size. In graphene nanoribbons, where the sublattice pseudospin plays the role of the real spin [193], Ghosh et al. [194] demonstrated that an oscillation *longitudinal* to the transport direction can be induced. To the best of our knowledge, Zitterbewegung has not been observed yet in 2DEG.

2.3.5 Quantum anomalous Hall and magnetoelectric effects

The ground state of topological materials is characterized by a topological invariant, the Chern number that quantifies the flux of Berry curvature experienced by itinerant electrons in the unit cell. In 2D Z_2 topological insulators, this flux can drive topologically protected spin-polarized edge states, leading to quantum spin Hall effect [50,173]. In 3D, this flux does not readily lead to 1D quantized spin Hall effect, but rather to the emergence of topologically protected spin-momentum locked surface states. Nevertheless if one applies a magnetic field perpendicular to the surface, time-reversal symmetry is broken and a gap opens at the surface Dirac point (see Fig. 2.2B). When the chemical potential lies in the gap, conductive 1D states build up at the edges of the surface, resulting in quantum anomalous Hall effect [54,195–197], which is the hallmark of 3D Z_2 topological insulators. In addition, because the velocity operator is directly related to the spin operator, $\hat{\mathbf{v}} = v\mathbf{z} \times \hat{\boldsymbol{\sigma}}$, the quantum anomalous Hall effect is automatically associated with quantized magnetoelectric effect, $\boldsymbol{S} = -\frac{e\hbar}{2\pi v}\mathbf{E}$ [54,198,199].

Because of its very large Rashba-like spin–orbit coupling, the coupling between spin currents and magnetic order at the magnetic surface of topological insulators has attracted substantial interest, both experimentally and theoretically. For instance, due to the presence of the induced gap perpendicular magnetic domain walls acquire an electrical charge $\rho = -(e\Delta/4\pi v)\nabla \cdot \mathbf{m}$ on the surface of the topological insulator [54,200,201]. The reciprocal effect of this electric charging is the induction of a magnetic monopole when approaching an electrical charge from the surface of a topological insulator [202]. Finally the nature of spin–orbit torque has been intensively studied theoretically [188,203–208] and experimentally [209–213], but a detailed discussion remains out of the scope of the present chapter.

2.3.6 Floquet physics in spin–orbit coupled systems

Floquet's theorem states that a system described by a Hamiltonian periodic in time displays quasistatic eigenstates [214]. These states, called Floquet states, are the analog of Bloch states in a crystal with spatial periodicity. In other words, when a material is submitted to *off-resonant* light that is an electromagnetic wave that does not drive direct interband transitions, the multiple photon absorption-emission processes "dress" the electronic band structure, resulting in a quasistatic Hamiltonian. The corresponding quasistatic states are referred to as Floquet–Bloch states [215]. A few years ago, Kitagawa et al. [216] and Lindner et al. [217] proposed to exploit such a scheme to induce topological phase transition in Dirac systems. More specifically, they showed that to the lowest order in light intensity the quasistatic Hamiltonian reads

$$\hat{H}_{\text{eff}} \approx \hat{H}_0 + \frac{1}{\Omega}\left[\hat{H}_{-1}, \hat{H}_1\right] + O(A^4) \tag{2.27}$$

Here $\hat{H}_n = (\Omega/2\pi)\int_0^{2\pi/\Omega} \hat{H}(t)e^{in\Omega t}dt$ is the nth order Fourier component of the Hamiltonian. Using the low-energy Hamiltonian of graphene in the presence of an oscillating vector potential, $\hat{H} = v\hat{\sigma}\cdot[(\mathbf{p}+e\mathbf{A}(t))\times\mathbf{z}]$, where $\mathbf{A}(t) = A(\pm\ \sin\ \Omega t, \cos\Omega t)$, and applying the above expansion, one obtains

$$\hat{H}_{\text{eff}} \approx v\hat{\sigma}\cdot(\hat{\mathbf{p}}\times\mathbf{z}) \pm \frac{(evA)^2}{\hbar\Omega}\hat{\sigma}_z \tag{2.28}$$

In other words, because of the direct coupling between the (pseudo)spin and linear momentum, that is, $\sim\hat{\sigma}_i\hat{p}_j$, shining circularly polarized light on a Dirac cone opens a gap at the second order in the field amplitude, whose sign depends on the light polarization. This gap opening can lead to quantum Hall effects even in the absence of applied magnetic field, but also to topological phase transition [216–218]. Such a light-induced gap opening was observed at the surface states of Bi_2Se_3 [219], and valley-dependent gap modulation was also predicted [220] and reported in WS_2 [221]. More recently, Hübener et al. demonstrated theoretically that a Dirac semimetal under off-resonant light can be turned into a Weyl semimetal [222]. The search for Floquet-driven topological transitions is currently an important topic expanding far away from Rashba physics [223].

2.4 Rashba devices

2.4.1 Aharonov–Casher interferometer

We mentioned above that electrons traveling through a loop pierced by a magnetic flux experience Aharonov–Bohm quantum oscillations [224] (see Fig. 2.8A). In a subsequent work, Aharonov and Casher demonstrated that charge neutral magnetic moments also experience a similar effect in the presence of an electric field [226]. Nitta et al. [227] proposed that such an effect can be observed in a quantum ring made of a Rashba 2DEG, and illustrated in Fig. 2.8B. Since the spin momenta flowing in opposite branches experience a different Rashba field, their state at the exit of the ring might differ, depending on the strength of Rashba spin–orbit coupling. Hence, modulating this strength through a gate voltage results in an oscillatory pattern, as observed in a HgTe [228] and InGaAs rings [225], and reported on Fig. 2.8C. The interference pattern can also be controlled by the bias voltage applied across the ring, as proposed in Ref. [229].

2.4.2 Datta–Das spin field-effect transistor

The most popular device concept utilizing Rashba spin–orbit coupling is undoubtedly the spin field-effect transistor proposed by Datta and Das [230], and illustrated in Fig. 2.9A. The device consists of two ferromagnetic leads connected through a Rashba channel. The Rashba field induces the flowing spins to precess, so that the state of the spin impinging on

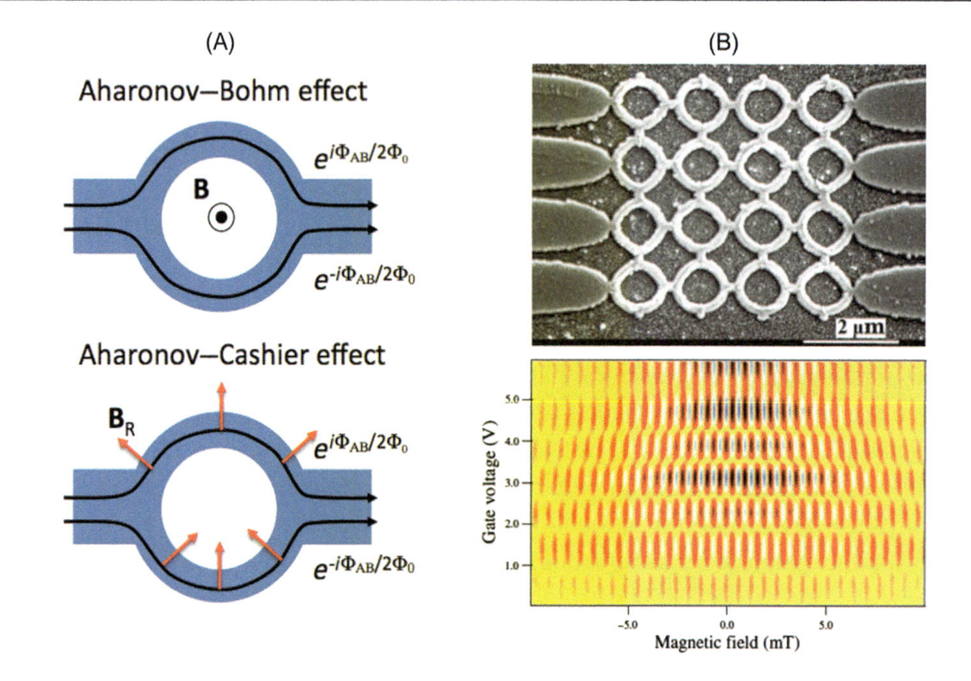

Figure 2.8

(A) Schematics of Aharonov–Bohm and (B) Aharonov–Casher effects. (C) Array of quantum rings made of InGaAs 2DEG, and corresponding interference pattern obtained upon varying the gate voltage and the external magnetic field [225].

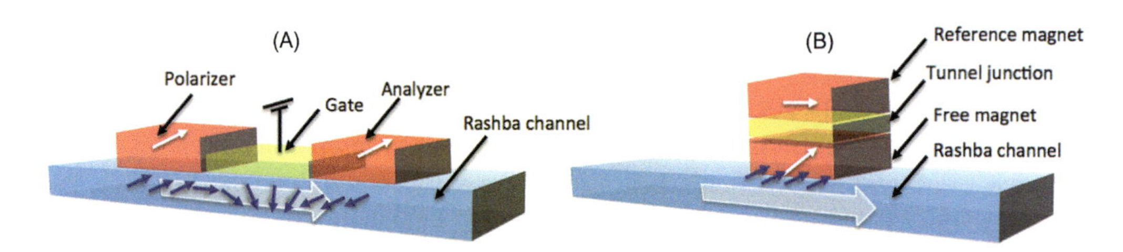

Figure 2.9

(A) Sketch of the Datta–Das transistor. Two magnets (*red layers*) are connected through a Rashba 2DEG (*blue layer*). The injected spins (*blue arrows*) experience a spin precession that is controlled by a gate voltage (*yellow layer*). The final state is detected through magnetoresistance. (B) Three-terminal spin–orbit torque device, as proposed by Manchon and Zhang [231]. The spin–orbit coupled channel promotes a torque that controls the magnetization direction of the overlayer. A magnetic tunnel junction is deposited on top to detect the magnetic state.

the ferromagnetic analyzer depends on the gate voltage. This first experimental demonstration of this device was performed by Koo et al. [141]. In a recent work, Chuang et al. [232] realized a magnet-free Datta–Das spin transistor, by replacing the ferromagnetic leads by spin-filtering point contacts with lateral gates. Another realization of magnet-free Datta-Das transistor was achieved last year by Choi et al. [233], where the spin current is injected and detected via spin Hall and inverse spin Hall effects. In this approach, the obtained signals are one to two orders of magnitude larger than that of the traditional Datta-Das transistor reported in [141].

2.4.3 Spin–orbit torques devices

The magnetic field generated by inverse spin galvanic effect can be used to torque the magnetization of an adjacent magnetic layer, a mechanism called spin–orbit torque [231,234,235,236]. The three-terminal device proposed by Manchon and Zhang [231] is sketched on Fig. 2.9B. The first observation of such a torque was reported on (Ga,Mn)As [237] and soon extended to a wide variety of systems involving asymmetric transition metal multilayers [238,239]. In the latter, another source of spin current, the spin Hall effect present in the heavy metal substrate, substantially contributes to the magnetization dynamics. The debate about the true origin of spin–orbit torques in these systems is not settled yet. That being said, Rashba-driven spin–orbit torque has been observed in several systems in which spin Hall effect is absent such as Ti/NiFe/AlO_x [240], $LaAlO_3$/$SrTiO_3$ superlattices [241], and MoS_2/CoFeB bilayers [242]. In most recent developments, current-driven magnetization switching driven by spin–orbit torque was achieved in heterostructures involving topological insulators [210,243,244]. In spite of the substantial bulk transport taking place in these systems, recent modeling suggests that the torque is dominated by the inverse spin galvanic effect of the interfacial Rashba–Dirac states [208]. The recent observation of room-temperature ultralow switching current density in *sputtered* Bi_xSe_{1-x}/Ta/CoFeB perpendicular heterostructures opens exciting perspectives for applications [245].

2.4.4 Spin–orbit Qubits

The EDSR predicted by Rashba [124,125] found its most elegant application with the demonstration of electrically driven spin–orbit Qubits [246,247]. This device is made of a double quantum dot, fabricated in a InAs nanowire (see Fig. 2.10). The first dot (QD1) hosts the Qubit, whose state can be manipulated via a gate voltage-controlled EDSR, while the second dot (QD2) serves as a reference. By tuning the double quantum dot to the Coulomb blockade regime (electrons cannot pass, no matter their spin state), the Qubit can be manipulated by modulating the gate voltage [248]. The spin state is then read out in the Pauli spin blockade regime (electrons can only pass if their spins are antiparallel) by measuring the current flowing through the double quantum dot.

2.5 Outlook and conclusion

Initially proposed to explain spin resonance experiments in semiconducting 2DEG, the concept of Rashba–Vas'ko spin–orbit coupling has been extended toward a wide variety

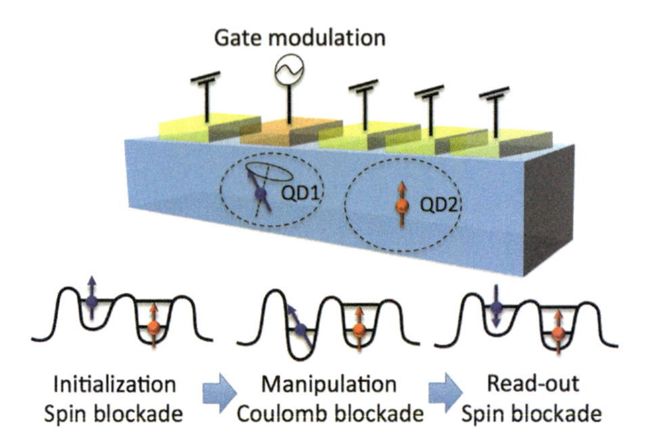

Figure 2.10
Schematic of the spin–orbit Qubit in a double quantum dot. The blue layer is the nanowire while the yellow layers are the different gates. The orange layer is the gate used to trigger EDSR. The spin–orbit Qubit (*blue arrow*) is initialized the first dot (QD1) in the spin blockade regime. Then the double quantum dot is set in the Coulomb regime, where the spin–orbit Qubit is manipulated via gate voltage driven EDSR. Once written in the desired state, the double quantum dot is set in the spin blockade regime and its conductance is read out.

of systems over the years, including bulk noncentrosymmetric crystals, metallic surfaces, and oxides interfaces. By enabling the direct coupling between linear momentum and (pseudo)spin degree of freedom, it has been successfully exploited to generate innovative device concepts, from the celebrated Datta–Das transistor to spin–orbit torque magnetic memories and spin–orbit Qubits. It is now a cornerstone of the vast field of topological materials, enabling the realization of massless quasiparticles in condensed matter (Dirac or Weyl fermions), and henceforth opening inspiring avenues for the exploration of high-energy physics in the lab.

Novel directions that could not be thoroughly addressed in the present overview are currently being pursued. For instance, the implementation of Rashba-like spin–orbit coupling in (topological or trivial) superconductors [249,250] enables helical-Cooper pairing, a key ingredient toward the realization Majorana fermions [120,251] that paves the way toward topological quantum computing [252]. In addition the possibility to synthetize Rashba-like spin–orbit coupling via laser fields [253,254] allows for the engineering and exploration of the impact of spin–orbit coupling on a wide variety of exotic situations such as Fermi gas [255,256], Bose gas [257], Bose–Einstein condensate [258], Bose–Hubbard model [259], etc. Finally the presence of Rashba-like spin–orbit coupling in noncentrosymmetric magnets promotes the emergence of antisymmetric magnetic exchange, known as Dzyaloshinskii–Moriya interaction [260,261]. This interaction stabilizes magnetic skyrmions, as well as homochiral spin spirals and domain walls [262,263], and has become a centerpiece for the realization of disruptive solutions for ultrahigh-density data storage [264,265]. New ideas rooted in the physics of Rashba spin–orbit coupling and applied to magnonic systems, such as magnon Hall effect [266,267], chiral magnetic damping [268], or Dzyaloshinskii–Moriya torque [269], are currently being explored. These recent developments demonstrate the continuous vitality and permanent creative power of this fascinating effect.

References

[1] J.J. Sakurai, Advanced Quantum Mechanics, Addison-Wesley, 1967.

[2] J.M. Luttinger, Quantum theory of cyclotron resonance in semiconductors: general theory, Phys. Rev. 102 (4) (1956) 1030.

[3] G. Dresselhaus, Spin-orbit coupling effects in zinc blende structures, Phys. Rev. 100 (2) (1955) 580.

[4] M. Dyakonov, V. Kachorovskii, Spin relaxation of two-dimensional electrons in noncentrosymmetric semiconductors, Sov. Phys. Semicond 20 (1986) 110.

[5] A. Manchon, H.C. Koo, J. Nitta, S.M. Frolov, R.A. Duine, New perspectives for Rashba spin-orbit coupling, Nat. Mater. 14 (2015) 871.

[6] E. Rashba, V. Sheka, Symmetry of energy bands in crystals of Wurtzite type II. symmetry of bands with spin-orbit interaction included, Fiz. Tverd. Tela: Collected Papers 2 (1959) 162–176.

[7] F.J. Ohkawa, Y. Uemura, Quantized surface states of a narrow-gap semiconductor, J. Phys. Soc. Jpn. 37 (5) (1974) 1325–1333. Available from: https://doi.org/10.1143/JPSJ.37.1325.

[8] F.T. Vas'ko, Spin splitting in the spectrum of two-dimensional electrons due to the surface potential, Pis'ma Zh. Eksp. Teor. Fiz 30 (9) (1979) 574.

[9] R. Lassnig, K.P theory, effective-mass approach, and spin splitting for two-dimensional electrons in GaAs-GaAlAs heterostructures, Phys. Rev. B 31 (12) (1985) 8076–8086.

[10] G. Engels, J. Lange, T. Schapers, H. Luth, Experimental and theoretical approach to spin splitting in modulation-doped InxGa1-xAs/InP quantum wells for B→0, Phys. Rev. B 55 (4) (1997) 1958. Available from: https://doi.org/10.1103/PhysRevB.55.R1958.

[11] T. Schapers, G. Engels, J. Lange, M. Hollfelder, H. Lu, I. Introduction, effect of the heterointerface on the spin splitting in modulation doped In_x Ga_1-x As/InP quantum wells for B→0, J. Appl. Phys. 83 (8) (1998) 4324–4333.

[12] D. Stein, K. Klitzing, G. Weimann, Electron spin resonance on G a A s-Al x Ga 1-x As heterostructures, Phys. Rev. Lett. 51 (2) (1983) 130–133. URL http://journals.aps.org/prl/abstract/10.1103/PhysRevLett.51.130.

[13] H. Stormer, Z. Schlesinger, A. Chang, D. Tsui, A. Gossard, W. Wiegmann, Energy structure and quantized Hall effect of two-dimensional holes, Phys. Rev. Lett. 51 (2) (1983) 126. Available from: https://doi.org/10.1103/PhysRevLett.51.126. URL http://link.aps.org/doi/10.1103/PhysRevLett.51.126.

[14] J. Luo, H. Munekata, F.F. Fang, P.J. Stiles, Observation of the zero-field spin splitting of the ground electron subband in GaSb-InaS-GaSb quantum wells, Phys. Rev. B 38 (14) (1988) 10142–10145. Available from: https://doi.org/10.1103/PhysRevB.38.10142.

[15] B. Das, D.C. Miller, S. Datta, R. Reifenberger, W.P. Hong, P.K. Bhattacharya, et al., Evidence for spin splitting in InxGa1-xAs/In0.52Al0.48As heterostructures as B0, Phys. Rev. B 39 (2) (1989) 1411. Available from: https://doi.org/10.1103/PhysRevB.39.1411.

[16] J.B. Miller, D.M. Zumbühl, C.M. Marcus, Y.B. Lyanda-Geller, D. Goldhaber-Gordon, K. Campman, et al., Gate-controlled spin-orbit quantum interference effects in lateral transport, Phys. Rev. Lett. 90 (7) (2003) 076807. Available from: https://doi.org/10.1103/PhysRevLett.90.076807. URL http://link.aps.org/doi/10.1103/PhysRevLett.90.076807.

[17] F.E. Meijer, A.F. Morpurgo, T.M. Klapwijk, J. Nitta, Universal spin-induced time reversal symmetry breaking in two-dimensional electron gases with Rashba spin-orbit interaction, Phys. Rev. Lett. 94 (18) (2005) 186805. Available from: https://doi.org/10.1103/PhysRevLett.94.186805. arXiv:0412731.

[18] J. Nitta, T. Akazaki, H. Takayanagi, T. Enoki, Gate control of spin-orbit interaction in an inverted InGaAs/InAlAs heterostructure, Phys. Rev. Lett. 78 (1997) 1335–1338.

[19] A.D. Caviglia, M. Gabay, S. Gariglio, N. Reyren, C. Cancellieri, J.-M. Triscone, Tunable Rashba spin-orbit interaction at oxide interfaces, Phys. Rev. Lett. 104 (12) (2010) 126803. Available from: https://doi.org/10.1103/PhysRevLett.104.126803. URL http://link.aps.org/doi/10.1103/PhysRevLett.104.126803.

[20] H. Nakamura, T. Koga, T. Kimura, Experimental evidence of cubic Rashba effect in an inversion-symmetric oxide, Phys. Rev. Lett. 108 (20) (2012) 206601. Available from: https://doi.org/10.1103/PhysRevLett.108.206601. URL http://link.aps.org/doi/10.1103/PhysRevLett.108.206601.

[21] L. Meier, G. Salis, I. Shorubalko, E. Gini, S. Schön, K. Ensslin, Measurement of Rashba and Dresselhaus spin-orbit magnetic fields, Nat. Phys. 3 (9) (2007) 650–654. Available from: https://doi.org/10.1038/nphys675. URL http://www.nature.com/doifinder/10.1038/nphys675.

[22] S.D. Stranks, H.J. Snaith, Metal-halide perovskites for photovoltaic and light-emitting devices, Nature Publishing Group 10 (5) (2015) 391–402. Available from: https://doi.org/10.1038/nnano.2015.90. URL https://doi.org/10.1038/nnano.2015.90.

[23] M. Kepenekian, R. Robles, C. Katan, D. Sapori, L. Pedesseau, J. Even, Rashba and Dresselhaus effects in hybrid organic–inorganic perovskites: from basics to devices, ACS Nano 9 (12) (2015) 11557–11567.

[24] J. Li, P.M. Haney, Optical spintronics in organic-inorganic perovskite photovoltaics, Phys. Rev. B 93 (2016) 155432. Available from: https://doi.org/10.1103/PhysRevB.93.155432.

[25] T. Etienne, E. Mosconi, F.D. Angelis, Dynamical origin of the Rashba effect in organohalide lead perovskites: a key to suppressed carrier recombination in perovskite solar cells? J. Phys. Chem. Lett. 7 (2016) 1638. Available from: https://doi.org/10.1021/acs.jpclett.6b00564.

[26] S. LaShell, B. McDougall, E. Jensen, Spin splitting of an Au(111) surface state band observed with angle resolved photoelectron spectroscopy, Phys. Rev. Lett. 77 (16) (1996) 3419–3422. Available from: https://doi.org/10.1103/PhysRevLett.77.3419. URL http://link.aps.org/doi/10.1103/PhysRevLett.77.3419.

[27] D. Popović, F. Reinert, S. Hüfner, V. Grigoryan, M. Springborg, H. Cercellier, et al., High-resolution photoemission on Ag/Au(111): spin-orbit splitting and electronic localization of the surface state, Phys. Rev. B 72 (4) (2005) 045419. Available from: https://doi.org/10.1103/PhysRevB.72.045419. URL http://link.aps.org/doi/10.1103/PhysRevB.72.045419.

[28] I. Gierz, B. Stadtmüller, J. Vuorinen, M. Lindroos, F. Meier, J.H. Dil, et al., Structural influence on the Rashba-type spin splitting in surface alloys, Phys. Rev. B 81 (24) (2010) 245430. Available from: https://doi.org/10.1103/PhysRevB.81.245430. URL http://link.aps.org/doi/10.1103/PhysRevB.81.245430.

[29] Y. Koroteev, G. Bihlmayer, J. Gayone, E. Chulkov, S. Blügel, P. Echenique, et al., Strong spin-orbit splitting on Bi surfaces, Phys. Rev. Lett. 93 (4) (2004) 046403. Available from: https://doi.org/10.1103/PhysRevLett.93.046403. URL http://link.aps.org/doi/10.1103/PhysRevLett.93.046403.

[30] C. Ast, J. Henk, A. Ernst, L. Moreschini, M. Falub, D. Pacilé, et al., Giant spin splitting through surface alloying, Phys. Rev. Lett. 98 (18) (2007) 186807. Available from: https://doi.org/10.1103/PhysRevLett.98.186807. URL http://link.aps.org/doi/10.1103/PhysRevLett.98.186807.

[31] O. Krupin, G. Bihlmayer, K. Starke, S. Gorovikov, J. Prieto, K. Döbrich, et al., Rashba effect at magnetic metal surfaces, Phys. Rev. B 71 (2005) 201403. Available from: https://doi.org/10.1103/PhysRevB.71.201403. URL http://link.aps.org/doi/10.1103/PhysRevB.71.201403.

[32] S. Mathias, A. Ruffing, F. Deicke, M. Wiesenmayer, I. Sakar, G. Bihlmayer, et al., Quantum-well-induced giant spin-orbit splitting, Phys. Rev. Lett. 104 (6) (2010) 066802. Available from: https://doi.org/10.1103/PhysRevLett.104.066802. URL http://link.aps.org/doi/10.1103/PhysRevLett. 104.066802.

[33] L. Moreschini, A. Bendounan, H. Bentmann, M. Assig, K. Kern, F. Reinert, et al., Influence of the substrate on the spin-orbit splitting in surface alloys on (111) noble-metal surfaces, Phys. Rev. B 80 (3) (2009) 035438. Available from: https://doi.org/10.1103/PhysRevB.80.035438. URL http://link.aps.org/doi/10.1103/PhysRevB.80.035438.

[34] G. Bihlmayer, Y. Koroteev, P. Echenique, E. Chulkov, S. Blügel, The Rashba-effect at metallic surfaces, Surface Sci. 600 (18) (2006) 3888–3891. Available from: https://doi.org/10.1016/j.susc.2006.01.098. URL http://linkinghub.elsevier.com/retrieve/pii/S0039602806004195.

[35] J.-H. Park, C.H. Kim, H.-W. Lee, J.H. Han, Orbital chirality and Rashba interaction in magnetic bands, Phys. Rev. B 87 (2013) 041301. Available from: https://doi.org/10.1103/PhysRevB.87.041301. URL http://link.aps.org/doi/10.1103/PhysRevB.87.041301.

[36] S. Grytsyuk, A. Belabbes, P.M. Haney, H.W. Lee, K.J. Lee, M.D. Stiles, et al., k-Asymmetric spin splitting at the interface between transition metal ferromagnets and heavy metals, Phys. Rev. B 93 (2016) 174421. Available from: https://doi.org/10.1103/PhysRevB.93.174421.

[37] V. Kashid, T. Schena, B. Zimmermann, Y. Mokrousov, S. Blügel, V. Shah, et al., Dzyaloshinskii-Moriya interaction and chiral magnetism in 3d-5d zigzag chains: tight-binding model and ab initio calculations, Phys. Rev. B 90 (5) (2014) 054412. Available from: https://doi.org/10.1103/PhysRevB.90.054412. URL http://link.aps.org/doi/10.1103/PhysRevB.90.054412.

[38] X. Xu, W. Yao, D. Xiao, T.F. Heinz, Spin and pseudospins in layered transition metal dichalcogenides, Nat. Phys. 10 (2014) 343. Available from: https://doi.org/10.1038/nphys2942. URL http://www.nature.com/doifinder/10.1038/nphys2942.

[39] D. Marchenko, A. Varykhalov, M.R. Scholz, G. Bihlmayer, E.I. Rashba, A. Rybkin, Giant Rashba splitting in graphene due to hybridization with gold., Nat. Commun. 3 (2012) 1232. Available from: https://doi.org/10.1038/ncomms2227. URL http://www.ncbi.nlm.nih.gov/pubmed/23187632.

[40] K. Lee, W.S. Yun, J.D. Lee, Giant Rashba-type splitting in molybdenum-driven bands of MoS2/Bi(111) heterostructure, Phys. Rev. B 91 (2015) 125420. Available from: https://doi.org/10.1103/PhysRevB.91.125420.

[41] Y.C. Cheng, Z.Y. Zhu, M. Tahir, U. Schwingenschlögl, Spin-orbit-induced spin splittings in polar transition metal dichalcogenide monolayers, Europhys. Lett. 102 (2013) 57001. Available from: https://doi.org/10.1209/0295-5075/102/57001. URL http://stacks.iop.org/0295-5075/102/i = 5/a = 57001.

[42] A.-y Lu, H. Zhu, J. Xiao, C.-p Chuu, Y. Han, M.-h Chiu, et al., Janus monolayers of transition metal dichalcogenides, Nat. Nanotechnol. 12 (2017) 744. Available from: https://doi.org/10.1038/nnano.2017.100 (May).

[43] M.Z. Hasan, C. Kane, Colloquium: topological insulators, Rev. Modern Phys. 82 (2010) 3045. Available from: https://doi.org/10.1103/RevModPhys.82.3045. URL http://link.aps.org/doi/10.1103/RevModPhys.82.3045.

[44] T. Wehling, A. Black-Schaffer, A. Balatsky, Dirac materials, Adv. Phys. 63 (1) (2014) 1–76. Available from: https://doi.org/10.1080/00018732.2014.927109. URL http://www.tandfonline.com/doi/abs/10.1080/00018732.2014.927109.

[45] Y. Ando, L. Fu, Topological crystalline insulators and topological superconductors: from concepts to materials, Annu. Rev. Condens. Matter Phys. 6 (2015) 361. Available from: https://doi.org/10.1146/annurev-conmatphys-031214-014501.

[46] F. Ortmann, S. Roche, S.O. Valenzuela, Topological Insulators, Wiley-VCH Verlag GmbH & Co, Weinheim, Germany, 2015.

[47] C.L. Kane, E.J. Mele, Quantum spin Hall effect in graphene, Phys. Rev. Lett. 95 (2005) 226801. Available from: https://doi.org/10.1103/PhysRevLett.95.226801. URL http://link.aps.org/doi/10.1103/PhysRevLett.95.226801.

[48] C.L. Kane, E.J. Mele, Z2 topological order and the quantum spin Hall effect, Phys. Rev. Lett. 95 (2005) 146802. Available from: https://doi.org/10.1103/PhysRevLett.95.146802. URL http://link.aps.org/doi/10.1103/PhysRevLett.95.146802.

[49] B.A. Bernevig, T.L. Hughes, S.-C. Zhang, Quantum spin Hall effect and topological phase transition in HgTe quantum wells, Science (New York, N.Y.) 314 (2006) 1757. Available from: https://doi.org/10.1126/science.1133734. URL http://www.ncbi.nlm.nih.gov/pubmed/17170299.

[50] M. König, S. Wiedmann, C. Brüne, A. Roth, H. Buhmann, L.W. Molenkamp, et al., Quantum spin Hall insulator state in HgTe quantum wells, Science (New York, N.Y.) 318 (5851) (2007) 766–770. Available from: https://doi.org/10.1126/science.1148047. URL http://www.ncbi.nlm.nih.gov/pubmed/17885096.

[51] L. Fu, C. Kane, E.J. Mele, Topological Insulators in three dimensions, Phys. Rev. Lett. 98 (10) (2007) 106803. Available from: https://doi.org/10.1103/PhysRevLett.98.106803. URL http://link.aps.org/doi/10.1103/PhysRevLett.98.106803.

[52] J.E. Moore, L. Balents, Topological invariants of time-reversal-invariant band structures, Phys. Rev. B 75 (2007) 121306. Available from: https://doi.org/10.1103/PhysRevB.75.121306. URL http://link.aps.org/doi/10.1103/PhysRevB.75.121306.

[53] S. Murakami, Phase transition between the quantum spin Hall and insulator phases in 3D: emergence of a topological gapless phase, New J. Phys. 9 (2007) 356. Available from: https://doi.org/10.1088/1367-2630/9/9/356. arXiv:0710.0930.

[54] X.-L. Qi, T.L. Hughes, S.-C. Zhang, Topological field theory of time-reversal invariant insulators, Phys. Rev. B 78 (2008) 195424. Available from: https://doi.org/10.1103/PhysRevB.78.195424. URL http://link.aps.org/doi/10.1103/PhysRevB.78.195424.

[55] H. Zhang, C.-X. Liu, X.-L. Qi, X. Dai, Z. Fang, S.-C. Zhang, Topological insulators in Bi2Se3, Bi2Te3 and Sb2Te3 with a single Dirac cone on the surface, Nat. Phys. 5 (2009) 438. Available from: https://doi.org/10.1038/nphys1270. URL http://www.nature.com/doifinder/10.1038/nphys1270.

[56] P. Gehring, H.M. Benia, Y. Weng, R. Dinnebier, C.R. Ast, M. Burghard, et al., A natural topological insulator, Nano Lett. 13 (3) (2013) 1179–1184. Available from: https://doi.org/10.1021/nl304583m. URL http://www.ncbi.nlm.nih.gov/pubmed/23438015.

[57] L. Fu, Topological crystalline insulators, Phys. Rev. Lett. 106 (10) (2011) 106802. Available from: https://doi.org/10.1103/PhysRevLett.106.106802. URL http://link.aps.org/doi/10.1103/PhysRevLett.106.106802.

[58] Q. Wang, F. Wang, J. Li, Z. Wang, X. Zhan, J. He, Low-dimensional topological crystalline insulators, Small 11 (2015) 4613. Available from: https://doi.org/10.1002/smll.201501381.

[59] T.H. Hsieh, H. Lin, J. Liu, W. Duan, A. Bansil, L. Fu, Topological crystalline insulators in the SnTe material class, Nat. Commun. 3 (2012) 982. Available from: https://doi.org/10.1038/ncomms1969. arXiv:1202.1003, URL http://www.ncbi.nlm.nih.gov/pubmed/22864575.

[60] Y. Tanaka, Z. Ren, T. Sato, K. Nakayama, S. Souma, T. Takahashi, et al., Experimental realization of a topological crystalline insulator in SnTe, Nat.Phys. 8 (2012) 800. Available from: https://doi.org/10.1038/nphys2442. arXiv:arXiv:1304.0430.

[61] S.-Y. Xu, C. Liu, N. Alidoust, M. Neupane, D. Qian, I. Belopolski, et al., Observation of a topological crystalline insulator phase and topological phase transition in Pb1-xSnxTe, Nat. Commun. 3 (2012) 1192. Available from: https://doi.org/10.1038/ncomms2191. arXiv:arXiv:1210.2917v1, URL http://www.nature.com/doifinder/10.1038/ncomms2191.

[62] P. Dziawa, B.J. Kowalski, K. Dybko, R. Buczko, A. Szczerbakow, M. Szot, et al., Topological crystalline insulator states in Pb(1-x)Sn(x)Se, Nat. Mater. 11 (12) (2012) 1023–1027. Available from: https://doi.org/10.1038/nmat3449. arXiv:1206.1705, URL https://doi.org/10.1038/nmat3449.

[63] Z. Wang, A. Alexandradinata, R.J. Cava, B.A. Bernevig, Hourglass fermions, Nature 532 (2016) 189. Available from: https://doi.org/10.1038/nature17410. URL https://doi.org/10.1038/nature17410.

[64] M. Kargarian, G.A. Fiete, Topological crystalline insulators in transition metal oxides, Phys. Rev. Lett. 110 (15) (2013) 156403. Available from: https://doi.org/10.1103/PhysRevLett.110.156403. arXiv: arXiv:1212.4162v2.

[65] Q. Zhang, Y. Cheng, U. Schwingenschlögl, Emergence of topological and topological crystalline phases in TlBiS2 and TlSbS2, Sci. Rep. 5 (2015) 8379. Available from: https://doi.org/10.1038/srep08379. URL http://www.nature.com/doifinder/10.1038/srep08379%5Cnhttp://www.nature.com/articles/srep08379.

[66] C.-K. Chiu, Y.H. Chan, X. Li, Y. Nohara, A.P. Schnyder, Type-II Dirac surface states in topological crystalline insulators, Phys. Rev. B 95 (2017) 035151. Available from: https://doi.org/10.1103/PhysRevB.95.035151. arXiv:1606.03456, URL http://arxiv.org/abs/1606.03456.

[67] N. Hao, F. Zheng, P. Zhang, S.-Q. Shen, Topological crystalline antiferromagnetic state in tetragonal FeS, Phys. Rev. B 96 (2017) 165102. Available from: https://doi.org/10.1103/PhysRevB.96.165102. arXiv:1702.01372, URL http://arxiv.org/abs/1702.01372%0Ahttps://doi.org/10.1103/PhysRevB.96.165102.

[68] A. Pham, S. Li, Unique topological surface states of full-Heusler topological crystalline insulators, Phys. Rev. B 95 (11) (2017) 115124. Available from: https://doi.org/10.1103/PhysRevB.95.115124. URL http://link.aps.org/doi/10.1103/PhysRevB.95.115124.

[69] Y.J. Wang, W.F. Tsai, H. Lin, S.Y. Xu, M. Neupane, M.Z. Hasan, et al., Nontrivial spin texture of the coaxial Dirac cones on the surface of topological crystalline insulator SnTe, Phys. Rev. B 87 (2013) 235317. Available from: https://doi.org/10.1103/PhysRevB.87.235317.

[70] J. Liu, W. Duan, L. Fu, Two types of surface states in topological crystalline insulators, Phys. Rev. B 88 (2013) 241303. Available from: https://doi.org/10.1103/PhysRevB.88.241303.

[71] I. Zeljkovic, Y. Okada, M. Serbyn, R. Sankar, D. Walkup, W. Zhou, et al., Dirac mass generation from crystal symmetry breaking on the surfaces of topological crystalline insulators, Nat. Mater. 14 (2015) 318. Available from: https://doi.org/10.1038/nmat4215. arXiv:1403.4906, URL https://doi.org/10.1038/nmat4215.

[72] P. Coleman, Heavy Fermions: electrons at the edge of magnetism, in: H. Kronmuller, S. Parkin (Eds.), Handbook of Magnetism and Advanced Magnetic Materials, John Wiley & Sons, Ltd, 2007, pp. 1–54. Ch. Fundamenta.

[73] M. Dzero, K. Sun, V. Galitski, P. Coleman, Topological Kondo insulators, Phys. Rev. Lett. 104 (2010) 106408. Available from: https://doi.org/10.1103/PhysRevLett.104.106408.

[74] M. Neupane, N. Alidoust, S.Y. Xu, T. Kondo, Y. Ishida, D.J. Kim, et al., Surface electronic structure of the topological Kondo-insulator candidate correlated electron system SmB6, Nat. Commun. 4 (2013) 2991. Available from: https://doi.org/10.1038/ncomms3991. arXiv:1312.1979.

[75] N. Xu, P.K. Biswas, J.H. Dil, R.S. Dhaka, G. Landolt, S. Muff, et al., Direct observation of the spin texture in SmB 6 as evidence of the topological Kondo insulator, Nat. Commun. 5 (2014) 4566. Available from: https://doi.org/10.1038/ncomms5566. arXiv:1407.8118.

[76] M. Dzero, J. Xia, V. Galitski, P. Coleman, Topological Kondo insulators, Annu. Rev. Con. Mat. Phys. 7 (2016) 249. Available from: https://doi.org/10.1146/annurev-conmatphys-031214-014749. arXiv:1506.05635, URL http://arxiv.org/abs/1406.3533%5Cnhttp://arxiv.org/abs/1506.05635.

[77] O. Erten, P.Y. Chang, P. Coleman, A.M. Tsvelik, Skyrme insulators: insulators at the brink of superconductivity, Phys. Rev. Lett. 119 (2017) 057603. Available from: https://doi.org/10.1103/PhysRevLett.119.057603.

[78] Z. Xiang, B. Lawson, T. Asaba, C. Tinsman, L. Chen, C. Shang, et al., Bulk rotational symmetry breaking in Kondo insulator SmB6, Phys. Rev. X 7 (3) (2017) 1–15. Available from: https://doi.org/10.1103/PhysRevX.7.031054. arXiv:1710.08945.

[79] N.J. Laurita, C.M. Morris, S.M. Koohpayeh, W.A. Phelan, T.M. McQueen, N.P. Armitage, Impurities or a neutral Fermi surface? A further examination of the low-energy ac optical conductivity of SmB6, Phys. B: Con. Mat. 536 (2018) 78. Available from: https://doi.org/10.1016/j.physb.2017.09.015. arXiv:1709.01508, URL https://doi.org/10.1016/j.physb.2017.09.015.

[80] X. Wan, A.M. Turner, A. Vishwanath, S.Y. Savrasov, Topological semimetal and Fermi-arc surface states in the electronic structure of pyrochlore iridates, Phys. Rev. B 83 (20) (2011) 205101. Available from: https://doi.org/10.1103/PhysRevB.83.205101. URL http://link.aps.org/doi/10.1103/PhysRevB.83.205101.

[81] S.M. Young, S. Zaheer, J.C.Y. Teo, C.L. Kane, E.J. Mele, A.M. Rappe, Dirac semimetal in three dimensions, Phys. Rev. Lett. 108 (2012) 140405. Available from: https://doi.org/10.1103/PhysRevLett.108.140405.

[82] N.P. Armitage, E.J. Mele, A. Vishwanath, Weyl and Dirac semimetals in three-dimensional solids, Rev. Mod. Phys. 90 (2018) 15001. Available from: https://doi.org/10.1103/RevModPhys.90.015001. arXiv:1705.01111, URL https://doi.org/10.1103/RevModPhys.90.015001.

[83] H. Weyl, Elektron und gravitation, I. Zeitschrift Physik 56 (1929) 330.

[84] G. Xu, H. Weng, Z. Wang, X. Dai, Z. Fang, Chern semimetal and the quantized anomalous Hall effect in HgCr 2Se4, Phys. Rev. Lett. 107 (2011) 186806. Available from: https://doi.org/10.1103/PhysRevLett.107.186806.

[85] A.A. Burkov, L. Balents, Weyl semimetal in a topological insulator multilayer, Phys. Rev. Lett. 107 (2011) 127205. Available from: https://doi.org/10.1103/PhysRevLett.107.127205.

[86] A.A. Zyuzin, S. Wu, A.A. Burkov, Weyl semimetal with broken time reversal and inversion symmetries, Phys. Rev. B Con. Mat. Mater. Phys. 85 (16) (2012) 165110. Available from: https://doi.org/10.1103/PhysRevB.85.165110.

[87] G.B. Halász, L. Balents, Time-reversal invariant realization of the Weyl semimetal phase, Phys. Rev. B 85 (2012) 035103. Available from: https://doi.org/10.1103/PhysRevB.85.035103.

[88] B. Singh, A. Sharma, H. Lin, M.Z. Hasan, R. Prasad, A. Bansil, Topological electronic structure and Weyl semimetal in the TlBiSe 2 class of semiconductors, Phys. Rev. B 86 (2012) 115208. Available from: https://doi.org/10.1103/PhysRevB.86.115208.

[89] S.-M. Huang, S.-Y. Xu, I. Belopolski, C.-C. Lee, G. Chang, B. Wang, et al., Fermion semimetal with surface Fermi arcs in the transition metal monopnictide TaAs class, Nat. Commun. 6 (2015) 7373. Available from: https://doi.org/10.1038/ncomms8373. arXiv:1501.00755, URL http://www.nature.com/ncomms/2015/150612/ncomms8373/full/ncomms8373.html.

[90] H. Weng, C. Fang, Z. Fang, B. Andrei Bernevig, X. Dai, Weyl semimetal phase in non-centrosymmetric transition-metal monophosphides, Phys. Rev. X 5 (2015) 011029. Available from: https://doi.org/10.1103/PhysRevX.5.011029.

[91] L. Balents, Weyl electrons kiss, Physics 4 (2011) 36. Available from: https://doi.org/10.1103/Physics.4.36. URL http://link.aps.org/doi/10.1103/Physics.4.36.

[92] S. Li, A.V. Andreev, Spiraling Fermi arcs in Weyl materials, Phys. Rev. B 92 (2015) 201107. Available from: https://doi.org/10.1103/PhysRevB.92.201107 (R).

[93] C. Fang, L. Lu, J. Liu, L. Fu, Topological semimetals with helicoid surface states, Nat. Phys. 12 (2016) 936. Available from: https://doi.org/10.1038/nphys3782. arXiv:1512.01552, URL http://arxiv.org/abs/1512.01552.

[94] B.Q. Lv, N. Xu, H.M. Weng, J.Z. Ma, P. Richard, X.C. Huang, et al., Observation of Weyl nodes in TaAs, Nat. Phys. 11 (2015) 724. Available from: https://doi.org/10.1038/NPHYS3426. arXiv:1503.09188, URL http://arxiv.org/abs/1503.09188.

[95] B.Q. Lv, H.M. Weng, B.B. Fu, X.P. Wang, H. Miao, J. Ma, et al., Experimental discovery of Weyl semimetal TaAs, Phys. Rev. X 5 (3) (2015) 031013. Available from: https://doi.org/10.1103/PhysRevX.5.031013. arXiv:1502.04684.

[96] S.-Y. Xu, I. Belopolski, N. Alidoust, M. Neupane, G. Bian, C. Zhang, et al., Discovery of a Weyl Fermion semimetal and topological Fermi arcs, Science 349 (6248) (2015) 613. Available from: https://doi.org/10.1126/science.aaa9297. arXiv:1502.03807, URL http://www.sciencemag.org/content/early/2015/07/15/science.aaa9297.

[97] L.X. Yang, Z.K. Liu, Y. Sun, H. Peng, H.F. Yang, T. Zhang, et al., Weyl semimetal phase in the non-centrosymmetric compound TaAs, Nat. Phys. 11 (2015) 728. Available from: https://doi.org/10.1038/nphys3425. URL http://www.nature.com/doifinder/10.1038/nphys3425.

[98] S.-Y. Xu, N. Alidoust, I. Belopolski, Z. Yuan, G. Bian, T.-R. Chang, et al., Discovery of a Weyl fermion state with Fermi arcs in niobium arsenide, Nat. Phys. 11 (9) (2015) 748−754. Available from: https://doi.org/10.1038/nphys3437. arXiv:1504.01350, URL http://arxiv.org/abs/1504.01350%5Cnhttp://www.nature.com/doifinder/10.1038/nphys3437.

[99] S.-Y. Xu, I. Belopolski, D.S. Sanchez, C. Zhang, G. Chang, C. Guo, et al., Experimental discovery of a topological Weyl semimetal state in TaP, Sci. Adv. 1 (10) (2015). Available from: https://doi.org/10.1126/sciadv.1501092. e1501092−e1501092, arXiv:1508.03102 URL http://arxiv.org/abs/1508.03102%5Cnhttp://advances.sciencemag.org/cgi/doi/10.1126/sciadv.1501092.

[100] F. Wilczek, Why are there analogies between condensed matter and particle theory? Phys. Today 51 (1998) 11.

[101] H.B. Nielsen, M. Ninomiya, The Adler-Bell-Jackiw anomaly and Weyl fermions in a crystal, Phys. Lett. B 130 (1983) 389−396. Available from: https://doi.org/10.1016/0370-2693(83)91529-0. arXiv: arXiv:1011.1669v3.

[102] A.A. Soluyanov, D. Gresch, Z. Wang, Q. Wu, M. Troyer, X. Dai, et al., Type-II Weyl semimetals, Nature 527 (7579) (2015) 495−498. Available from: https://doi.org/10.1038/nature15768. arXiv:1507.01603, URL https://doi.org/10.1038/nature15768.

[103] Z. Wang, D. Gresch, A.A. Soluyanov, et al., MoTe2: a Type-II Weyl topological metal, Phys. Rev. Lett. 117 (2016) 056805. Available from: https://doi.org/10.1103/PhysRevLett.117.056805.

[104] Y. Qi, P.G. Naumov, M.N. Ali, C.R. Rajamathi, W. Schnelle, O. Barkalov, et al., Superconductivity in Weyl semimetal candidate MoTe2, Nat. Commun. 7 (2016) 11038. Available from: https://doi.org/10.1038/ncomms11038. URL http://www.nature.com/doifinder/10.1038/ncomms11038.
[105] G. Autes, D. Gresch, M. Troyer, A.A. Soluyanov, O.V. Yazyev, Robust type-II Weyl semimetal phase in transition metal diphosphides X P2 (X = Mo, W), Phys. Rev. Lett. 117 (6) (2016) 066402. Available from: https://doi.org/10.1103/PhysRevLett.117.066402. arXiv:1603.04624.
[106] J. Sanchez-Barriga, M.G. Vergniory, D. Evtushinsky, I. Aguilera, A. Varykhalov, S. Blugel, et al., Surface Fermi arc connectivity in the type-II Weyl semimetal candidate WTe2, Phys. Rev. B – Con. Mat. Mater. Phys. 94 (16) (2016) 161401. Available from: https://doi.org/10.1103/PhysRevB.94.161401. arXiv:1608.05633.
[107] F.Y. Bruno, A. Tamai, Q.S. Wu, I. Cucchi, C. Barreteau, A. De La Torre, et al., Observation of large topologically trivial Fermi arcs in the candidate type-II Weyl semimetal WTe2, Phys. Rev. B 94 (2016) 121112. Available from: https://doi.org/10.1103/PhysRevB.94.121112.
[108] B. Feng, Y.H. Chan, Y. Feng, R.Y. Liu, M.Y. Chou, K. Kuroda, et al., Spin texture in type-II Weyl semimetal WTe2, Phys. Rev. B – Con. Mat.Mater. Phys. 94 (19) (2016) 195134. Available from: https://doi.org/10.1103/PhysRevB.94.195134. arXiv:1606.00085.
[109] A. Tamai, Q.S. Wu, I. Cucchi, F.Y. Bruno, S. Ricco, T.K. Kim, et al., Fermi arcs and their topological character in the candidate type-II Weyl semimetal $MoTe_2$, Phys. Rev. X 6 (2016) 031021. Available from: https://doi.org/10.1103/PhysRevX.6.031021.
[110] L. Huang, T.M. McCormick, M. Ochi, Z. Zhao, M.-T. Suzuki, R. Arita, et al., Spectroscopic evidence for a type II Weyl semimetallic state in MoTe2, Nat. Mater. 15 (July) (2016) 1155. Available from: https://doi.org/10.1038/nmat4685. arXiv:1603.06482, URL http://arxiv.org/abs/1603.06482%5Cnhttp://www.nature.com/doifinder/10.1038/nmat4685.
[111] J. Jiang, Z. Liu, Y. Sun, H. Yang, C. Rajamathi, Y. Qi, et al., Signature of type-II Weyl semimetal phase in MoTe2, Nat. Commun. 8 (2017) 13973. Available from: https://doi.org/10.1038/ncomms13973. URL http://www.nature.com/doifinder/10.1038/ncomms13973.
[112] K. Deng, G. Wan, P. Deng, K. Zhang, S. Ding, E. Wang, et al., Experimental observation of topological Fermi arcs in type-II Weyl semimetal MoTe2, Nat. Phys. 12 (2016) 1105. Available from: https://doi.org/10.1038/nphys3871. arXiv:1603.08508, URL http://arxiv.org/abs/1603.08508.
[113] I. Belopolski, S.Y. Xu, Y. Ishida, X. Pan, P. Yu, D.S. Sanchez, et al., Fermi arc electronic structure and Chern numbers in the type-II Weyl semimetal candidate Mox W1-xTe2, Phys. Rev. B 94 (2016) 085127. Available from: https://doi.org/10.1103/PhysRevB.94.085127.
[114] M. Kargarian, M. Randeria, Y.-M. Lu, Are the double Fermi arcs of Dirac semimetals topologically protected? Proc.Natl Acad. Sci. 113 (31) (2015) 8648. Available from: https://doi.org/10.1073/pnas.1524787113. arXiv:1509.02180, URL http://arxiv.org/abs/1509.02180.
[115] Z.K. Liu, B. Zhou, Y. Zhang, Z.J. Wang, H.M. Weng, D. Prabhakaran, et al., Discovery of a three-dimensional topological Dirac semimetal, Na3Bi, Science 343 (2014) 864–867.
[116] M. Neupane, S.-y Xu, R. Sankar, N. Alidoust, G. Bian, C. Liu, et al., Observation of a three-dimensional topological Dirac semimetal phase in high-mobility Cd3As2, Nat. Commun. 5 (2014) 3786. Available from: https://doi.org/10.1038/ncomms4786.
[117] P. Tang, Q. Zhou, G. Xu, S.-C. Zhang, Dirac Fermions in antiferromagnetic semimetal, Nat. Phys. 12 (2016) 1100. Available from: https://doi.org/10.1038/nphys3839. URL http://arxiv.org/abs/1603.08060.
[118] F. Schindler, A.M. Cook, M.G. Vergniory, Z. Wang, S.S.P. Parkin, B.A. Bernevig, et al., Higher-order topological insulators, Sci. Adv. 4 (6) (2018) eaat0346. Available from: https://doi.org/10.1126/sciadv.aat0346. URL http://advances.sciencemag.org/lookup/doi/10.1126/sciadv.aat0346.
[119] M. Ezawa, Higher-order topological insulators and semimetals on the breathing kagome and pyrochlore lattices, Phys. Rev. Lett. 120 (2018) 26801. Available from: https://doi.org/10.1103/PhysRevLett.120.026801. arXiv:1709.08425.
[120] V. Mourik, K. Zuo, S.M. Frolov, S.R. Plissard, E.P.A.M. Bakkers, L.P. Kouwenhoven, Signatures of Majorana fermions in hybrid superconductor-semiconductor nanowire devices, Science (New York, N.

Y.) 336 (2012) 1003. Available from: https://doi.org/10.1126/science.1222360. URL http://www.ncbi.nlm.nih.gov/pubmed/22499805.

[121] A. Das, Y. Ronen, Y. Most, Y. Oreg, M. Heiblum, H. Shtrikman, Zero-bias peaks and splitting in an Al-InAs nanowire topological superconductor as a signature of Majorana fermions, Nat. Phys. 8 (12) (2012) 887–895. Available from: https://doi.org/10.1038/nphys2479. URL https://doi.org/10.1038/nphys2479.

[122] C.H.L. Quay, T.L. Hughes, J.A. Sulpizio, L.N. Pfeiffer, K.W. Baldwin, K.W. West, Observation of a one-dimensional spin-orbit gap in a quantum wire, Nat. Phys. 6 (5) (2010) 336–339. Available from: https://doi.org/10.1038/nphys1626. URL https://doi.org/10.1038/nphys1626.

[123] S. Heedt, N.T. Ziani, F. Crépin, W. Prost, S. Trellenkamp, J. Schubert, et al., Signatures of interaction-induced helical gaps in nanowire quantum point contacts, Nat. Phys. 13 (June) (2017) 563–568. Available from: https://doi.org/10.1038/NPHYS4070.

[124] E. Rashba, Properties of semiconductors with an extremum loop. 1. Cyclotron and combinational resonance in a magnetic field perpendicular to the plane of the loop, Sov. Phys. Solid State 2 (1960) 1109–1122.

[125] E.I. Rashba, V.I. Sheka, Electric-dipole spin resonances, in: G. Landwehr, E. Rashba (Eds.), Landau Level Spectroscopy, Elsevier Science Publishers B.V, 1991, pp. 133–206. Ch. 4 URL https://doi.org/10.1016/B978-0-444-88535-7.50011-X.

[126] M. Duckheim, D. Loss, Electric-dipole-induced spin resonance in disordered semiconductors, Nat. Phys. 2 (3) (2006) 195–199. Available from: https://doi.org/10.1038/nphys238. arXiv:0605735, URL http://www.nature.com/doifinder/10.1038/nphys238.

[127] M. Dobrowolska, Y. Chen, J.K. Furdyna, S. Rodriguez, Effects of photon-momentum and magnetic-field reversal on the far-infrared electric-dipole spin resonance in InSb, Phys. 51 (2) (1983) 134.

[128] R.L. Bell, Electric dipole spin translations in InSb, Phys. Rev. Lett. 9 (2) (1962) 52–54. Available from: https://doi.org/10.1103/PhysRevLett.9.52.

[129] M. Schulte, J.G.S. Lok, G. Denninger, W. Dietsche, Electron spin resonance on a two-dimensional electron gas in a single AlAs quantum well, Phys. Rev. Lett. 94 (13) (2005) 137601. Available from: https://doi.org/10.1103/PhysRevLett.94.137601.

[130] Z. Wilamowski, H. Malissa, F. Schäffler, W. Jantsch, G-actor tuning and manipulation of spins by an electric current, Phys. Rev. Lett. 98 (18) (2007) 187203. Available from: https://doi.org/10.1103/PhysRevLett.98.187203. arXiv:0610046.

[131] R. Hanson, D.D. Awschalom, Coherent manipulation of single spins in semiconductors, Nature 453 (7198) (2008) 1043–1049. Available from: https://doi.org/10.1038/nature07129.

[132] S.M. Frolov, S. Lüscher, W. Yu, Y. Ren, J.A. Folk, W. Wegscheider, Ballistic spin resonance, Nature 458 (2009) 868–871. Available from: https://doi.org/10.1038/nature07873. URL http://www.nature.com/doifinder/10.1038/nature07873.

[133] Y. Kato, R.C. Myers, D. Driscoll, A.C. Gossard, J. Levy, D.D. Awschalom, Gigahertz electron spin manipulation using voltage-controlled g-tensor modulation, Science 299 (2003) 1201.

[134] K. Nowack, F.H.L. Koppens, Y.V. Nazarov, L.M.K. Vandersypen, Coherent control of a single electron spin with electric fields, Science 318 (2007) 1430–1434.

[135] J. Luo, H. Munekata, F.F. Fang, P.J. Stiles, Effects of inversion asymmetry on electron-energy band structures in Gasb/InAs/GaSb quantum-wells, Phys. Rev. B 41 (11) (1990) 7685. Available from: https://doi.org/10.1103/PhysRevB.41.7685.

[136] Ya Bychkov, E.I. Rashba, Oscillatory effects and the magnetic susceptibility of carriers in inversion layers, J. Phys. C: Solid State Phys. 17 (1984) 6039. Available from: https://doi.org/10.1088/0022-3719/17/33/015.

[137] C.R. Whitsett, Oscillatory magnetoresistance in mercuric selenide, Phys. Rev. 138 (1965) A829. Available from: https://doi.org/10.1103/PhysRev.138.A829.

[138] D. Seiler, W. Becker, L. Roth, Inversion-asymmetry splitting of the conduction band in GaSb from Shubnikov-de Haas measurements, Phys. Rev. B 1 (2) (1970) 764.

[139] J.P. Eisenstein, H.L. Stormer, V. Narayanamurti, A.C. Gossard, W. Wiegmann, Effect of inversion symmetry on the band structure of semiconductor heterostructures, Phys. Rev. Lett. 53 (27) (1984) 2579–2582. Available from: https://doi.org/10.1103/PhysRevLett.53.2579.

[140] T. Ando, A.B. Fowler, F. Stern, Electronic properties of two-dimensional systems, Rev. Mod. Phys. 54 (2) (1982) 437–672. Available from: https://doi.org/10.1103/RevModPhys.54.437.

[141] H.C. Koo, J.H. Kwon, J. Eom, J. Chang, S.H. Han, M. Johnson, Control of spin precession in a spin-injected field effect transistor, Science (New York, N.Y.) 325 (5947) (2009) 1515–1518. Available from: https://doi.org/10.1126/science.1173667. URL http://www.ncbi.nlm.nih.gov/pubmed/19762637.

[142] G. Bergmann, Weak Localization in thin films a time-of-flight experiment with conduction electrons, Phys. Rep. 107 (1984) 1–58.

[143] L.E. Golub, I.V. Gornyi, V.Y. Kachorovskii, Weak antilocalization in two-dimensional systems with large Rashba splitting, Phys. Rev. B 93 (24) (2016). Available from: https://doi.org/10.1103/PhysRevB.93.245306. arXiv:1512.05800.

[144] W. Knap, C. Skierbiszewski, A. Zduniak, D. Bertho, F. Kobbi, J.L. Robert, et al., Weak antilocalization and spin precession in quantum wells, Phys. Rev. B 53 (7) (1996) 3912–3924.

[145] J. Chen, H.J. Qin, F. Yang, J. Liu, T. Guan, F.M. Qu, et al., Gate-voltage control of chemical potential and weak antilocalization in Bi2Se3, Phys. Rev. Lett. 105 (2010) 176602. Available from: https://doi.org/10.1103/PhysRevLett.105.176602 (October).

[146] E.L. Ivchenko, G.E. Pikus, New photogalvanic effect in gyrotropic crystals, Pis'ma Zh. Eksp. Teor. Fiz 27 (1978) 604.

[147] V.M. Edelstein, Spin polarization of conduction electrons induced by electric current in two-dimensional asymmetric electron systems, Sol. Stat. Commun. 73 (3) (1990) 233–235.

[148] L.E. Vorobev, E.L. Ivchenko, G.E. Pikus, I.I. Farbshtein, V.A. Shalygin, A.V. Shturbin, Optical activity in tellurium induced by a current, JETP Lett. 29 (1979) 485.

[149] A.Y. Silov, P.A. Blajnov, J.H. Wolter, R. Hey, K.H. Ploog, N.S. Averkiev, Current-induced spin polarization at a single heterojunction, Appl. Phys. Lett. 85 (24) (2004) 5929. Available from: https://doi.org/10.1063/1.1833565. URL http://scitation.aip.org/content/aip/journal/apl/85/24/10.1063/1.1833565.

[150] Y.K. Kato, R. Myers, A. Gossard, D.D. Awschalom, Current-induced spin polarization in strained semiconductors, Phys. Rev. Lett. 93 (2004) 176601. Available from: https://doi.org/10.1103/PhysRevLett.93.176601. URL http://link.aps.org/doi/10.1103/PhysRevLett.93.176601.

[151] C.H. Li, O.M.J. van't Erve, J.T. Robinson, Y. Liu, L. Li, B.T. Jonker, Electrical detection of charge-current-induced spin polarization due to spin-momentum locking in Bi2Se3, Nat. Nanotechnol. 9 (2014) 218. Available from: https://doi.org/10.1038/nnano.2014.16. URL http://www.ncbi.nlm.nih.gov/pubmed/24561354.

[152] E.L. Ivchenko, Y.B. Lyanda-Geller, G.E. Pikus, Photocurrent in structures with quantum wells with an optical orientation of free carriers, JETP Lett. 50 (3) (1989) 175–177.

[153] S.D. Ganichev, E.L. Ivchenko, V.V. Bel'kov, S.A. Tarasenko, M. Sollinger, D. Weiss, et al., Spin-galvanic effect, Nature 417 (2002) 153–156. Available from: https://doi.org/10.1038/nature747.1.

[154] J.C. Rojas-Sánchez, L. Vila, G. Desfonds, S. Gambarelli, J.-P. Attane, J.M. De Teresa, et al., Spin-to-charge conversion using Rashba coupling at the interface between non-magnetic materials, Nat. Commun. 4 (2013) 2944. Available from: https://doi.org/10.1038/ncomms3944. URL http://www.ncbi.nlm.nih.gov/pubmed/24343336.

[155] J.C. Rojas-Sánchez, S. Oyarzún, Y. Fu, A. Marty, C. Vergnaud, S. Gambarelli, et al., Spin to charge conversion at room temperature by spin pumping into a new type of topological insulator: α-Sn films, Phys. Rev. Lett. 116 (2016) 096602. Available from: https://doi.org/10.1103/PhysRevLett.116.096602.

[156] J. Sinova, S.O. Valenzuela, J. Wunderlich, C.H. Back, T. Jungwirth, Spin Hall effect, Rev. Mod. Phys. 87 (2015) 1213.

[157] J. Sinova, D. Culcer, Q. Niu, N.A. Sinitsyn, T. Jungwirth, A.H. MacDonald, Universal intrinsic spin Hall effect, Phys. Rev. Lett. 92 (2004) 126603. Available from: https://doi.org/10.1103/PhysRevLett.92.126603. URL http://link.aps.org/doi/10.1103/PhysRevLett.92.126603.

[158] J.-i Inoue, G.E.W. Bauer, L. Molenkamp, Suppression of the persistent spin Hall current by defect scattering, Phys. Rev. B 70 (2004) 041303. Available from: https://doi.org/10.1103/PhysRevB.70.041303. URL http://link.aps.org/doi/10.1103/PhysRevB.70.041303.

[159] I. Adagideli, G.E.W. Bauer, Intrinsic spin Hall edges, Phys. Rev. Lett. 95 (25) (2005) 256602. Available from: https://doi.org/10.1103/PhysRevLett.95.256602. URL http://link.aps.org/doi/10.1103/PhysRevLett.95.256602.

[160] X. Wang, A. Manchon, Diffusive spin dynamics in ferromagnetic thin films with a Rashba interaction, Phys. Rev. Lett. 108 (2012) 117201. Available from: https://doi.org/10.1103/PhysRevLett.108.117201. URL http://link.aps.org/doi/10.1103/PhysRevLett.108.117201.

[161] M. Dyakonov, V. Perel, Spin relaxation of conduction electrons in noncentrosymmetric semiconductors, Soviet Phys. Solid State 13 (12) (1972) 3023.

[162] A.A. Burkov, D.G. Hawthorn, Spin and charge transport on the surface of a topological insulator, Phys. Rev. Lett. 105 (2010) 066802. Available from: https://doi.org/10.1103/PhysRevLett.105.066802. URL http://link.aps.org/doi/10.1103/PhysRevLett.105.066802.

[163] A. Johansson, J. Henk, I. Mertig, Edelstein effect in Weyl semimetals, Phys. Rev. B 97 (2018) 085417. Available from: https://doi.org/10.1103/PhysRevB.97.085417.

[164] D.T. Son, B.Z. Spivak, Chiral anomaly and classical negative magnetoresistance of Weyl metals, Phys. Rev. B Con. Mat. Mater. Phys. 88 (2013) 104412. Available from: https://doi.org/10.1103/PhysRevB.88.104412. arXiv:1206.1627.

[165] A.A. Burkov, Chiral anomaly and diffusive magnetotransport in Weyl metals, Phys. Rev. Lett. 113 (24) (2014) 247203. Available from: https://doi.org/10.1103/PhysRevLett.113.247203. arXiv:1409.0013v2.

[166] A.A. Burkov, Negative longitudinal magnetoresistance in Dirac and Weyl metals, Phys. Rev. B 91 (2015) 245157. Available from: https://doi.org/10.1103/PhysRevB.91.245157.

[167] X. Huang, L. Zhao, Y. Long, P. Wang, D. Chen, Z. Yang, et al., Observation of the chiral-anomaly-induced negative magnetoresistance: In 3D Weyl semimetal TaAs, Phys. Rev. X 5 (3) (2015) 1–9. Available from: https://doi.org/10.1103/PhysRevX.5.031023. arXiv:1503.01304.

[168] C.-L. Zhang, S.-Y. Xu, I. Belopolski, Z. Yuan, Z. Lin, B. Tong, et al., Signatures of the Adler–Bell–Jackiw chiral anomaly in a Weyl fermion semimetal, Nat. Commun. 7 (2016) 10735. Available from: https://doi.org/10.1038/ncomms10735. URL http://www.nature.com/doifinder/10.1038/ncomms10735.

[169] C.-z Li, L.-X. Wang, H. Liu, J. Wang, Z.-M. Liao, D.-P. Yu, Giant negative magnetoresistance induced by the chiral anomaly in individual Cd3As2 nanowires, Nat. Commun. 6 (2015) 10137. Available from: https://doi.org/10.1038/ncomms10137. URL https://doi.org/10.1038/ncomms10137%5Cnhttp://www.nature.com/ncomms/2015/151217/ncomms10137/full/nc.

[170] J. Xiong, S.K. Kushwaha, T. Liang, J.W. Krizan, M. Hirschberger, W. Wang, et al., Evidence for the chiral anomaly in the Dirac semimetal Na3Bi, Sci. Exp. 350 (6259) (2015) 1–8. Available from: https://doi.org/10.1126/science.aac6089. arXiv:arXiv:1503.08179v1 URL http://www.sciencemag.org/content/350/6259/413.long.

[171] T. Liang, Q. Gibson, M.N. Ali, M. Liu, R.J. Cava, N.P. Ong, Ultrahigh mobility and giant magnetoresistance in the Dirac semimetal Cd3As2, Nat. Mater. 14 (3) (2014) 280–284. Available from: https://doi.org/10.1038/nmat4143. arXiv:1404.7794 URL http://www.nature.com/doifinder/10.1038/nmat4143.

[172] J. Schliemann, J.C. Egues, D. Loss, Nonballistic spin-field-effect transistor, Phys. Rev. Lett. 90 (2003) 146801.

[173] B.A. Bernevig, J. Orenstein, S.-C. Zhang, Exact SU(2) symmetry and persistent spin helix in a spin-orbit coupled system, Phys. Rev. Lett. 97 (2006) 236601. Available from: https://doi.org/10.1103/PhysRevLett.97.236601. URL http://link.aps.org/doi/10.1103/PhysRevLett.97.236601.

[174] J. Schliemann, Colloquium: persistent spin textures in semiconductor nanostructures, Rev. Mod. Phys. 89 (March) (2017) 011001. Available from: https://doi.org/10.1103/RevModPhys.89.011001.

[175] J.D. Koralek, C.P. Weber, J. Orenstein, B.A. Bernevig, S.-C. Zhang, S. Mack, et al., Emergence of the persistent spin helix in semiconductor quantum wells, Nature 458 (7238) (2009) 610–613. Available from: https://doi.org/10.1038/nature07871URL. Available from: http://www.ncbi.nlm.nih.gov/pubmed/19340077.

[176] A. Sasaki, S. Nonaka, Y. Kunihashi, M. Kohda, T. Bauernfeind, T. Dollinger, et al., Direct determination of spin-orbit interaction coefficients and realization of the persistent spin helix symmetry, Nat. Nanotechnol. 9 (2014) 703–709. Available from: https://doi.org/10.1038/nnano.2014.128. URL https://doi.org/10.1038/nnano.2014.128.

[177] M. Kohda, V. Lechner, Y. Kunihashi, T. Dollinger, P. Olbrich, C. Sch, et al., Gate-controlled persistent spin helix state in (In, Ga) As quantum wells, Phys. Rev. B 86 (2012) 081306. Available from: https://doi.org/10.1103/PhysRevB.86.081306.

[178] P.W. Brouwer, Scattering approach to parametric pumping, Phys. Rev. B 58 (1998) R10135.

[179] Y. Tserkovnyak, A. Brataas, G.E.W. Bauer, Enhanced Gilbert damping in thin ferromagnetic films, Phys. Rev. Lett. 88 (2002) 117601. Available from: https://doi.org/10.1103/PhysRevLett.88.117601. URL http://link.aps.org/doi/10.1103/PhysRevLett.88.117601.

[180] A. Brataas, Y. Tserkovnyak, G.E.W. Bauer, P.J. Kelly, Spin pumping and spin transfer, in: S. Maekawa, S. Valenzuela, E. Saitoh, Y. Kimura (Eds.), Spin Current, Oxford University Press, New York, 2012.

[181] A.G. Mal'Shukov, C.S. Tang, C.S. Chu, K.A. Chao, Spin-current generation and detection in the presence of an ac gate, Phys. Rev. B 68 (2003) 233307. Available from: https://doi.org/10.1103/PhysRevB.68.233307.

[182] P. Sharma, P.W. Brouwer, Mesoscopic effects in adiabatic spin pumping, Phys. Rev. Lett. 91 (October) (2003) 166801. Available from: https://doi.org/10.1103/PhysRevLett.91.166801.

[183] M. Governale, F. Taddei, R. Fazio, Pumping spin with electrical fields, Phys. Rev. B 68 (2003) 155324. Available from: https://doi.org/10.1103/PhysRevB.68.155324.

[184] K. Hals, A. Brataas, Y. Tserkovnyak, Scattering theory of charge-current–induced magnetization dynamics, Europhys. Lett. 90 (2010) 47002. Available from: https://doi.org/10.1209/0295-5075/90/47002. URL http://stacks.iop.org/0295-5075/90/i = 4/a = 47002? key = crossref.6f16244b8eddddb354d517e92b5e0597.

[185] C. Ciccarelli, K.M.D. Hals, A. Irvine, V. Novak, Y. Tserkovnyak, H. Kurebayashi, et al., Magnonic charge pumping via spin–orbit coupling, Nat. Nanotechnol. 10 (2014) 50–54. Available from: https://doi.org/10.1038/nnano.2014.252. URL https://doi.org/10.1038/nnano.2014.252.

[186] H.T. Ueda, A. Takeuchi, G. Tatara, T. Yokoyama, Topological charge pumping effect by the magnetization dynamics on the surface of three-dimensional topological insulators, Phys. Rev. B 85 (2012) 115110. Available from: https://doi.org/10.1103/PhysRevB.85.115110.

[187] F. Mahfouzi, N. Nagaosa, B.K. Nikolić, Spin-to-charge conversion in lateral and vertical topological-insulator/ferromagnet heterostructures with microwave-driven precessing magnetization, Phys. Rev. B 90 (2014) 115432. Available from: https://doi.org/10.1103/PhysRevB.90.115432.

[188] P.B. Ndiaye, C.A. Akosa, M.H. Fischer, A. Vaezi, E.A. Kim, A. Manchon, Dirac spin-orbit torques and charge pumping at the surface of topological insulators, Phys. Rev. B 96 (2017) 014408. Available from: https://doi.org/10.1103/PhysRevB.96.014408.

[189] T. Yokoyama, J. Zang, N. Nagaosa, Theoretical study of the dynamics of magnetization on the topological surface, Phys. Rev. B 81 (2010) 241410. Available from: https://doi.org/10.1103/PhysRevB.81.241410. URL http://link.aps.org/doi/10.1103/PhysRevB.81.241410.

[190] O.A. Barut, N. Zanghi, Classical model of the Dirac electron, Phys. Rev. Lett. 52 (1984) 2009–2012.

[191] J. Schliemann, D. Loss, R.M. Westervelt, Zitterbewegung of electronic wave packets in III-V zinc-blende semiconductor quantum wells, Phys. Rev. Lett. 206801 (May) (2005) 1–4. Available from: https://doi.org/10.1103/PhysRevLett.94.206801.

[192] J. Cserti, G. David, Unified description of Zitterbewegung for spintronic, graphene, and superconducting systems, Phys. Rev. B 74 (2006) 172305. Available from: https://doi.org/10.1103/PhysRevB.74.172305.

[193] T.M. Rusin, W. Zawadzki, Zitterbewegung of electrons in graphene in a magnetic field, Phys. Rev. B 78 (2008) 125419. Available from: https://doi.org/10.1103/PhysRevB.78.125419.

[194] S. Ghosh, U. Schwingenschl, A. Manchon, Resonant longitudinal Zitterbewegung in zigzag graphene nanoribbons, Phys. Rev. B 91 (2015) 045409. Available from: https://doi.org/10.1103/PhysRevB.91.045409.

[195] C.X. Liu, X.L. Qi, X. Dai, Z. Fang, S.C. Zhang, Quantum anomalous hall effect in Hg1-yMnyTe quantum wells, Phys. Rev. Lett. 101 (2008) 146802. Available from: https://doi.org/10.1103/PhysRevLett.101.146802.

[196] R. Yu, W. Zhang, H.-J. Zhang, S.-C. Zhang, X. Dai, Z. Fang, Quantized anomalous Hall effect in magnetic topological insulators, Science 329 (2010) 61–64.

[197] C.-Z. Chang, J. Zhang, X. Feng, J. Shen, Z. Zhang, M. Guo, et al., Experimental observation of the quantum anomalous Hall effect in a magnetic topological insulator, Science 340 (2013) 167. Available from: https://doi.org/10.1126/science.1234414. URL http://www.ncbi.nlm.nih.gov/pubmed/23493424.

[198] K. Nomura, N. Nagaosa, Surface-quantized anomalous Hall current and the magnetoelectric effect in magnetically disordered topological insulators, Phys. Rev. Lett. 106 (2011) 166802. Available from: https://doi.org/10.1103/PhysRevLett.106.166802. URL http://link.aps.org/doi/10.1103/PhysRevLett.106.166802.

[199] I. Garate, M. Franz, Inverse spin-galvanic effect in the interface between a topological insulator and a ferromagnet, Phys. Rev. Lett. 104 (2010) 146802. Available from: https://doi.org/10.1103/PhysRevLett.104.146802. URL http://link.aps.org/doi/10.1103/PhysRevLett.104.146802.

[200] K. Nomura, N. Nagaosa, Electric charging of magnetic textures on the surface of a topological insulator, Phys. Rev. B 82 (2010) 161401. Available from: https://doi.org/10.1103/PhysRevB.82.161401. URL http://link.aps.org/doi/10.1103/PhysRevB.82.161401.

[201] Y. Tserkovnyak, D. Loss, Thin-film magnetization dynamics on the surface of a topological insulator, Phys. Rev. Lett. 108 (2012) 187201. Available from: https://doi.org/10.1103/PhysRevLett.108.187201. URL http://link.aps.org/doi/10.1103/PhysRevLett.108.187201.

[202] X.-L. Qi, R. Li, J. Zang, S.-C. Zhang, Inducing a magnetic monopole with topological surface states, Science 323 (2009) 1184. Available from: https://doi.org/10.1126/science.1167747.

[203] K. Taguchi, K. Shintani, Y. Tanaka, Spin-charge transport driven by magnetization dynamics on the disordered surface of doped topological insulators, Phys. Rev. B 92 (2015) 035425. Available from: https://doi.org/10.1103/PhysRevB.92.035425.

[204] A. Sakai, H. Kohno, Spin torques and charge transport on the surface of topological insulator, Phys. Rev. B 89 (2014) 165307. Available from: https://doi.org/10.1103/PhysRevB.89.165307URL. Available from: http://link.aps.org/doi/10.1103/PhysRevB.89.165307.

[205] J. Fujimoto, H. Kohno, Transport properties of Dirac ferromagnet, Phys. Rev. B 90 (2014) 214418. Available from: https://doi.org/10.1103/PhysRevB.90.214418.

[206] Y. Tserkovnyak, Da Pesin, D. Loss, Spin and orbital magnetic response on the surface of a topological insulator, Phys. Rev. B 91 (2015) 041121. Available from: https://doi.org/10.1103/PhysRevB.91.041121. URL http://link.aps.org/doi/10.1103/PhysRevB.91.041121.

[207] F. Mahfouzi, B.K. Nikolić, N. Kioussis, Antidamping spin-orbit torque driven by spin-flip reflection mechanism on the surface of a topological insulator: a time-dependent nonequilibrium green function approach, Phys. Rev. B 93 (2016) 115419. Available from: https://doi.org/10.1103/PhysRevB.93.115419.

[208] S. Ghosh, A. Manchon, Spin-orbit torque in 3D topological insulator-ferromagnet heterostructure: crossover between bulk and surface transport, Phys. Rev. B 97 (2018) 134402. Available from: https://doi.org/10.1103/PhysRevB.97.134402. URL http://arxiv.org/abs/1711.11016.

[209] A.R. Mellnik, J.S. Lee, A. Richardella, J.L. Grab, P.J. Mintun, M.H. Fischer, et al., Spin-transfer torque generated by a topological insulator, Nature 511 (2014) 449. Available from: https://doi.org/10.1038/nature13534. URL http://www.nature.com/doifinder/10.1038/nature13534.

[210] Y. Fan, P. Upadhyaya, X. Kou, M. Lang, S. Takei, Z. Wang, et al., Magnetization switching through giant spin-orbit torque in a magnetically doped topological insulator heterostructure, Nat. Mater. 13 (2014) 699–704. Available from: https://doi.org/10.1038/nmat3973. URL http://www.ncbi.nlm.nih.gov/pubmed/24776536.

[211] Y. Shiomi, K. Nomura, Y. Kajiwara, K. Eto, M. Novak, K. Segawa, et al., Spin-electricity conversion induced by spin injection into topological insulators, Phys. Rev. Lett. 113 (2014) 196601. Available from: https://doi.org/10.1103/PhysRevLett.113.196601. URL http://link.aps.org/doi/10.1103/PhysRevLett.113.196601.

[212] Y. Wang, P. Deorani, K. Banerjee, N. Koirala, M. Brahlek, S. Oh, et al., Topological surface states originated spin-orbit torques in Bi2Se3, Phys. Rev. Lett. 114 (2015) 257202. Available from: https://doi.org/10.1103/PhysRevLett.114.257202.

[213] M. Jamali, J.S. Lee, J.S. Jeong, F. Mahfouzi, Y. Lv, Z. Zhao, et al., Giant spin pumping and inverse spin Hall effect in the presence of surface spin-orbit coupling of topological insulator Bi2Se3, Nano Lett. 15 (2015) 7126. Available from: https://doi.org/10.1021/acs.nanolett.5b03274URL. Available from: http://pubsdc3.acs.org/doi/10.1021/acs.nanolett.5b03274.

[214] G. Floquet, Sur les equations differentielles lineaires a coefficients presque-periodiques, Annales scientifiques de l'ENS 12 (1883) 47. Available from: https://doi.org/10.1007/BF02545660.

[215] V.M. Galitskii, S. Goreslavkii, V.F. Elesin, Electric and magnetic properties of a semiconductor in the field of a strong electromagnetic wave, JEPT. 30 (1970) 117–122.

[216] T. Kitagawa, T. Oka, A. Brataas, L. Fu, E. Demler, Transport properties of nonequilibrium systems under the application of light: photoinduced quantum Hall insulators without Landau levels, Phys. Rev. B 84 (2011) 235108. Available from: https://doi.org/10.1103/PhysRevB.84.235108. arXiv:1104.4636v3.

[217] N.H. Lindner, G. Refael, V. Galitski, Floquet topological insulator in semiconductor quantum wells, Nat. Phys. 7 (2011) 490. Available from: https://doi.org/10.1038/nphys1926.

[218] M. Ezawa, Photoinduced topological phase transition and a single Dirac-cone state in silicene, Phys. Rev. Lett. 110 (2013) 026603. Available from: https://doi.org/10.1103/PhysRevLett.110.026603. arXiv: arXiv:1207.6694v2.

[219] Y.H. Wang, H. Steinberg, P. Jarillo-Herrero, N. Gedik, Observation of Floquet-Bloch states on the surface of a topological insulator, Science 342 (2013) 453. Available from: https://doi.org/10.1126/science.1239834. URL http://www.ncbi.nlm.nih.gov/pubmed/24159040.

[220] M. Tahir, A. Manchon, U. Schwingenschlögl, Photoinduced quantum spin and valley Hall effects, and orbital magnetization in monolayer MoS2, Phys. Rev. B Con. Mat. Mater. Phys. 90 (12) (2014). Available from: https://doi.org/10.1103/PhysRevB.90.125438.

[221] E.J. Sie, J.W. McLver, Y.H. Lee, L. Fu, J. Kong, N. Gedik, Valley-selective optical Stark effect in monolayer WS2, Nat. Mater. 14 (2015) 290. Available from: https://doi.org/10.1038/nmat4156. arXiv:1407.1825.

[222] H. Hübener, M.A. Sentef, U. de Giovannini, A.F. Kemper, A. Rubio, Creating stable Floquet-Weyl semimetals by laser-driving of 3D Dirac materials, Nat. Commun. 8 (2017) 13940. Available from: https://doi.org/10.1038/ncomms13940. arXiv:1604.03399, URL http://arxiv.org/abs/1604.03399.

[223] J. Cayssol, B. Dóra, F. Simon, R. Moessner, Floquet topological insulators, Phys. Status Solidi Rapid Res. Lett. 7 (2013) 101. Available from: https://doi.org/10.1002/pssr.201206451. arXiv:1211.5623.

[224] Y. Aharonov, D. Bohm, Significance of electromagnetic potentials in the quantum theory, Phys. Rev. 115 (3) (1959) 485–491. Available from: https://doi.org/10.1103/PhysRev.115.485. URL http://link.aps.org/doi/10.1103/PhysRev. 115.485.

[225] T. Bergsten, T. Kobayashi, Y. Sekine, J. Nitta, Experimental demonstration of the time reversal Aharonov-Casher effect, Phys. Rev. Lett. 97 (19) (2006) 196803. Available from: https://doi.org/10.1103/PhysRevLett.97.196803. URL http://link.aps.org/doi/10.1103/PhysRevLett.97.196803.

[226] Y. Aharonov, A. Casher, Topological quantum effects for neutral particles, Phys. Rev. Lett. 53 (4) (1984) 319–321. Available from: https://doi.org/10.1103/PhysRevLett.53.319. URL http://link.aps.org/doi/10.1103/PhysRevLett.53.319.

[227] J. Nitta, F.E. Meijer, H. Takayanagi, Spin-interference device, Appl. Phys. Lett. 75 (5) (1999) 695. Available from: https://doi.org/10.1063/1.124485. URL http://scitation.aip.org/content/aip/journal/apl/75/5/10.1063/1.124485.

[228] M. König, A. Tschetschetkin, E.M. Hankiewicz, J. Sinova, V. Hock, V. Daumer, et al., Direct observation of the Aharonov-Casher phase, Phys. Rev. Lett. 96 (7) (2006) 076804. Available from: https://doi.org/10.1103/PhysRevLett.96.076804. URL http://link.aps.org/doi/10.1103/PhysRevLett.96.076804.

[229] X. Wang, A. Manchon, Rashba diamond in an Aharonov-Casher ring, Appl. Phys. Lett. 99 (2011) 142507.

[230] S. Datta, B. Das, Electronic analog of the electro-optic modulator, Appl. Phys. Lett. 56 (1990) 665. Available from: https://doi.org/10.1063/1.102730. URL http://scitation.aip.org/content/aip/journal/apl/56/7/10.1063/1.102730.

[231] A. Manchon, S. Zhang, Theory of nonequilibrium intrinsic spin torque in a single nanomagnet, Phys. Rev. B 78 (2008) 212405.

[232] P. Chuang, S.-C. Ho, L.W. Smith, F. Sfigakis, M. Pepper, C.-H. Chen, et al., All-electric all-semiconductor spin field-effect transistors, Nat. Nanotechnol. 10 (2015) 35. Available from: https://doi.org/10.1038/nnano.2014.296. arXiv:1506.06507, URL http://www.nature.com/doifinder/10.1038/nnano.2014.296.

[233] W.Y. Choi, Hyung-jun Kim, J. Chang, S.H. Han, A. Abbout, H.B. Mohamed Saidaoui, et al., Ferromagnet-Free All-Electric Spin Hall Transistors, Nano Lett. 18 (12) (2018) 7998–8002.

[234] B.A. Bernevig, O. Vafek, Piezo-magnetoelectric effects in p-doped semiconductors, Phys. Rev. B 72 (2005) 033203. Available from: https://doi.org/10.1103/PhysRevB.72.033203. URL http://link.aps.org/doi/10.1103/PhysRevB.72.033203.

[235] I. Garate, A.H. MacDonald, Influence of a transport current on magnetic anisotropy in gyrotropic ferromagnets, Phys. Rev. B 80 (2009) 134403. Available from: https://doi.org/10.1103/PhysRevB.80.134403. URL http://link.aps.org/doi/10.1103/PhysRevB.80.134403.

[236] A. Manchon, et al., Current-induced spin-orbit torques in ferromagnetic and antiferromagnetic systems, Rev. Mod. Phys. 91 (2019) 035004. Available from: https://doi.org/10.1103/RevModPhys.91.0350040.

[237] A. Chernyshov, M. Overby, X. Liu, J.K. Furdyna, Y. Lyanda-Geller, L.P. Rokhinson, Evidence for reversible control of magnetization in a ferromagnetic material by means of spin–orbit magnetic field, Nat. Phys. 5 (9) (2009) 656–659. Available from: https://doi.org/10.1038/nphys1362. URL http://www.nature.com/doifinder/10.1038/nphys1362.

[238] I.M. Miron, K. Garello, G. Gaudin, P.-J. Zermatten, M.V. Costache, S. Auffret, et al., Perpendicular switching of a single ferromagnetic layer induced by in-plane current injection, Nature 476 (2011) 189. Available from: https://doi.org/10.1038/nature10309. URL http://www.ncbi.nlm.nih.gov/pubmed/21804568.

[239] X. Liu, J. Sinova, Unified theory of spin dynamics in a two-dimensional electron gas with arbitrary spin-orbit coupling strength at finite temperature, Phys. Rev. B 86 (2012) 174301. Available from: https://doi.org/10.1103/PhysRevB.86.174301. URL http://link.aps.org/doi/10.1103/PhysRevB.86.174301.

[240] S. Emori, T. Nan, A.M. Belkessam, X. Wang, A.D. Matyushov, C.J. Babroski, et al., Interfacial spin-orbit torque without bulk spin-orbit coupling, Phys. Rev. B 93 (2016) 180402. Available from: https://doi.org/10.1103/PhysRevB.93.180402. URL http://link.aps.org/doi/10.1103/PhysRevB.93.180402.

[241] K. Narayanapillai, K. Gopinadhan, X. Qiu, A. Annadi, Ariando, T. Venkatesan, et al., Current-driven spin orbit field in LaAlO3/SrTiO3 heterostructures, Appl. Phys. Lett. 105 (16) (2014) 162405. Available from: https://doi.org/10.1063/1.4899122. arXiv:1410.0456.

[242] Q. Shao, G. Yu, Y.-W. Lan, Y. Shi, M.-Y. Li, C. Zheng, et al., Strong Rashba-Edelstein effect-induced spin-orbit torques in monolayer transition metal dichalcogenide/ferromagnet bilayers, Nano Lett. 16 (2016) 7514. Available from: https://doi.org/10.1021/acs.nanolett.6b03300. URL http://pubs.acs.org/doi/abs/10.1021/acs.nanolett.6b03300.

[243] J. Han, A. Richardella, S.A. Siddiqui, J. Finley, N. Samarth, L. Liu, Room-temperature spin-orbit torque switching induced by a topological insulator, Phys. Rev. Lett. 119 (2017) 077702. Available from: https://doi.org/10.1103/PhysRevLett.119.077702.

[244] Y. Wang, D. Zhu, Y. Wu, Y. Yang, J. Yu, R. Ramaswamy, et al., Room temperature magnetization switching in topological insulator-ferromagnet heterostructures by spin-orbit torques, Nat. Commun. 8 (2017) 1364. Available from: https://doi.org/10.1038/s41467-017-01583-4. URL https://doi.org/10.1038/s41467-017-01583-4.

[245] D. Mahendra, R. Grassi, J.-Y. Chen, M. Jamali, D. Reifsnyder Hickey, D. Zhang, et al., Room temperature giant spin-orbit torque due to quantum confinement in sputtered BixSe(1-x) films, Nature Materials 17, (2018) 800. https://doi.org/10.1038/s41563-018-0136-z.
[246] S. Nadj-Perge, S.M. Frolov, E.P.A.M. Bakkers, L.P. Kouwenhoven, Spin-orbit qubit in a semiconductor nanowire, Nature 468 (7327) (2010) 1084–1087. Available from: https://doi.org/10.1038/nature09682. Available from: http://www.nature.com/nature/journal/v468/n7327/abs/nature09682.
[247] J.W.G. Van Den Berg, S. Nadj-Perge, V.S. Pribiag, S.R. Plissard, E.P.A.M. Bakkers, S.M. Frolov, et al., Fast spin-orbit qubit in an indium antimonide nanowire, Phys. Rev. Lett. 110 (6) (2013) 066806. Available from: https://doi.org/10.1103/PhysRevLett.110.066806. arXiv:1210.7229.
[248] E.I. Rashba, A.L. Efros, Orbital mechanisms of electron-spin manipulation by an electric field, Phys. Rev. Lett. 91 (12) (2003) 126405. Available from: https://doi.org/10.1103/PhysRevLett.91.126405.
[249] L.P. Gor'kov, E.I. Rashba, Superconducting 2D system with lifted spin degeneracy: mixed singlet-triplet state, Phys. Rev. Lett. 87 (2001) 37004. Available from: https://doi.org/10.1103/PhysRevLett.87.037004. arXiv:0103449.
[250] M. Sato, Y. Ando, Topological superconductors: a review, Rep. Prog. Phys. 80 (2017) 076501. Available from: https://doi.org/10.1088/1361-6633/aa6ac7. arXiv:1608.03395, URL https://doi.org/10.1088/1361-6633/aa6ac7.
[251] S. Nadj-Perge, I. Drozdov, J. Li, H. Chen, S. Jeon, Observation of Majorana fermions in ferromagnetic atomic chains on a superconductor, Science 346 (6209) (2014) 602. URL http://www.sciencemag.org/content/346/6209/602.short.
[252] A. Stern, N.H. Lindner, Topological quantum computation—from basic concepts to first experiments, Science 339 (2013) 1179. Available from: https://doi.org/10.1126/science.1172133.
[253] X.-j Liu, M.F. Borunda, X. Liu, J. Sinova, Effect of induced spin-orbit coupling for atoms via laser fields, Phys. Rev. Lett. 102 (2009) 046402. Available from: https://doi.org/10.1103/PhysRevLett.102.046402.
[254] V. Galitski, I.B. Spielman, Spin-orbit coupling in quantum gases, Nature 494 (2013) 49. Available from: https://doi.org/10.1038/nature11841. arXiv:1312.3292, URL https://doi.org/10.1038/nature11841.
[255] P. Wang, Z.-q Yu, Z. Fu, J. Miao, L. Huang, Spin-orbit coupled degenerate Fermi gases, Phys. Rev. Lett. 109 (2012) 095301. Available from: https://doi.org/10.1103/PhysRevLett.109.095301.
[256] L.W. Cheuk, A.T. Sommer, Z. Hadzibabic, T. Yefsah, W.S. Bakr, M.W. Zwierlein, Spin-injection spectroscopy of a spin-orbit coupled Fermi gas, Phys. Rev. Lett. 109 (9) (2012) 095302. Available from: https://doi.org/10.1103/PhysRevLett.109.095302. URL http://link.aps.org/doi/10.1103/PhysRevLett.109.095302.
[257] S.C. Ji, J.Y. Zhang, L. Zhang, Z.D. Du, W. Zheng, Y.J. Deng, et al., Experimental determination of the finite-temperature phase diagram of a spin-orbit coupled Bose gas, Nat. Phys. 10 (2014) 314. Available from: https://doi.org/10.1038/nphys2905.
[258] Y.-J. Lin, K. Jimenez-GarcÄśa, I.B. Spielman, Spin-orbit-coupled Bose-Einstein condensates, Nature 471 (2011) 83. Available from: https://doi.org/10.1038/nature09887.
[259] W.S. Cole, S. Zhang, A. Paramekanti, N. Trivedi, Bose-Hubbard models with synthetic spin-orbit coupling: Mott insulators, spin textures, and superfluidity, Phys. Rev. Lett. 109 (2012) 085302. Available from: https://doi.org/10.1103/PhysRevLett.109.085302. URL http://link.aps.org/doi/10.1103/PhysRevLett.109.085302.
[260] I. Dzyaloshinskii, Thermodynamic theory of weak ferromagnetism in antiferromagnetic substances, Sov. Phys. JETP 5 (1957) 1259.
[261] T. Moriya, Anisotropic superexchange interaction and weak ferromagnetism, Phys. Rev. 120 (1960) 91. Available from: https://doi.org/10.1103/PhysRev.120.91. URL https://journals.aps.org/pr/pdf/10.1103/PhysRev.120.91 http://prola.aps.org/abstract/PR/v120/i1/p91_1.

[262] N. Nagaosa, Y. Tokura, Topological properties and dynamics of magnetic skyrmions, Nat. Nanotechnol. 8 (12) (2013) 899–911. Available from: https://doi.org/10.1038/nnano.2013.243. URL http://www.ncbi.nlm.nih.gov/pubmed/24302027.

[263] A. Soumyanarayanan, N. Reyren, A. Fert, C. Panagopoulos, Spin-orbit coupling induced emergent phenomena at surfaces and interfaces, Nature 539 (2016) 509. Available from: https://doi.org/10.1038/nature19820. arXiv:1611.09521.

[264] S.S.P. Parkin, M. Hayashi, L. Thomas, Magnetic domain-wall racetrack memory, Science 320 (2008) 190. Available from: https://doi.org/10.1126/science.1145799. URL http://www.ncbi.nlm.nih.gov/pubmed/18403702.

[265] S.-H. Yang, K.-S. Ryu, S. Parkin, Domain-wall velocities of up to 750 m/s driven by exchange-coupling torque in synthetic antiferromagnets, Nat. Nanotechnol. 10 (2015) 221. Available from: https://doi.org/10.1038/nnano.2014.324. URL http://www.nature.com/doifinder/10.1038/nnano.2014.324.

[266] Y. Onose, T. Ideue, H. Katsura, Y. Shiomi, N. Nagaosa, Y. Tokura, Observation of the magnon Hall effect, Science 329 (2010) 297. Available from: https://doi.org/10.1126/science.1188260. URL http://www.ncbi.nlm.nih.gov/pubmed/20647460.

[267] R. Matsumoto, S. Murakami, Theoretical prediction of a rotating magnon wave packet in ferromagnets, Phys. Rev. Lett. 106 (19) (2011) 197202. Available from: https://doi.org/10.1103/PhysRevLett.106.197202. URL http://journals.aps.org/prl/abstract/10.1103/PhysRevLett.106.197202.

[268] E. Jué, C.K. Safeer, M. Drouard, A. Lopez, P. Balint, Chiral damping of magnetic domain walls, Nat. Mater. 15 (2016) 272. Available from: https://doi.org/10.1038/nmat4518.

[269] A. Manchon, P. Ndiaye, J.-H. Moon, H.-W. Lee, K.-J. Lee, Magnon-mediated Dzyaloshinskii-Moriya torque in homogeneous ferromagnets, Phys. Rev. B 90 (2014) 224403. Available from: https://doi.org/10.1103/PhysRevB.90.224403.

CHAPTER 3

Two-dimensional ferrovalley materials

Xin-Wei Shen[1], He Hu[1] and Chun-Gang Duan[1,2]

[1]*State Key Laboratory of Precision Spectroscopy and Key Laboratory of Polar Materials and Devices, Ministry of Education, Department of Optoelectronics, East China Normal University, Shanghai, P.R. China,* [2]*Collaborative Innovation Center of Extreme Optics, Shanxi University, Taiyuan, P.R. China*

Chapter Outline

3.1 Valleytronics in 2D hexagonal lattices

Conventional electronics and spintronics are based on the intrinsic degrees of freedom of an electron, namely its charge and spin. The large-scale semiconductor integrated circuits are currently rooted in the manipulation of electric charge. With the discovery of giant magnetoresistance effect [1], the spin degree of freedom shows its great potential in the information processing with the advantages of ultrahigh integration density and ultrahigh speed [2,3]. In recent years, an additional degree of freedom, valley [4], has attracted intensive attention due to the emergence of novel two-dimensional (2D) graphene-related materials [5–8] with hexagonal lattice symmetry, as shown in Fig. 3.1A. In the momentum space of Bloch electrons, the local energy extreme point in the conduction band or valence band (VB) is referred to as a valley, as illustrated in Fig. 3.1B. In analogy to charge and spin, the valley degree of freedom (VDF) constitutes the similar binary logic states in solids, leading to the possibilities for information applications. Till now, various functional devices such as valley separators [10], valley filters, and valves [11–13] have been achieved in valleytronics.

However, the pristine graphene is protected by both space and time-reversal symmetry. Apart from this none other experimentally measurable physical quantity can be identified

Spintronic 2D Materials.
DOI: https://doi.org/10.1016/B978-0-08-102154-5.00003-5

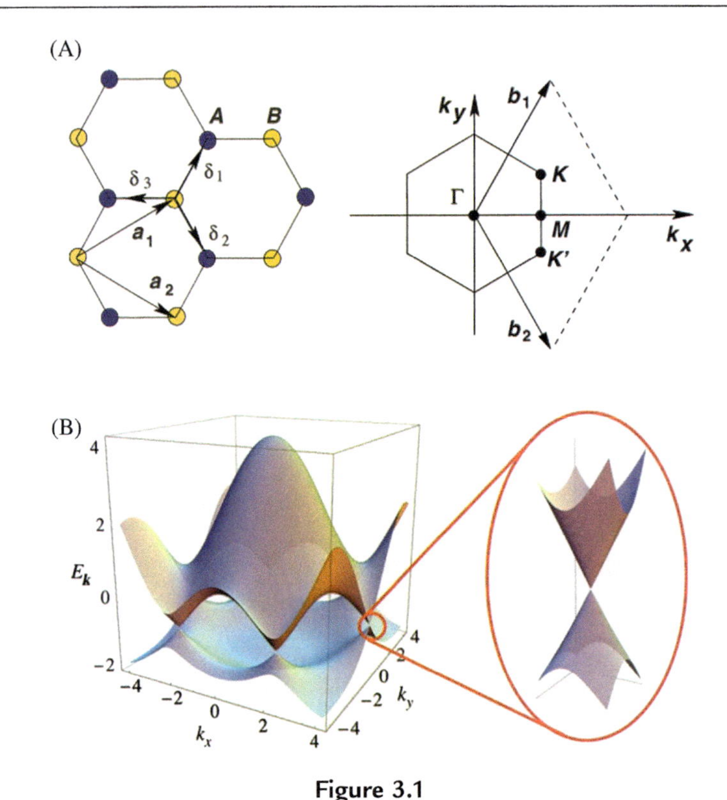

Figure 3.1
(A) Honeycomb lattice and its BZ of graphene. (B) Schematic diagram of electronic dispersion, showing the valley index. *BZ*, Brillouin zone. Source: *Permission from A.H. Castro Neto, et al., The electronic properties of graphene. Rev. Mod. Phys. 81(1) (2009) 109–162 [9].*

to distinguish the inequivalent valleys and manipulate this binary VDF. Thus it is not an ideal valleytronic material that can be used for designing the practical devices in industrial applications. On account of this, immense attention has gradually been devoted to 2H-phase group-VI transition metal dichalcogenides (TMDs), which have the unique potential for utilizing the valley index effectively [14–17]. Similar to graphene, these materials exhibit honeycomb lattice structures with the hexagonal Brillouin zone (BZ), as shown in Fig. 3.2A. Moreover the intrinsic spin–orbit coupling (SOC) derived from the d orbitals of heavy transition metal [20] will give rise to significant but opposite spin splitting at valleys K_+ and K_-, as shown in Fig. 3.2B, inducing strong coupled between spin and VDF. Note that a pair of inequivalent valleys is expected at the K_+ and K_- points located at the corners of BZ. With respect to centrosymmetric graphene, these 2H-phase TMDs lack space inversion symmetry, leading to the existence of the valley Hall effect (VHE) [18] and the valley-dependent optical selection rules [19], as shown in Fig. 3.2C and D. Therefore 2H-TMDs monolayers are generally regarded as the promising platform to study the fundamental physics in spintronics, valleytronics, and crossing areas.

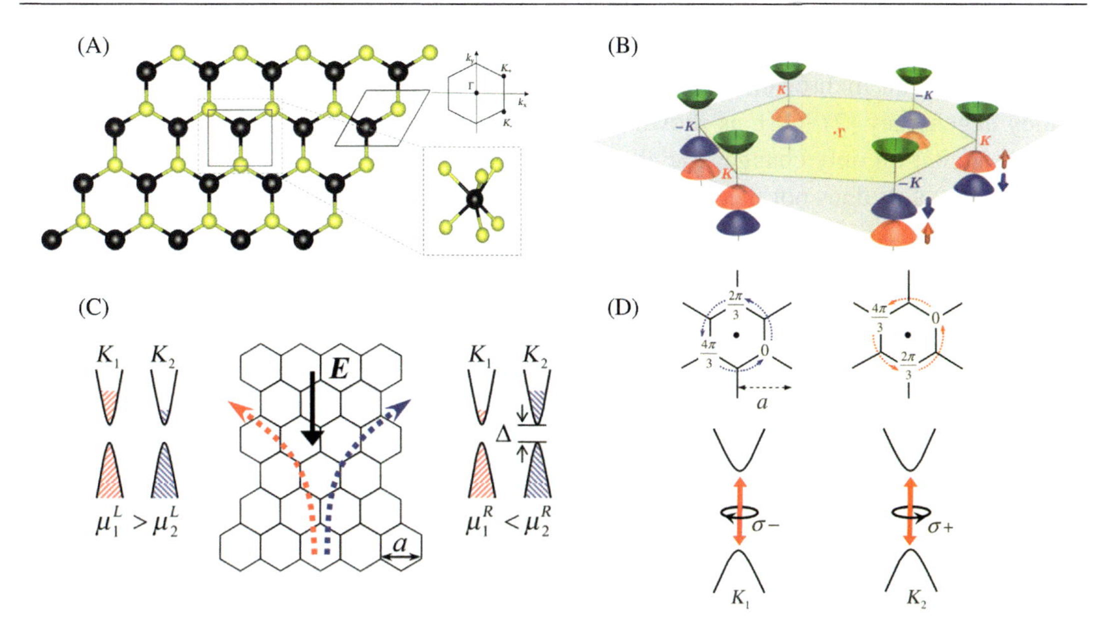

Figure 3.2
(A) The 2D hexagonal crystal structure of a monolayer TMDs composed of transition metal atoms (*black*) and chalcogen atoms (*yellow*) resembles that of graphene but with broken inversion symmetry. (B) Schematic drawing of the band structure at the band edges located at the K points. (C) The mechanism of VHE, where the electrons at two inequivalent valleys will travel to the opposite edges under an in-plane electric field. (D) Valley-dependent optical selection rules: the optical transitions at K_1 (K_2) valley could only be excited by σ_- (σ_+) circularly polarized light. *TMDs*, transition metal dichalcogenides; *VHE*, valley Hall effect. Source: *(B) Permission from D. Xiao, et al., Coupled spin and valley physics in monolayers of MoS2 and other group-VI dichalcogenides, Phys. Rev. Lett. 108(19) (2012) 196802 [16]; (C) Permission from D. Xiao, W. Yao, Q. Niu, Valley-contrasting physics in graphene: magnetic moment and topological transport, Phys. Rev. Lett. 99(23) (2007) 236809; (D) Permission from W. Yao, D. Xiao, Q. Niu, Valley-dependent optoelectronics from inversion symmetry breaking, Phys. Rev. B, 77(23) (2008) 235406.*

3.2 Valley polarization induced by external fields

In spintronics, two opposite spin polarizations constitute binary logic states for data encoding, processing, and storage. From the point of view on long-term information storage, if we want to manipulate the valley index in valleytronics, the major challenge is to break the degeneracy between two prominent valleys that is achieving the valley polarization. In this regard, the pristine TMDs monolayer are not suitable for the valleytronic applications, as the valleys in these systems still undergo energy degeneration without valley polarization. In analogy with paraelectric and paramagnetic materials, they can be called paravalley materials. Consequently, the achievement of valley polarization is highly desirable in valleytronics. At the present stage, enormous attempts have been adopted to induce the valley polarization. In consideration of valley-selective circular dichroism, the optical pumping is the principal mechanism to break the valley degeneracy [21–23], as shown in Fig. 3.3A. However as a dynamic process, optical excitation only changes the chemical potential rather than the energy degeneracy in two valleys. It does not meet the requirement of long-term information storage. The electron–electron interaction has been proposed to break time-reversal symmetry, which is considered as an effective way to induce the valley polarization [27]. Unfortunately, an external electric field is necessary to guarantee the stability of the system. The Zeeman effect rooted in an external magnetic field [25,28,29] offers an additional way to induce the valley polarization, as illustrated in Fig. 3.3B. However a sizable valley splitting requires an extreme field strength, which is difficult to achieve in the practical applications. Fig. 3.3C demonstrates that the optical stark effect [26] is another approach to control valley splitting in the valleytronic materials. A similar technical problem appears here, that is a huge amplitude of oscillating electric field is required for the valley polarization.

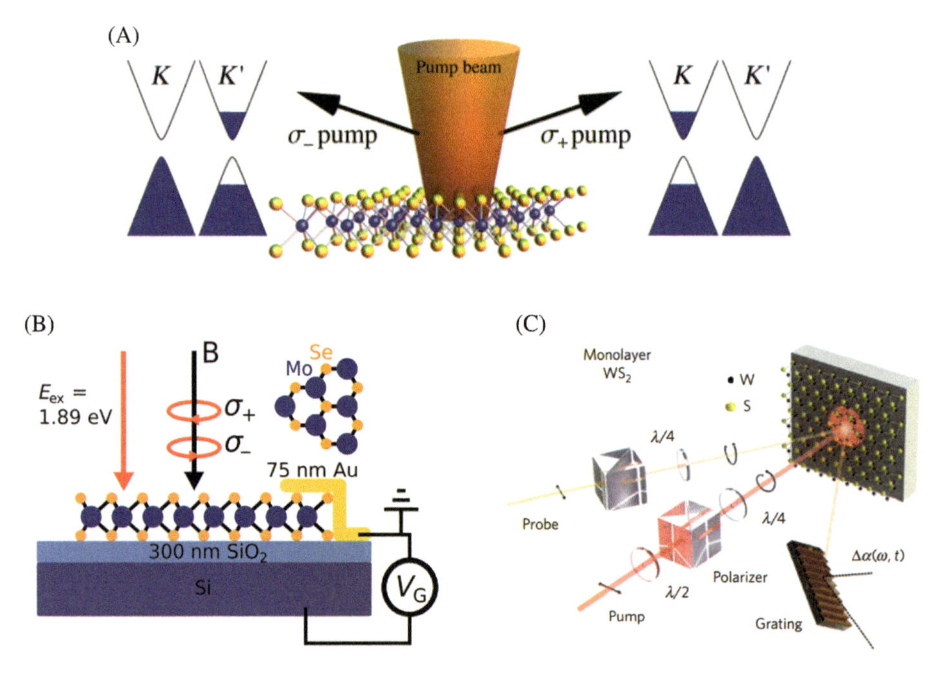

Figure 3.3

Valley polarization induced by various external fields: (A) Optical pumping with circularly polarized light; (B) out-of-plane magnetic fields. (C) Valley-selective optical Stark effect. *(A) Permission from A. Kumar, et al., Chiral plasmon in gapped Dirac systems, Phys. Rev. B, 93(4) (2016) 041413 [24]; (B) Permission from D. MacNeill, et al., Breaking of valley degeneracy by magnetic field in monolayer MoSe2, Phys. Rev. Lett. 114(3) (2015) 037401 [25]; (C) Permission from E.J. Sie, et al., Valley-selective optical Stark effect in monolayer WS2, Nat. Mater. 14 (2014) 290.*

Although all the physical attempts mentioned above have profound connotations, they are limited for the data devices. When the valley degeneracy is generally lifted through external means, the valley polarization is extrinsic and volatile as the applied external fields are removed. The system with valley polarization state will degenerate into the initial nonpolarization state, bringing up disadvantages of practical applications. The volatility restricts the next-generation memory applications based on valleytronic products. Compared with the external magnetic field, magnetic doping [30–32] and magnetic proximity effect [10,33] seem to be a more intelligent approach to meet the requirement of nonvolatility, as shown in Fig. 3.4. However they still belong to external means in essence. It is then natural to raise the following questions: how to realize intrinsic and nonvolatile valley polarization?; and could the robust valley polarization persist in the valleytronic materials without any external fields? In the following Sections,

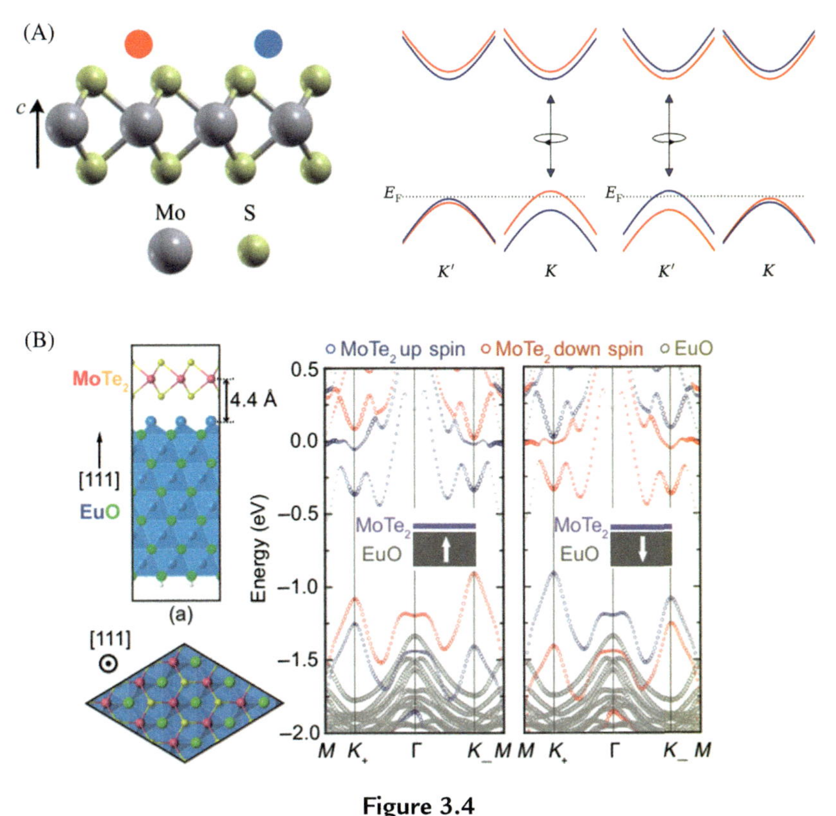

Figure 3.4

Valley splitting is tunable through: (A) Mn atom magnetic doping in monolayer MoS2. Right panel is the band structures with exchange field; (B) proximity-induced Zeeman effect in MoT_2/EuO heterostructure. Right panel is the band structures with opposite magnetism of EuO. Source: *(A) Permission from Y. Cheng, Q. Zhang, U. Schwingenschlögl, Valley polarization in magnetically doped single-layer transition-metal dichalcogenides, Phys. Rev. B, 89(15) (2014) 155429; (B) Permission from J. Qi, et al., Giant and tunable valley degeneracy splitting in MoTe2. Phys. Rev. B, 92(12) (2015) 121403.*

we will introduce a new member of the ferroic family that is ferrovalley materials. In addition to ferroelectric and ferromagnetic materials that have been routinely explored, the ferrovalley materials with spontaneous valley polarization are for the first time unveiled in a series of 2D valleytronic materials. These ferrovalley materials have the potential to avoid the limitation of volatility as seen in the conventional valleytronic materials, thus is of great importance in paving the way for the practical applications of valleytronics.

3.3 Ferrovalley materials with spontaneous valley polarization

3.3.1 VSe_2-Ferrovalley materials induced by ferromagnetism

Taking hexagonal TMDs monolayer as an example, the mechanism of spontaneous valley polarization induced by ferromagnetism can be represented using a two-band $\mathbf{k}\cdot\mathbf{p}$ model. The total Hamiltonian of the system can be written in the following form:

$$H(\boldsymbol{k}) = I_2 \otimes H_0(\boldsymbol{k}) + H_{\text{SOC}}(\boldsymbol{k}) + H_{\text{ex}}(\boldsymbol{k}) \tag{3.1}$$

The SOC effect is equivalent to a momentum space-dependent pseudomagnetic field, which moves the spin-up energy band upward and the spin-down band downward at K_+ valley, while the spin splitting at K_-valley is opposite. The additional term $H_{\text{ex}}(\boldsymbol{k})$ is used to describe the intrinsic exchange interaction, which is equivalent to an intrinsic magnetic field, tending to split the spin-majority and spin-minority states. Combining the SOC effect with the valley-independent exchange interaction, the degeneracy of the two valleys is broken, leading to the valley polarization. According to the total Hamiltonian, the band structure near the valleys $K_\pm$ of TMDs monolayers can be easily deduced. Fig. 3.5 shows the band structures at valleys K_+ and K_- of representative 2H-phase TMD monolayers referenced to monolayer MoS_2. The Fermi level lies between the upper band (UB) and lower band (LB). The spin splitting of UB induced by SOC effect has the opposite sign to the one of LB.

The valley-dependent optical selection rules, as well as the impact of valley polarization on optical properties can be demonstrated systematically using group theory analysis. The allowed interband transitions excited by circularly polarized light near the band edges have been plotted as A_+, B_+, A_- and B_-, and σ_+ and σ_- represent the left-handed and right-handed radiation, respectively. Fig. 3.5A shows energy band with absence of SOC effect, for K_+ valley, the irreducible representations (IRs) of LB and UB are A' and $2E'$ respectively. However they are A' and $1E'$ for K_-. Obviously, although the IRs of LB is the same for the two valleys, the symmetry of UB is different between K_+ and K_-. Using the great orthogonality theorem of group theory, the electric-dipole transition is allowed only if the reduced direct product representation between the initial state IRs and the incident light IRs contains the representation of the final state. The one-to-one correspondence

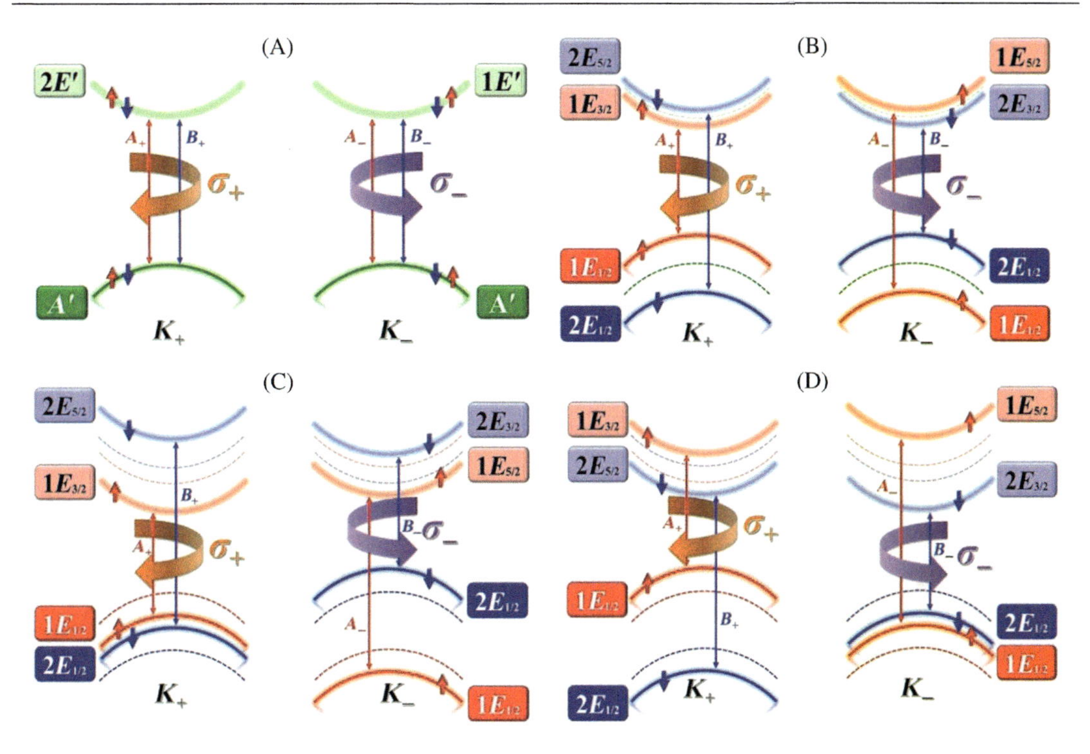

Figure 3.5
The schematic band structures at valleys K_+ and K_- of representative 2H-phase TMD monolayers. (A) Without SOC effect, (B) with SOC effect, (C) with SOC effect and a positive exchange field which is the valley-polarized case, and (D) is same as (C) but with negative exchange field.

between the energy valley index and the rotational light excitation can be obtained. When the left-handed (right-handed) incident light with $2E'$ ($1E'$) symmetry acts on the two valleys, we can note that:

$$A' \otimes 2E'(1E') = 2E'(1E') \tag{3.2}$$

Apparently, the optical absorption at K_+ (K_-) could only be excited by the left-handed (right-handed) light, which reveals the valley-dependent optical selection rules. At present, the optical band gaps (E_g^{opt}) of the two valleys are identical that can be demonstrated by:

$$E_g^{opt}(A_+) = E_g^{opt}(B_+) = E_g^{opt}(A_-) = E_g^{opt}(B_-) = \Delta \tag{3.3}$$

When the SOC effect is considered, as shown in Fig. 3.5B, the symmetry for valleys needs to be described by the double group C_{3h}^{D}. When the spin index is taken into account, the identical representation A' degenerates to $1E_{1/2}$ and $2E_{1/2}$ for spin-up and spin-down components. Meanwhile, $2E'$ and $1E'$ change to $1E_{3/2}$ (spin-up), $2E_{5/2}$ (spin-down), $1E_{5/2}$ (spin-up), and $2E_{3/2}$ (spin-down), accordingly. By applying direct product between IRs of the ground state and circularly polarized light, the allowed interband transitions can be easily obtained, as labeled in Fig. 3.5B. Note that the chirality is locked in each valley. Specifically, at the valley K_+, both spin-up and spin-down components can only be excited by left-handed light, while the valley K_- only corresponds to right-handed light. The valley-dependent SOC effect splits the previously degenerated A_+ (A_-) and B_+ (B_-), and makes the E_g^{opt} in K_+ and K_- stemming from different spin states. The spin splitting energy at the bottom of UB (the top of LB) is labeled as $2\lambda_u$ ($2\lambda_l$), which is defined by the energy difference $E_{u(l)\uparrow} - E_{u(l)\downarrow}$ at the K_+ points. Although the electron transitions of K_+ and K_- corresponding to different spin states, E_g^{opt} of the two valleys are the same, which means the energy levels are still degenerate:

$$E_g^{opt}(A_+) = E_g^{opt}(B_-) = \Delta - \lambda_l + \lambda_u \tag{3.4}$$

Once the intrinsic exchange field exists (see Fig. 3.5C and D), time-inversion symmetry of the system is broken, which decouples the energetically degenerated valleys, elucidating the occurrence of valley polarization. Taking the positive exchange field as an example (Fig. 3.5C), the optical band gap of valley K_+ corresponds to the transition of spin-up electrons excited by left-handed radiation, given by:

$$E_g^{opt}(A_+) = \Delta - \lambda_l + \lambda_u + m_l - m_u \tag{3.5}$$

For the K_-valley,

$$E_g^{opt}(B_-) = \Delta - \lambda_l + \lambda_u - m_l + m_u \tag{3.6}$$

where $m_u(m_l) = E_{u(l)\uparrow} - E_{u(l)\downarrow}$ represents the effective exchange splitting in the band edge of UB (LB). The effect of the intrinsic exchange field leads to the difference of the optical band gap between the two valleys with the magnitude of $2|m_l - m_u|$. For such a valley-polarized system, once the polarity of circularly polarized light is reversed, different optical band gaps can be seen directly, indicating the possibility to judge the valley polarization utilizing noncontact and nondestructive circularly polarized optical means.

The above two-band $\mathbf{k} \cdot \mathbf{p}$ model and group theory analysis clearly show the general rule to hunt for ferrovalley materials with spontaneous valley polarization, which is the coexistence of SOC effect with intrinsic exchange interaction. Fortunately, 2H-VSe_2 monolayer [34–36] has been predicted to be such a certain ferrovalley material [37]. As a peculiar ferromagnetic semiconductor among TMDs, the unpaired 3d electrons of vanadium atom lead to the existence of intrinsic magnetism. The intrinsic magnetic moment with the magnitude of 1.01 μ_B implies remarkable exchange interaction, which can cause significant spontaneous valley polarization.

When the magnetism in monolayer VSe_2 is ignored, as shown in Fig. 3.6A, the band structure is essentially similar to the representative one for TMDs (see Fig. 3.5B). It is a metal with the

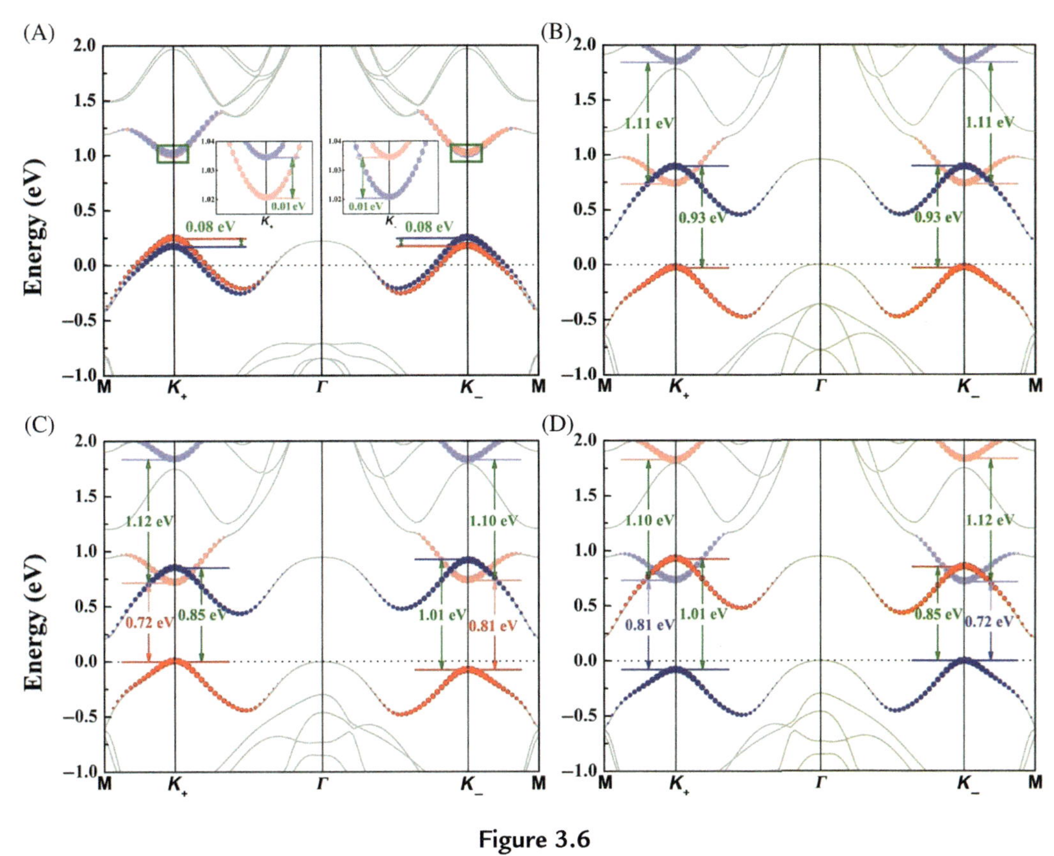

Figure 3.6

The band structures of the 2H-VSe2 monolayer. (A) With SOC effect but without ferromagnetism, and (B) with magnetic moment but without SOC effect. (C) The real case including both the magnetism and SOC effect, and (D) is the same as (C) but with opposite magnetic moment. The insets in (A) amplify the spin splitting at the bottom of the UB. The radius of dots is proportional to its population in the corresponding state near valleys: red and blue ones for spin-up and spin-down components of $d_{x^2-y^2}$ and d_{xy} orbitals on cation-V, and light red and light blue symbols represent spin-up and spin-down states for d_{z^2} characters. The Fermi level EF is set to zero.

Fermi level passing through the states predominantly comprising $d_{x^2-y^2}$ and d_{xy} orbitals on cation-V. Once the intrinsic exchange interaction of unpaired d electrons is taken into account, this equivalent pseudomagnetic field completely splits the degenerated spin-up and spin-down components of the states occupied near the Fermi level, as shown in Fig. 3.6B. Inspiringly, the spin-splitting energy of LB is up to 0.93 eV. As a result, the system becomes a ferromagnetic semiconductor with an indirect band gap. Although the top of the VB is located at the center of the BZ, the direct band gap still corresponds to the valley $K_{\pm}$. As shown in Figs. 3.6C and 3.2D, when considering the nonnegligible SOC effect and the strong exchange interaction, the spontaneous valley polarization can be then induced.

When the magnetic moment is positive (see Fig. 3.6C), the spin splitting at LB states equals to $|2m_l-2\lambda_l|\sim0.85$ eV in the valley K_+, and this is much smaller than $|2m_l+2\lambda_l|\sim1.01$ eV in the valley K_-. Conversely, that of UB states is with a relatively greater value at the point $K_+(|2m_u-2\lambda_u|\sim1.12$ eV) than at $K_-(|2m_u+2\lambda_u|\sim1.10$ eV) because of the opposite sign between m_u and λ_u. More importantly, the band gap with energy difference $|2\lambda_l-2\lambda_u|\sim0.09$ eV will directly reflect in the optical properties excited by circularly polarized light. Compared with the E_g^{opt} related to the left-handed radiation, the right-handed one experiences a blue shift (Fig. 3.7A). When the magnetic moment is inverted, as clearly displayed in Fig. 3.6D, valley polarization possesses reversed polarity. As a result, in comparison with the left-handed one, the red shift of E_g^{opt} excited by right-handed light happens (Fig. 3.7B) The above chirality-dependent optical band gap indicates that circularly polarized light can be used to characterize the occurrence of valley polarization and its chiral characteristics.

It is also interesting to inspect the Berry curvature that has a crucial influence on the electronic transport properties and is the kernel parameter to various Hall effects. The expression of spin-resolved nonzero z-component Berry curvature from the Kubo formula derivation [38]:

$$\Omega_{n,z}^{\uparrow(\downarrow)}(\boldsymbol{k}) = -\sum_{n'\neq n}\frac{2\mathrm{Im}[p_{nn'}^{x\uparrow(\downarrow)}(\boldsymbol{k})p_{n'n}^{y\uparrow(\downarrow)}(\boldsymbol{k})]}{[(E_{n'}^{\uparrow(\downarrow)}(\boldsymbol{k})-E_{n}^{\uparrow(\downarrow)}(\boldsymbol{k})]^2} \tag{3.7}$$

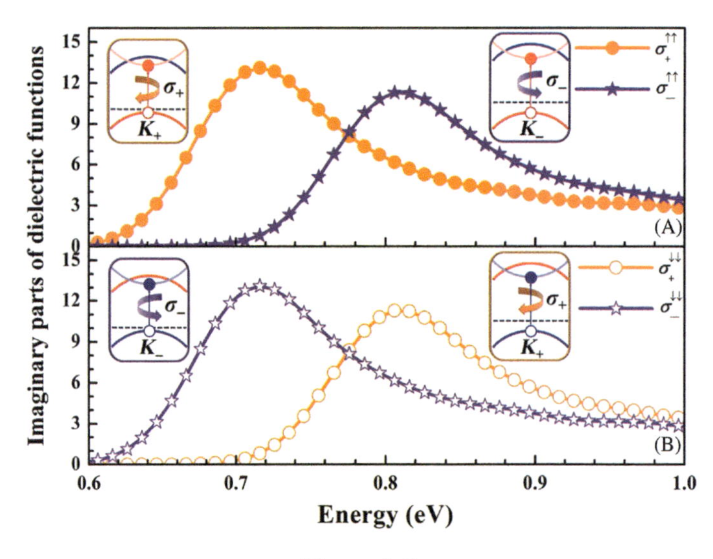

Figure 3.7

The imaginary parts of complex dielectric function ε_2 for monolayer VSe_2. Cases excited by left-handed radiation σ_+ and right-handed radiation σ_- (A) with positive magnetic moment, and (B) with negative one are presented. Insets are the schematic interband transitions related to certain E_g^{opt}.

For the bands with a major contribution from the $d_{x^2-y^2}$ and d_{xy} orbitals of V atoms, the Berry curvature can be calculated as:

$$\Omega_{l,z}(\boldsymbol{k}) = \Omega_{l,z}^{\uparrow}(\boldsymbol{k}) + \Omega_{l,z}^{\downarrow}(\boldsymbol{k}) \quad (3.8)$$

For a paravalley system, $\Omega_{l,z}(\boldsymbol{k})$ is an odd function in the momentum space because of time-reversal symmetry and broken space inversion symmetry. Although the absolute values in opposite valleys are no longer identical in the ferrovalley material, Berry curvatures still have opposite sign, as displayed in Fig. 3.8A. It is important to emphasize that reversal of valley polarization would not change the sign of Berry curvatures in the two valleys, only exchanging the absolute values of Berry curvature. In order to explore the difference between $\Omega_{l,z}(\boldsymbol{k})$ and $\Omega_{l,z}(_{-}\boldsymbol{k})$, they are summated. When valley polarization is induced by the positive magnetic moment, the absolute value of Berry curvature in K_+ valley is greater than the one in the valley K_-, giving rise to a positive summation, as shown in Fig. 3.8B. As expected, when the valley polarization is reversed, the Berry curvatures are obtained with the same values but in the opposite sign. Consequently, Berry curvatures, as circularly polarized radiations, are another effective method to determine the occurrence of valley polarization and its polarity reversal.

The sign change of Berry curvature in different valleys will lead to the new type of Hall effect, called VHE that has been widely found in 2D honeycomb lattice [12,16,39–42]. The accumulation of spin and valley information at the boundary of the sample has brought about a series of interesting phenomena such as emission of photons with opposite circularly polarized light on the two boundaries. The existence of the VHEmakes it possible for the cross study of spintronics and valleytronics.

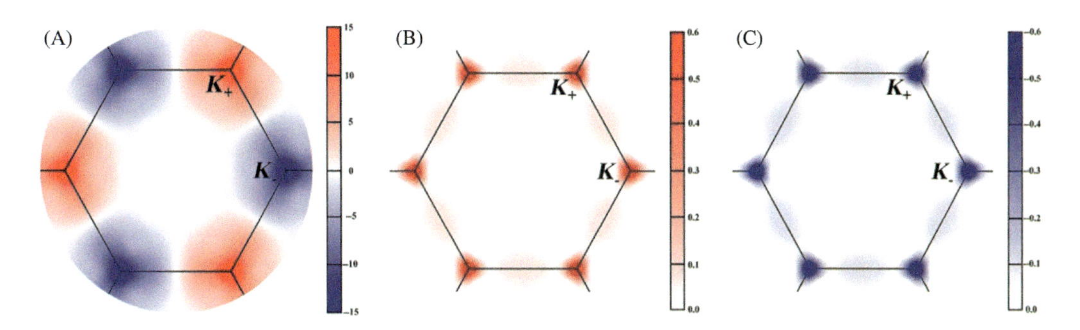

Figure 3.8
Berry curvatures of ferrovalley 2H-VSe_2 monolayer. (A) Contour maps of Berry curvatures in the $\boldsymbol{k}$ space for bands mainly occupied by $d_{x^2-y^2}$ and d_{xy} states in units of $\mathring{A}^2$. Summation of Berry curvatures in the point K and its space inversion are shown in (B) and (C) for positive and negative valley polarization cases, accordingly.

In ferrovalley materials, the VHE exhibit a more interesting feature, which is the presence of additional charge Hall current originating from the spontaneous valley polarization. Analogous to the anomalous Hall effect in ferromagnetic materials, this effect in ferrovalley materials is named as anomalous VHE (AVHE). The AVHE offers a possible way to realize data storage utilizing ferrovalley materials. An example for moderate p-type VSe_2 achieved by slight hole doping with Fermi energy lying between the VB tops of K_+ and K_- valleys is displayed in Fig. 3.9. The p-type VSe_2 possesses 100% spin polarizability around the Fermi level. Because of the almost zero Berry curvature in the center of BZ, the carries from the Γ point and its neighbors pass through the ribbon directly without transverse deflection. When the p-type VSe_2 possess positive valley polarization, the majority carriers, which are spin-down holes from K_+ valley, gain transverse velocities towards left side in the presence of an external electric field. The accumulation of holes in the left boundary of the ribbon generate a charge Hall current that can be detected as a positive voltage. When the polarity of valley polarization reversed, spin-up holes from K_- valley will accumulate in the right side of the sample because of the negative Berry curvature, resulting in measurable transverse voltage with opposite sign.

Based on the AVHE in ferrovalley materials, the electrically reading and magnetically writing memory devices are coming up. The binary logic states are stored by the valley polarization, which can be manipulated by external magnetic field. It can be then easily read out by measuring the sign of the transverse Hall voltage. Furthermore due to the intrinsic valley polarization, the memory applications based on the ferrovalley materials is nonvolatile, which is of great significance to the innovation of storage technology and the development of next-generation of data storage.

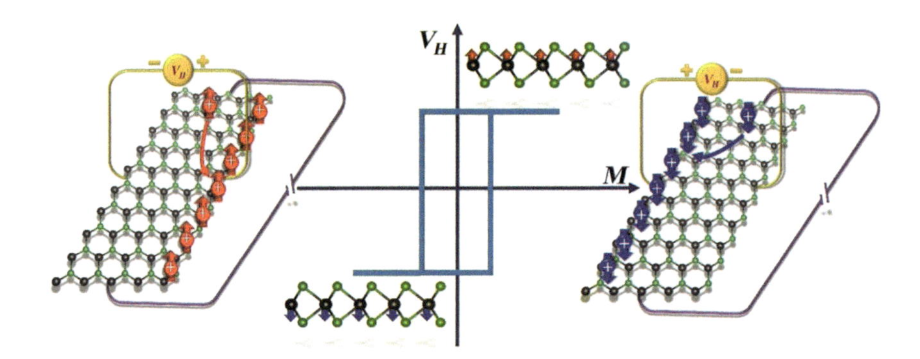

Figure 3.9
Sketch of data storage utilizing hole-doped ferrovalley materials based on AVHE. The carriers that are denoted by white " + " symbol are holes. Upward arrows in red color and downward arrows in blue color represent spin-up and spin-down carriers, respectively. *AVHE*, anomalous valley Hall effect.

3.3.2 Bilayer VSe_2-antiferrovalley materials

The spontaneous valley polarization has been presented in the hexagonal systems, such as ferromagnetic semiconductor 2H-VSe_2 monolayer. However the reversal of valley polarization originating from inherent exchange interaction strongly relies on the external magnetic field, which is a relatively energy-consuming way. Compared with the magnetic field, all-electric tuning with advantages of ultrahigh speed, and ultra-low power consumption is highly desirable for information processing devices. As we all know, in the pristine TMDs bilayer system, an applied gate voltage will break the spatial symmetry and leads to a potential difference between the two monolayers. All of the subbands from one layer move upward with respect to those from the other, lifting the valley degeneracy energetically. According to such structural design [43], valley degeneracy couples to the electrically controlled inversion symmetry, rather than the intrinsic exchange interaction in ferrovalley monolayers. Therefore energetically splitting between valleys strongly depends on the applied electric field. If we can combine electrically detectable AVHE with electrical control of valley polarization, such a bilayer system is possible to achieve all-electrically valleytronic memory devices. Following the strategy, firstly, we consider an antiferrovalley bilayer of typical TMDs in H-type stacking, where one monolayer sits on the other with 180-degree rotation (see Fig. 3.10). Monolayers are in ferrovalley state with opposite magnetic moment to guarantee the presence of both space and time-inversion symmetry. In analogy to the Section 2.3.1, through adding interlayer hopping term $H_\perp$ to the two-band model of ferrovalley monolayers, a minimal $\mathbf{k}\cdot\mathbf{p}$ model is used to describe electron behavior near points $K_\pm$. An external electric field is taken into account as the term H_E. Then the total Hamiltonian is expressed as follows:

$$H(\boldsymbol{k}) = \begin{bmatrix} H_0^u(\boldsymbol{k}) - H_{\mathrm{SOC}} - H_{\mathrm{ex}} - H_E & H_\perp \\ H_\perp & H_0^l(\boldsymbol{k}) + H_{\mathrm{SOC}} + H_{\mathrm{ex}} + H_E \end{bmatrix} \tag{3.9}$$

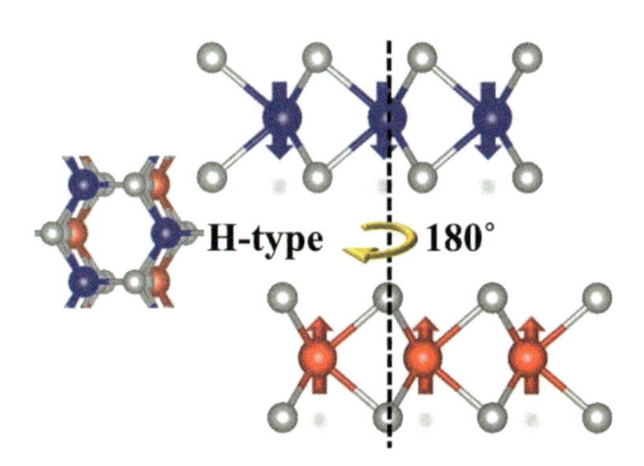

Figure 3.10
Schematic views of the antiferrovalley bilayer with H-type stacking. Silvery, red, and blue spheres represent chalcogen, spin-up, and spin-down transition-metal atoms, respectively.

Note that signs of the terms originating from the SOC effect H_{SOC}, as well as that of the ones from exchange interaction H_{ex}, are reversed for the upper and lower monolayers. The former is due to the exchanged valley index, imported by the H-type valley-alignment. The antiferrovalley coupling causes the opposite magnetic moment between two single layers which means the $K_+(K_-)$ of the lower layer is equivalent to the $K_+(K_-)$ of the upper layer and then leads to the reversed sign of H_{ex}.

In consideration of the circularly polarized optical selectivity for valleys, the two valleys are distinguished by using the symbol $K\sigma_+$ and $K\sigma_-$ rather than K_+ or K_- itself. In other words, if the states at a certain valley that can only be excited by the left-handed light, they are situated at the valley $K\sigma_+$. For the case where the optical transition is locked with right-handed light, the valley is defined as $K\sigma_-$.

As shown in Fig. 3.11A, without an electric field, spatial inversion symmetry accompanies with antiferrovalley layer coupling in the pristine H-type bilayer with D_{3d} point group, resulting in energetically degenerated band structures between two monolayers. However their spin states and valley indexes are opposite as a result of the reversal of the magnetic moment. Optical selections rules show that for the inversion symmetric bilayer, electrons occupying the double-degenerated VB maxima can be excited by both the right-handed and left-handed lights, indicating the energy equivalency between two valley indexes.

For the Hall effect in the antiferrovalley system, a moderate hole-doping bilayer is adopted. The majority carriers are spin-down holes from the valley $K\sigma_+$ of the upper monolayer and spin-up $K\sigma_-$ holes from the lower layer. When a longitudinal bias voltage is applied, the spin-up and spin-down holes gain transverse velocities in opposite directions, and accumulate toward the right and left side of the ribbon, accordingly, as shown in Fig. 3.12A. Although the accumulation of carriers in the boundary of the upper and lower layer come from the same K_- point, they respectively correspond to the left and right circularly polarized radiations. Obviously, the valley index is opposite as the same as the spin ones between the upper layer and lower layer. Such a phenomenon can be called the VHE. Interestingly, in typical bilayer TMDs, the VHE should be induced by an electric field [23,44]. Here even without any external fields, it directly happens in the antiferrovalley bilayer, which is consistent with the presence of valley degeneracy in such system.

When a vertical external bias is applied, the bilayer loses its inherent inversion symmetry and now holds the C_{3v} point group. As illustrated in Fig. 3.11B and C, a positive (negative) electric field moves the subbands from the lower (upper) layer upward, with respect to those from the upper (lower) monolayer. Therefore the spin and valley degeneracy between two monolayers is decoupled with an interlayer potential difference. Since the external

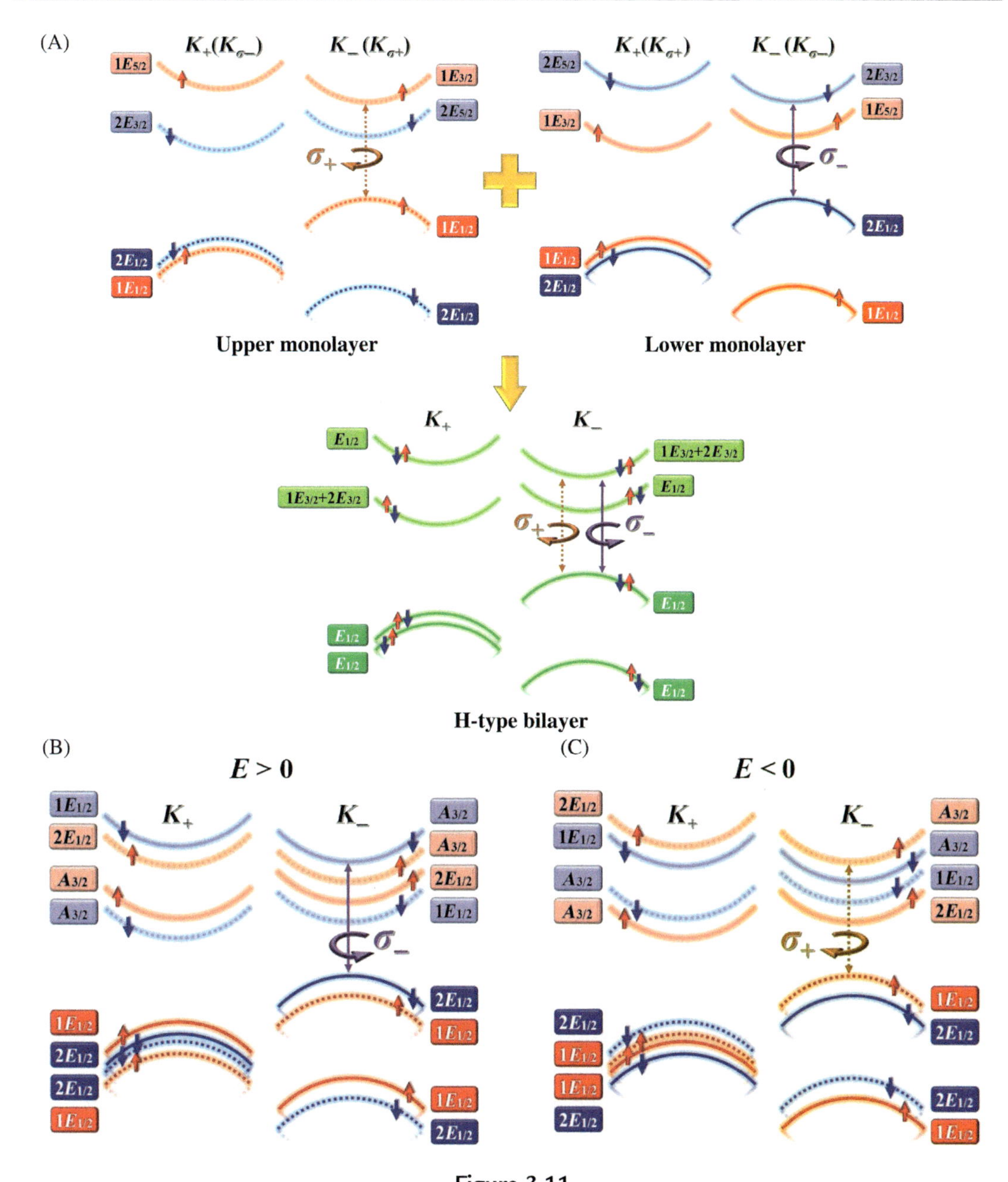

Figure 3.11

The band structures near points $K\pm$. (A) Without electric field, under a(B) positive, and (C) negative perpendicular electric field. Optical selection rules for the top VB are plotted. σ_+ and σ_-correspond to the left-handed and right-handed radiations, accordingly. *VB*, valence band.

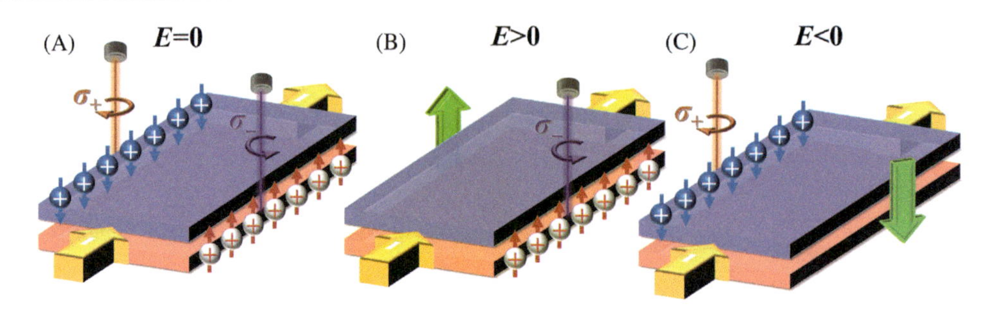

Figure 3.12

Illustration of the VHE and field-controlled AVHE in moderately hole-doped antiferrovalley bilayers. (A) Without electric field, (B) positive, and (C) negative perpendicular electric field are applied. The carriers denoted by " + " symbol are holes. Upward (downward) arrows with red (*blue*) color indicate spin-up (spin-down) states. Green arrows show the direction of the vertical *E* field.

bias is effective to break and optionally tune the valley degeneracy, its manipulation for AVHE is obvious.

To explore the Hall effect in the bilayer antiferrovalley system with the existence of a perpendicular electric field, a slightly p-type doping bilayer with Fermi energy lying between the VB edges of two monolayers is adopted. When the applied *E* field is positive (shown as Fig. 3.12B), the majority carriers, that is spin-up holes from $K\sigma_-$ valley of the lower monolayer, acquire anomalous velocities proportional to the positive Berry curvatures and then accumulate in the right side of the sample. By contraries, When the applied *E* field is negative (shown as Fig. 3.12C), the majority carriers, that is spin-down holes from $K\sigma_+$ valley of the lower monolayer accumulate in the left side of the sample, which obviously leads to the measurable transverse voltage with opposite sign. The combination of valley, spin, and charge accumulations in such case, similar to magnetic controlled ferrovalley 2H-VSe_2 monolayer, manifesting the presence of AVHE.

The previous minimal $\mathbf{k} \cdot \mathbf{p}$ model proposed the feasibility of governing the energetically valley degeneracy and eventually the AVHE utilizing a vertical electric field in the antiferrovalley bilayer. However in the actual VSe_2 system, the relatively small but nonnegligible SOC effect couples with strong exchange interaction make its band structures quite different from the representative ones displayed in Fig. 3.11. Taking bilayer VSe_2 as an example, density-functional theory calculations are performed to demonstrate it.

First of all, in the absence of the external electric field, the band structure of bilayer antiferrovalley VSe_2 are displayed in middle panel of Fig. 3.13B. Due to the interlayer antiferromagnetic configuration, spin-up and spin-down subbands are originally overlapped.

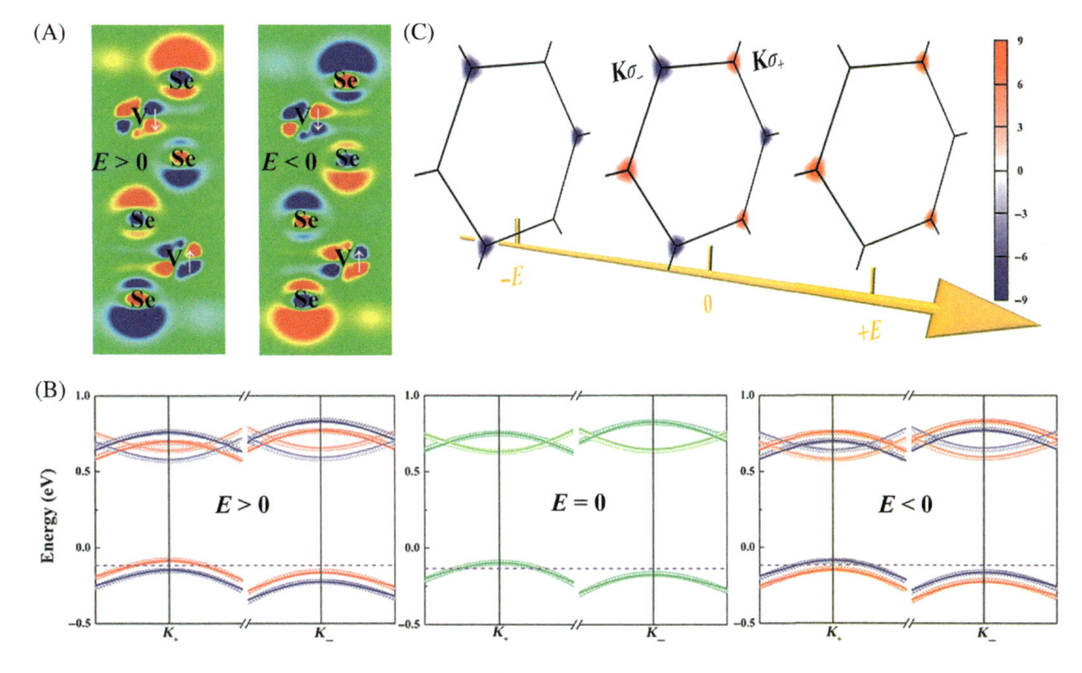

Figure 3.13

Charge density difference, band structures, and Berry curvatures. (A) Charge density difference of H-stacking VSe_2 bilayer between nonzero and zero external field with electron depletion shown in blue, and electron accumulation in red. (B) Electric field-dependent band structures near points $K\pm$. Solid lines are gained from first-principles calculations. Open triangles are the results from minimal $\mathbf{k}\cdot\mathbf{p}$ model. Upward-pointing (downward-pointing) triangles correspond to electronic states from lower (upper) monolayer. (C) Contour maps of Berry curvatures at valleys near the top VB. *VB*, valence band.

The VB maxima located at the valley $\boldsymbol{K}_+$ are comprised by $(|d_{x^2-y^2}> + i|d_{xy}>)/\sqrt{2}$ states of cation in lower layer and $(|d_{x^2-y^2}> - i|d_{xy}>)/\sqrt{2}$ ones from upper layer's V atom. Although these two states occupied the same $\boldsymbol{K}_+$ point in momentum space, they are with different symmetry and thus correspond to opposite valley index.

Based on the conservation of overall azimuthal quantum number, for the states in lower layer, it can only be excited by the left-handed light, namely its $K\sigma_+$ valley index. While the upper layer ones are related to right-handed light, indicating the valley $K\sigma_-$ states. Therefore in the absence of an electric field, band structures demonstrate the spin degeneracy, as well as the valley doublet.

When a positive vertical electric field with the magnitude of 0.1 V/Å is applied to the surface of the film, electronic charge transfer between two layers occurs, as shown in the left panel of Fig. 3.13A. In addition, the electric potential difference makes the states from one monolayer moves upward (see right panel of Fig. 3.13B), while the other layer atoms contribute to the opposite behavior. The direct band gap is located at K_+, and the VB top is occupied by the spin-up electrons from the lower layer, corresponding to $K\sigma_+$ valley. When the electric field is negative, as shown in right panel of Fig. 3.13A, the direction of the potential difference is opposite to that of positive electric field. The direct band gap is located at K_+, which is the same as $E>0$ case. However the VB top is occupied by the spin-down electrons from the upper layer, corresponding to $K\sigma_-$ valley. As a result, the degeneracy of band structure is broken, further lifting the degeneracy of spin and the valley in the bilayer.

Through calculating nonzero z-component Berry curvatures from the Kubo-formula [38], the influence of a perpendicular electric field on the Hall effect for a p-type bilayer (with Fermi level shown as violet dashed lines in Fig. 3.13B) is inspected. Although the global top of the VB is located in the Γ point, the carriers around there possess almost zero Berry curvatures which means no contribution to the Hall effect and therefore are not considered in our analysis below.

As shown in Fig. 3.13C, in the absence of electric fields, the majority carriers, that is spin-down holes from the valley $K\sigma_+$ of the lower layer and $K\sigma_-$ spin-up holes of the upper monolayer, acquire opposite transverse velocities because of their opposite Berry curvatures in the same magnitude, then moving to different sides. The long lived spin and valley accumulations on sample boundaries in the Hall bar demonstrate the existence of the VHE. When $E>0$, the majority carriers are spin-down holes from valley $K\sigma_+$, while $E<0$, is the $K\sigma_-$ spin-up holes. Obviously, a perpendicular electric field introduces additional charge Hall current. Net carriers derived from an individual valley accumulate in a single side of the ribbon, leading to a measurable transverse voltage. The combination of spin, valley, and charge accumulations here implies the AVHE induced by electric fields. Furthermore the sign of the Hall voltage depends strongly on the direction of the applied field. When the applied electric field is reversed, the valley index of the majority carriers would be opposite. The reverse sign of Berry curvatures causes their accumulations on the other boundary. Then the transverse Hall voltage would be opposite. Antiferrovalley VSe_2 bilayer thus presents an electrically tunable system for the AVHE.

This purely electrical driven and tuning AVHE demonstrates the feasibility for realizing all-electric valleytronic devices in antiferrovalley bilayers, where the valley information can be controlled by the electric field, and easily read out through the sign of the Hall voltage. The bit information is written by the vertical electric field. The absence of the field refresh leads to the extremely low power consumption. And the vertical electric recording method is beneficial to reduce the size of the information bit and finally improve the recording density; the reading of information is realized by measuring transverse voltage of the Hall bar, which is nondestructive. Furthermore when the volatile electric field is replaced by some ferroelectric substrates or even construct the antiferrovalley bilayer using polar TMDs

[45], it is possible to realize the information functional devices with advantage in nonvolatility, which will undoubtedly promote the application of valley electronics in the next-generation.

3.3.3 GeSe-Ferrovalley materials induced by ferroelectricity

At present stage, the spontaneous valley polarization in hexagonal ferrovalley materials can originate from intrinsic ferromagnetism, and be manipulated by electric field through the design of antiferrovalley bilayer. If we could combine ferrovalley with ferroelectricity, such an additional mechanism would provide the additional opportunity to realize the attractive electrical control of ferrovalley states in a more straightforward way. Very recently, through both experimental works and first-principles calculations, a number of intrinsic ferroelectrics based on 2D van der Waals materials with switchable in-plane, or out-of-plane ferroelectricity have been reported, as shown in Fig. 3.14. Therefore the emergence of 2D ferroelectrics offers exciting prospects for the next-generation technological revolution.

In particular, as one of the promising 2D ferroelectrics, monolayer group-IV monochalcogenides (MXs, M = Ge, Sn; X = S, Se, Te) have been predicted to be a new member of ferrovalley materials [51], makes the realization of tuning VDF through electric means possible. In analogy to black phosphorus [52], these waved materials exhibit orthorhombic lattice with 2D rectangular BZ [53], as presented in Fig. 3.15A. A relative displacement between cations and anions will occur upon applying an in-plane electric field, as shown in Fig. 3.15B, driving the nonpolar phase to convert into two equivalent polar phases [54,57]. Based on the first-principles calculations, the double-well potential energy plots have strongly demonstrated the ferroelectric behavior as a function of the atomic displacement [55], as obtained in Fig. 3.15B. The robust in-plane ferroelectricity has also been confirmed in atomic-thick SnTe film by the experimental work [58]. The valleytronics-related research generally focused on 2D hexagonal lattice systems. More importantly, multi-valley band structures are proved to exist in these orthorhombic structures [56], as shown in Fig. 3.15C, providing a promising and entirely new platform for the studies of the fundamental physics in valleytronics. The combination of intrinsic ferroelectricity, and valley physics makes the realization of tuning VDF in monolayer group-IV monochalcogenides through electric means possible.

Taking the monolayer GeSe as an example, the primitive cell without relative displacement between cations and anions is plotted in Fig. 3.16A. The pristine monolayer GeSe with inversion symmetry indicates an in-plane paraelectric phase, which can be labeled as P_0. The multi-valley characteristics can be obviously confirmed by the band structure of the nonpolar GeSe monolayer, as displayed in Fig. 3.16B. Instead of the symmetry corners of the BZ in the honeycomb systems, two pairs of valleys locate in the X–Γ and Y–Γ

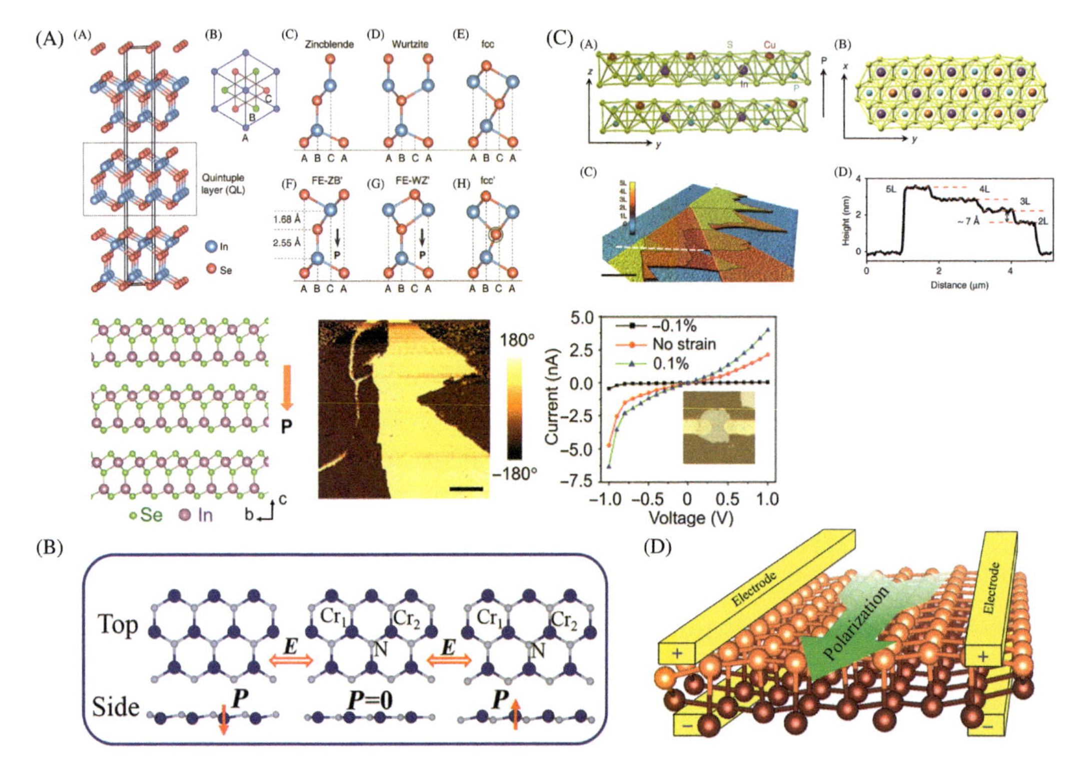

Figure 3.14

2D ferroelectrics in (A) In2Se3 and other III_2-VI_3 van der Waals multilayer; (B) buckled monolayer CrN; (C) $CuInP_2S_6$ thin film; (D) atomic-thick phosphorene nanoribbons. Source: *(A) Permission from W. Ding, et al., Prediction of intrinsic two-dimensional ferroelectrics in In2Se3 and other III2-VI3 van der Waals materials, Nat. Commun., 8 (2017) 14956; Y. Zhou, et al., Out-of-plane piezoelectricity and ferroelectricity in layered α-In2Se3 nanoflakes, Nano Lett. 17(9) (2017) 5508–5513 [46,47]; (B) Permission from W. Luo, K. Xu, H. Xiang, Two-dimensional hyperferroelectric metals: a different route to ferromagnetic-ferroelectric multiferroics, Phys. Rev. B, 96(23) (2017) 235415 [48]; (C) Permission from F. Liu, et al., Room-temperature ferroelectricity in CuInP2S6 ultrathin flakes, Nat. Commun. 7 (2016) 12357 [49]; (D) Permission from T. Hu, et al., New ferroelectric phase in atomic-thick phosphorene nanoribbons: existence of in-plane electric polarization, Nano Lett. 16(12) (2016) 8015–8020 [50].*

high-symmetry paths, which are referred to as valleys V_x and V_y. According to the paraelectric GeSe monolayer with space group *Cmcm* [54], these two valleys are related by the fourfold rotational symmetry and energetically degenerate with the C_{2v} point group symmetry. For the V_x (V_y) valley, the occupied states in the VB maximum (VBM), and the conduction band minimum (CBM) are purely dominated by p_x (p_y) atomic orbitals of Se and Ge atoms, respectively. The band gaps of V_x and V_y valleys are identical (about 0.83 eV), implying the degeneracy between valleys. On account of this, the paraelectric

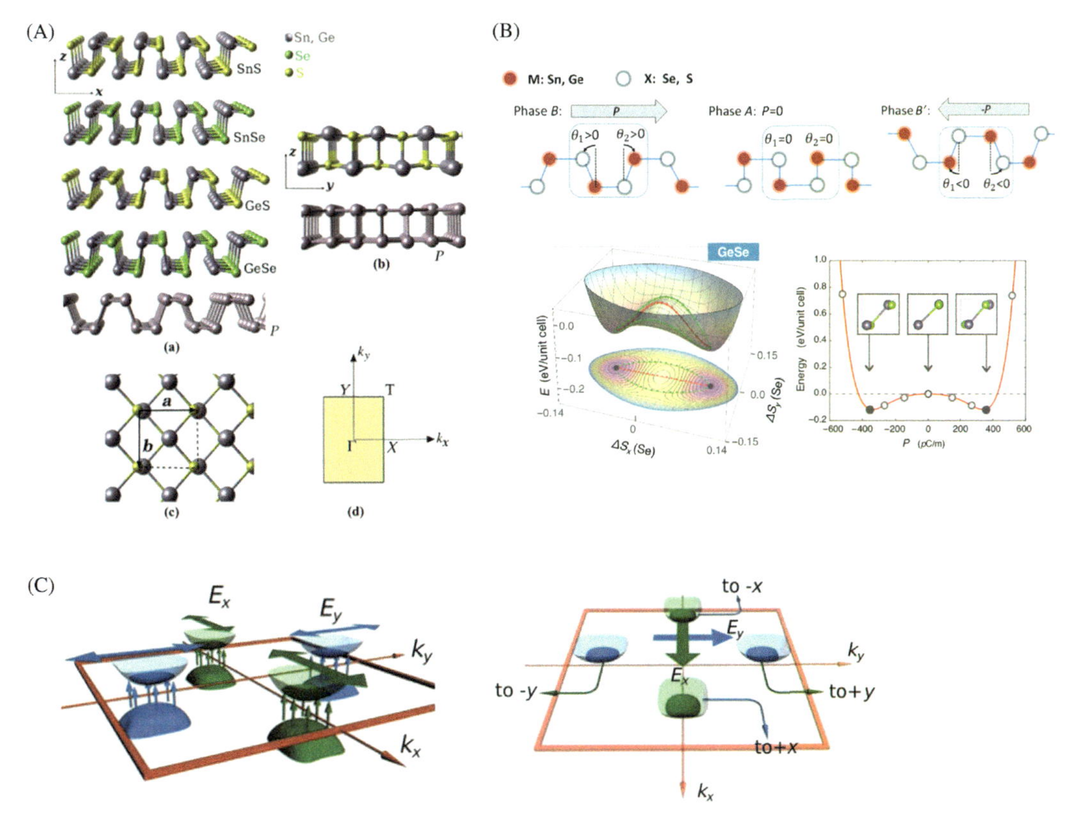

Figure 3.15

(A) Schematic diagrams of monolayers group-IV monochalcogenides and respective high-symmetry paths of the BZ. (B) Geometric distortion and double-well potential energy plots. (C) Valley characteristics displayed in the 2D BZ. *BZ*, Brillouin zone. Source: *(A) Permission from L.C. Gomes, A. Carvalho, Phosphorene analogues: isoelectronic two-dimensional group-IV monochalcogenides with orthorhombic structure, Phys. Rev. B, 92(8) (2015) 085406; (B) Permission from R. Fei, W. Kang, L. Yang, Ferroelectricity and phase transitions in monolayer group-IV monochalcogenides, Phys. Rev. Lett. 117(9) (2016) 097601; H. Wang, X. Qian, Two-dimensional multiferroics in monolayer group IV monochalcogenides. 2D Mater. 4(1) (2017) 015042 [54,55]; (C) Permission from A. Rodin, et al., Valley physics in tin (II) sulfide. Phys. Rev. B, 93(4) (2016) 045431.*

GeSe monolayer demonstrates the paravalley state that is state without spontaneous valley polarization.

Different from the valley-selective circular dichroism in hexagonal lattice, distinctive valley-related optical properties for valleys exists in the new member of valleytronic materials. The valley-dependent optical selection rules in monolayer GeSe can be demonstrated from the conservation of overall azimuthal quantum number [23].

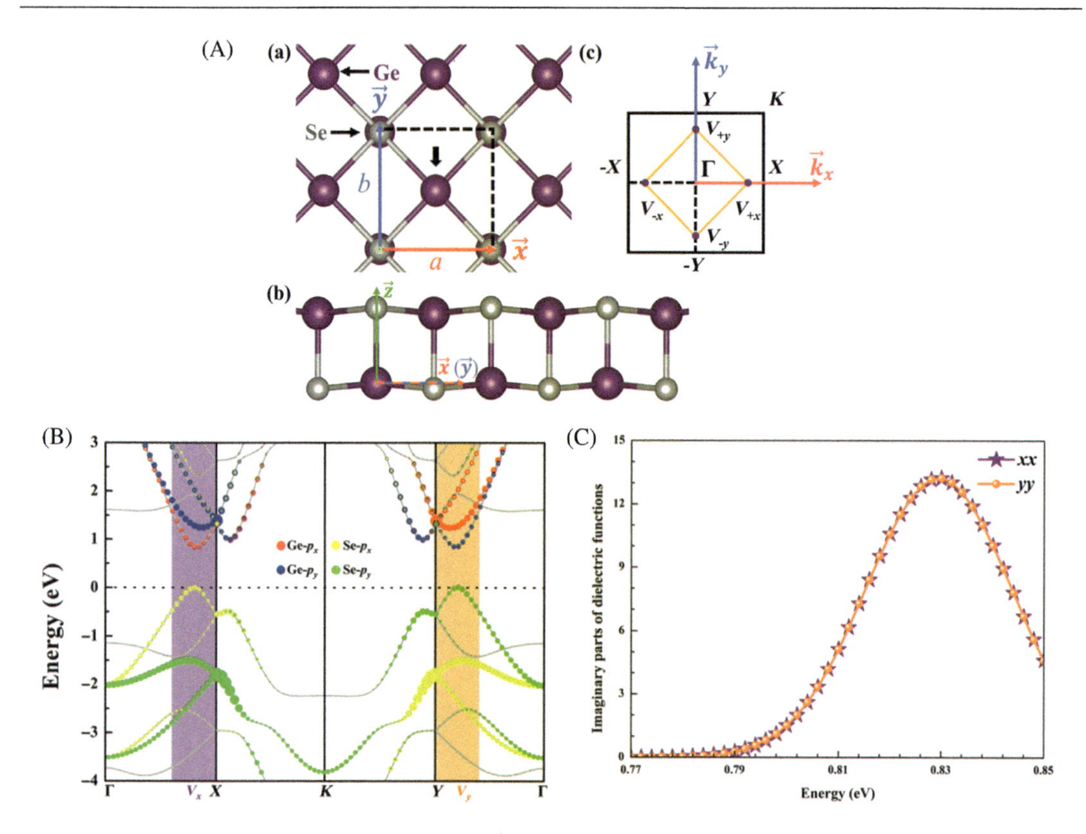

Figure 3.16

(A) Top and side views of the lattice structure of paraelectric GeSe monolayer with the rectangular BZ. Two pairs of valleys $V_{\pm x}$ and $V_{\pm y}$ locate in the $\pm X$–Γ and $\pm Y$–Γ symmetry lines, respectively. (B) Band structure, and (C) the imaginary parts of complex dielectric function ε_2 of paraelectric GeSe monolayer. The radius of solid color circles is proportional to the contributions of specific orbitals: the red (*yellow*) and blue (*green*) one represent p_x and p_y of Ge (Se) atoms. The Fermi level is set to zero. The optical characters excited by both linearly *x*- and *y*-polarized light are presented. *BZ*, Brillouin zone.

Assuming that the local atomic states bear different principal quantum number n, azimuthal quantum number l, and magnetic quantum number m_l, then the local wavefunctions can be expressed as:

$$\phi_{n,l,m_l}(r - R_j) = R_{n,l}(r - R_j)Y_{l,m_l}(\theta, \varphi) \tag{3.10}$$

Here (r, θ, φ) are the standard spherical coordinates. Wavefunctions can be constructed as Bloch functions in terms of atomic orbitals $\phi(r - R_j)$ localized at the lattice site R_j:

$$\psi = \frac{1}{\sqrt{N}}\sum_j e^{ik \cdot R_j}\phi(r - R_j) \tag{3.11}$$

where N is the number of unit cells for the crystal, and k is the wavevector. For the V_x (V_y) valley in the monolayer GeSe, the basic functions of conduction and VB are $|\psi_c\rangle = |p_x\rangle(|p_y\rangle)$ on Ge and $|\psi_v\rangle = |p_x\rangle(|p_y\rangle)$ on Se atoms, respectively. When the twofold rotation (C_2) is applied, the phase factor $e^{ik\cdot R_j}$ in the wave functions guarantees the Bloch phase to shift from one lattice site to another. Simultaneously, due to the symmetry operation, the rotation of local atomic coordinates of the spherical harmonics gives rise to the desynchronization of the azimuthal phase. After taking these two distinct contributions into account, we can deduce the transformation of electronic states associated with the twofold rotation easily. For the adapted symmetry C_{2x} at V_x valley, after tedious yet straightforward calculations, we have

$$\hat{C}_{2x}|\psi_i(\mathbf{V}_x)\rangle = |\psi_i(\mathbf{V}_x)\rangle, i = c, v \tag{3.12}$$

Here the symbol c and v refer to the CBM and VBM, respectively. We notice that basis functions with linear combination of atomic orbitals are invariant through rotational symmetry. Then the overall azimuthal quantum number M_x is eventually calculable,

$$\exp[iM_x(\varphi + \pi)] = \exp[iM_x\varphi] \tag{3.13}$$

It is obviously that the M_x is equal to zero for both the VBM and CBM, then, $\Delta M_x = 0$. As a consequence, the optical transition from VBM to CBM at V_x valley could only be excited by the x-polarized light, corresponding to the symmetry of $\cos(\theta)$ in the case. The similar approach could be adopted for the V_y valley. When the twofold rotation C_{2y} applied, $\Delta M_y = 0$ at V_y valley can be deduced, implying the absorption of photons polarized along y-direction. Therefore the conservation of overall angular momentum selection rule shows that, linearly polarized optical selectivity for valleys exists in such orthorhombic system, which is distinct from that of hexagonal materials.

On account of the specific valley-selective linear dichroism, the calculated optical properties excited by x- and y-polarized radiation for the paravalley GeSe monolayer is shown in Fig. 3.16C. Due to the identical optical band gap at valleys V_x and V_y, the optical curves are completely overlapped. These degenerate optical properties are due to the paravalley state in such monolayer group-IV monochalcogenides.

Compared with the paraelectric state, the ferroelectric phase of such orthorhombic systems can be confirmed as the ground state by both the theoretical and experimental works, which is dynamically and thermally stable at high transition temperature [59–61]. Fig. 3.17A shows the ferroelectric structures of monolayer GeSe, where the inherent inversion symmetry has been broken. The relative displacement between Ge and Se atoms will occur along x and y axes, making the ferroelectric states with relative polarization direction labeled as P_x and P_y states, accordingly.

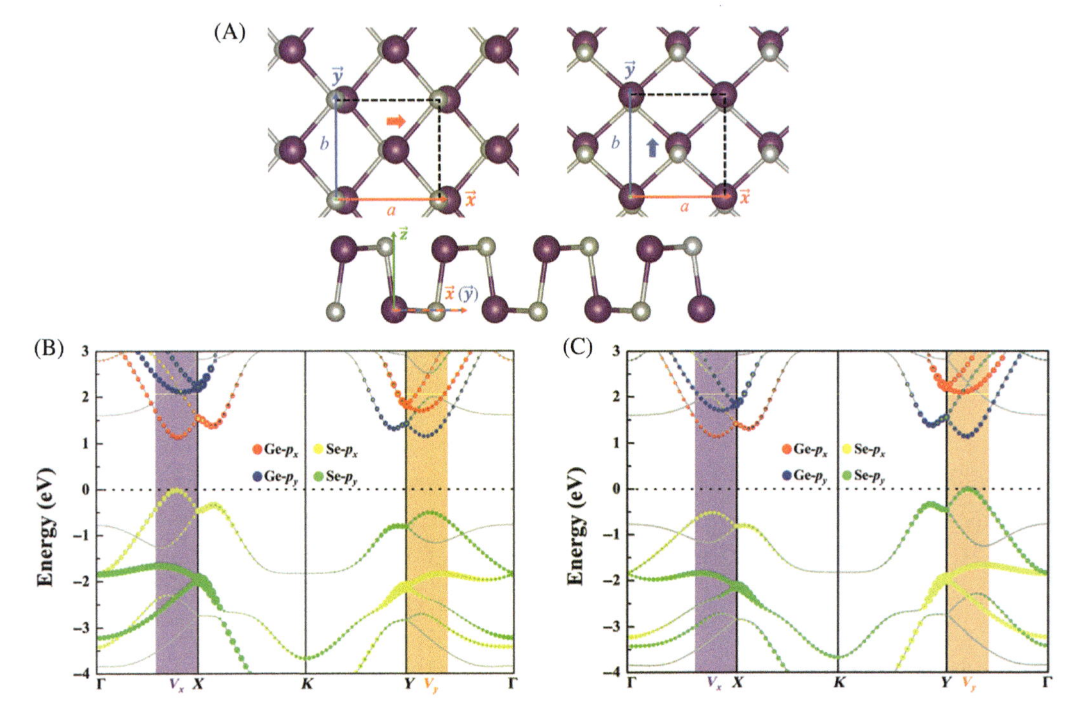

Figure 3.17
(A) Top and side views of the lattice structure for monolayer GeSe with ferroelectric polarization P_x and P_y states. The arrows point out the orientation of polarization along x and y axes, respectively. The band structure of ferroelectric phase GeSe monolayer in (B) P_x and (C) P_y states.

The band structures of the ferroelectric GeSe monolayer are plotted in Fig. 3.17B and C. We can note that the occupied states of VBM and CBM at valleys V_x and V_y are almost the same as the paraelectric phase. However due to the lack of fourfold rotational symmetry, the band gaps of two pairs of valleys are no longer identical. When the ferroelectric GeSe monolayer under P_x case, the V_x valley owns the global band gap with the magnitude of 1.14 eV, however, a larger band gap around 1.67 eV now locates in V_y valley.
The nondegenerate valleys induced by ferroelectricity suggest the emergence of ferrovalley state with spontaneous valley polarization. Through the group theory analysis, the degeneracy of point group in two pairs of valleys has also been lifted. The V_x valley still belongs to C_{2v} points group with optical excitation locked by x-polarized light. Whereas the V_y valley with lower symmetry holds the C_s point group, which is induced by ferroelectricity along x-direction. Now, the optical transition at V_y valley can be excited by both the x- and y-polarized light.

The calculated optical properties excited by linearly x-polarized light and y-polarized light in ferrovalley GeSe monolayer are shown in Fig. 3.18. As the occurrence of energetically

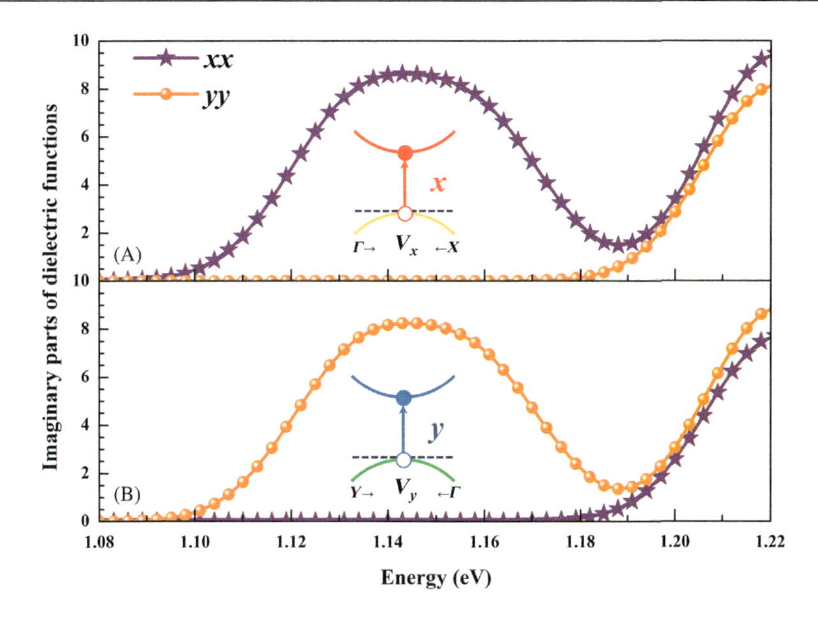

Figure 3.18
The imaginary parts of complex dielectric function ε_2 excited by linearly x-polarized light and y-polarized light of ferrovalley GeSe monolayer in (A) P_x and (B) P_y states.

nondegenerate valleys, the optical band gaps of linearly polarized light are decoupled. The optical curve of yy component experiences a blue shift compared with that of the xx component. The first peak with the photon energy of about 1.14 eV appears on the curve related to x-polarized light. Due to the Gaussian smearing, it implies the optical absorption starts from the energy of 1.14 eV, which is consistent with the direct optical band gap at V_x valley of ferroelectric GeSe monolayer. For the ferroelectric polarization along y-direction, the global band gap now locates in V_y valley with the same magnitude as P_x state, implying the polarity of valley polarization is reversed in the meantime. In such P_y state, the V_y valley reverts to C_{2v} point group with the optical transition can only be excited by the y-polarized light. Similarly the interband transitions at V_x valley with the symmetry reduced C_s point group now couple to the double-degenerate in-plane linearly polarized radiations. The energetical nondegeneracy between valleys is obviously verified as unequal optical band gaps between xx and yy components.

An idea for the unique valleytronic device application is thus inspired. Based on the monolayer GeSe, the prototype of electrically controllable polarizer is displayed in Fig. 3.19. When the in-plane electric field is applied along the x-axis, the valley polarization could be induced with the global band gap locating in V_x valley. Then a laser beam with the energy of 1.14 eV is incident on the monolayer GeSe. After absorbing the laser beam, only the electrons at V_x valley will jump to the excited state. In consideration of the photoluminescence process, the excited electrons finally re-emit radiations when they

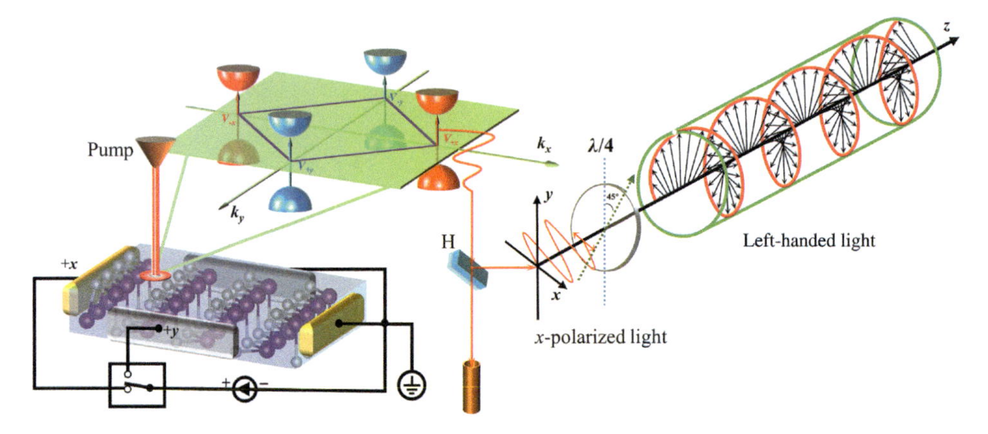

Figure 3.19
Proposed valleytronic device based on the ferrovalley GeSe monolayer. In the electrically tunable polarizer, x- or y-polarized light can be optionally generated, depending on the direction of external electric fields. Specific chirality of circularly polarized light is available and reversible via a place of quarter-wave plate. H is a half mirror to transmit the polarized light for different demands.

drop back to the ground state. Due to the optical selectivity for the V_x valley, the radiation along x-direction will be excited. Through a double-throw switch, the direction of electric field can be converted to y axis, resulting in the reversal of the polarity of valley polarization. Now, the V_y valley possesses the smallest band gap. Its symmetry indicates the emission of y-polarized radiation. Furthermore the polarizer could electrically manipulate the chirality of circularly polarized light. Only if a set of quarter-wave plate is placed perpendicular to the propagation of linearly polarized light, with its optical axis deviating 45 degrees anticlockwise from the x-direction. The x- or y- polarized light then transfers to the left- or right-handed radiations. Hence, through controlling the direction of external electric fields, a laser beam with specific wavelength can be electrically polarized to linearly, and even circularly polarized radiation. Compared with traditional mechanically driven ones, electrical control of the polarization state of polarized radiation in this ferrovalley-based polarizer is a more accurate and rapid way. Additionally, the proposed device based on 2D materials meets the requirement of next-generation electronic products towards miniaturization and multi-functionality.

3.4 Summary and outlook

In conclusion, combining the theory with first-principles calculations, the concept of ferrovalley materials with spontaneous valley polarization have been recently proposed as a new member of ferroic family. In the hexagonal systems, 2H-VSe_2 monolayer is predicted to be such a ferrovalley material, due to the coexistence of spin- orbit coupling and intrinsic exchange interaction. The polarity of valley polarization in this ferromagnetic semiconductor can be tunable via external magnetic field, and further detected by the circularly polarized light. Through elaborate structural design, antiferrovalley bilayer VSe_2, where the valley information can be controlled by the electric field, is promising to offer an all-electric approach to realize valleytronic devices. In addition, monolayer group-IV monochalcogenides in ferroelectric state is an entirely new ferrovalley member, extending the concept of ferrovalley materials from hexagonal to orthorhombic systems. The spontaneous valley polarization in this new type of ferrovalley materials can be induced by the intrinsic ferroelectricity, instead of ferromagnetism. Moreover the optical selectivity for valleys is now coupled to the linearly polarized light. The polarity of valley polarization is optionally switchable via external electric fields, making the realization of the electrical control of ferrovalley states.

In terms of future research, we can focus on the following potentially interesting directions: For the hexagonal honeycomb low-dimensional systems, the underlying physical mechanisms of the interaction among valley, spin, and charge degree of freedoms is of great importance to explore. Although many prototypes of functional devices based on the ferrovalley materials have been designed, there is still a long way for their use in the practical applications. Therefore experimental efforts on the proposed ferrovalley materials, to confirm the abundant valley physics and achieve some related valleytronic devices are necessary. Finally recent valleytronic researches almost rely on the 2D materials, however the concept of valley index originates from the bulk systems. If we could predict new ferrovalley materials in the three-dimensional systems through systematic screening, it may provide an interesting and entirely new platform for the future study in valleytronics.

References

[1] M.N. Baibich, et al., Giant magnetoresistance of (001)Fe/(001)Cr magnetic superlattices, Phys. Rev. Lett. 61 (21) (1988) 2472–2475.

[2] S. Wolf, et al., Spintronics: a spin-based electronics vision for the future, Science 294 (5546) (2001) 1488–1495.

[3] I. Žutić, J. Fabian, S.D. Sarma, Spintronics: fundamentals and applications, Rev. Mod. Phys. 76 (2) (2004) 323.

[4] O. Gunawan, et al., Valley susceptibility of an interacting two-dimensional electron system, Phys. Rev. Lett. 97 (18) (2006) 186404.

[5] A.K. Geim, K.S. Novoselov, The rise of graphene, Nat. Mater. 6 (2007) 183.

[6] K.F. Mak, et al., Atomically thin MoS_2: a new direct-gap emiconductor, Phys. Rev. Lett. 105 (13) (2010) 136805.

[7] K.S. Novoselov, et al., Electric field effect in atomically thin carbon films, Science 306 (5696) (2004) 666–669.

[8] K.S. Novoselov, et al., Two-dimensional gas of massless Dirac fermions in graphene, Nature 438 (2005) 197.

[9] A.H. Castro Neto, et al., The electronic properties of graphene, Rev. Mod. Phys. 81 (1) (2009) 109–162.

[10] Q. Zhang, et al., Large spin-valley polarization in monolayer $MoTe_2$ on top of EuO (111), Adv. Mater. 28 (5) (2016) 959–966.
[11] D. Pesin, A.H. MacDonald, Spintronics and pseudospintronics in graphene and topological insulators, Nat. Mater. 11 (2012) 409.
[12] A. Rycerz, J. Tworzydlo, C.W.J. Beenakker, Valley filter and valley valve in graphene, Nat. Phys. 3 (3) (2007) 172–175.
[13] P. San-Jose, et al., Pseudospin valve in bilayer graphene: towards graphene-based pseudospintronics, Phys. Rev. Lett. 102 (24) (2009) 247204.
[14] X. Xu, et al., Spin and pseudospins in layered transition metal dichalcogenides, Nat. Phys. 10 (5) (2014) 343–350.
[15] A. Kuc, T. Heine, The electronic structure calculations of two-dimensional transition-metal dichalcogenides in the presence of external electric and magnetic fields, Chem. Soc. Rev. 44 (9) (2015) 2603–2614.
[16] D. Xiao, et al., Coupled spin and valley physics in monolayers of MoS_2 and other group-VI dichalcogenides, Phys. Rev. Lett. 108 (19) (2012) 196802.
[17] A.M. Jones, et al., Optical generation of excitonic valley coherence in monolayer WSe_2, Nat. Nanotechnol. 8 (9) (2013) 634.
[18] D. Xiao, W. Yao, Q. Niu, Valley-contrasting physics in graphene: magnetic moment and topological transport, Phys. Rev. Lett. 99 (23) (2007) 236809.
[19] W. Yao, D. Xiao, Q. Niu, Valley-dependent optoelectronics from inversion symmetry breaking, Phys. Rev. B 77 (23) (2008) 235406.
[20] Z. Zhu, Y. Cheng, U. Schwingenschlögl, Giant spin-orbit-induced spin splitting in two-dimensional transition-metal dichalcogenide semiconductors, Phys. Rev. B 84 (15) (2011) 153402.
[21] H. Zeng, et al., Valley polarization in MoS_2 monolayers by optical pumping, Nat. Nanotechnol. 7 (8) (2012) 490.
[22] K.F. Mak, et al., Control of valley polarization in monolayer MoS_2 by optical helicity, Nat. Nanotechnol. 7 (8) (2012) 494.
[23] T. Cao, et al., Valley-selective circular dichroism of monolayer molybdenum disulphide, Nat. Commun. 3 (2012) 887.
[24] A. Kumar, et al., Chiral plasmon in gapped Dirac systems, Phys. Rev. B 93 (4) (2016) 041413.
[25] D. MacNeill, et al., Breaking of valley degeneracy by magnetic field in monolayer $MoSe_2$, Phys. Rev. Lett. 114 (3) (2015) 037401.
[26] E.J. Sie, et al., Valley-selective optical Stark effect in monolayer WS_2, Nat. Mater. 14 (2014) 290.
[27] F. Zhang, et al., Spontaneous quantum Hall states in chirally stacked few-layer graphene systems, Phys. Rev. Lett. 106 (15) (2011) 156801.
[28] A. Srivastava, et al., Valley Zeeman effect in elementary optical excitations of monolayer WSe_2, Nat. Phys. 11 (2) (2015) 141.
[29] G. Aivazian, et al., Magnetic control of valley pseudospin in monolayer WSe2, Nat. Phys. 11 (2) (2015) 148.
[30] A.N. Andriotis, M. Menon, Tunable magnetic properties of transition metal doped MoS_2, Phys. Rev. B 90 (12) (2014) 125304.
[31] Y. Cheng, Q. Zhang, U. Schwingenschlögl, Valley polarization in magnetically doped single-layer transition-metal dichalcogenides, Phys. Rev. B 89 (15) (2014) 155429.
[32] A. Ramasubramaniam, D. Naveh, Mn-doped monolayer MoS_2: an atomically thin dilute magnetic semiconductor, Phys. Rev. B 87 (19) (2013) 195201.
[33] J. Qi, et al., Giant and tunable valley degeneracy splitting in $MoTe_2$, Phys. Rev. B 92 (12) (2015) 121403.
[34] F. Li, K. Tu, Z. Chen, Versatile electronic properties of VSe_2 bulk, few-layers, monolayer, nanoribbons, and nanotubes: a computational exploration, J. Phys. Chem. C 118 (36) (2014) 21264–21274.
[35] H. Pan, Electronic and magnetic properties of vanadium dichalcogenides monolayers tuned by hydrogenation, J. Phys. Chem. C 118 (24) (2014) 13248–13253.
[36] M. Priyanka, S. Ralph, 2D transition-metal diselenides: phase segregation, electronic structure, and magnetism, J. Phys.: Cond. Matt. 28 (6) (2016) 064002.

[37] W.-Y. Tong, et al., Concepts of ferrovalley material and anomalous valley Hall effect, Nat. Commun. 7 (2016) 13612.
[38] D.J. Thouless, et al., Quantized Hall conductance in a two-dimensional periodic potential, Phys. Rev. Lett. 49 (6) (1982) 405–408.
[39] K.F. Mak, et al., The valley Hall effect in MoS_2 transistors, Science 344 (6191) (2014) 1489–1492.
[40] H. Pan, et al., Valley-polarized quantum anomalous Hall effect in silicene, Phys. Rev. Lett. 112 (10) (2014) 106802.
[41] W. Feng, et al., Intrinsic spin Hall effect in monolayers of group-VI dichalcogenides: a first-principles study, Phys. Rev. B 86 (16) (2012) 5391–5397.
[42] M. Ezawa, Valley-polarized metals and quantum anomalous Hall effect in silicene, Phys. Rev. Lett. 109 (5) (2012) 515–565.
[43] W.-Y. Tong, C.-G. Duan, Electrical control of the anomalous valley Hall effect in antiferrovalley bilayers, npj Quantum Mater. 2 (1) (2017) 47.
[44] J. Lee, K.F. Mak, J. Shan, Electrical control of the valley Hall effect in bilayer MoS_2 transistors, Nat. Nanotechnol. 11 (5) (2016) 421–425.
[45] Q.-F. Yao, et al., Manipulation of the large Rashba spin splitting in polar two-dimensional transition-metal dichalcogenides, Phys. Rev. B 95 (16) (2017) 165401.
[46] W. Ding, et al., Prediction of intrinsic two-dimensional ferroelectrics in In2Se3 and other III2-VI3 van der Waals materials., Nat. Commun. 8 (2017) 14956.
[47] Y. Zhou, et al., Out-of-plane piezoelectricity and ferroelectricity in layered α-In2Se3 nanoflakes, Nano Lett. 17 (9) (2017) 5508–5513.
[48] W. Luo, K. Xu, H. Xiang, Two-dimensional hyperferroelectric metals: a different route to ferromagnetic-ferroelectric multiferroics, Phys. Rev. B 96 (23) (2017) 235415.
[49] F. Liu, et al., Room-temperature ferroelectricity in CuInP2S6 ultrathin flakes, Nat. Commun. 7 (2016) 12357.
[50] T. Hu, et al., New ferroelectric phase in atomic-thick phosphorene nanoribbons: existence of in-plane electric polarization, Nano Lett. 16 (12) (2016) 8015–8020.
[51] S. Xin-Wei, et al., Electrically tunable polarizer based on 2D orthorhombic ferrovalley materials, 2D Mater. 5 (1) (2018) 011001.
[52] H. Liu, et al., Phosphorene: an unexplored 2D semiconductor with a high hole mobility, ACS nano 8 (4) (2014) 4033–4041.
[53] L.C. Gomes, A. Carvalho, Phosphorene analogues: isoelectronic two-dimensional group-IV monochalcogenides with orthorhombic structure, Phys. Rev. B 92 (8) (2015) 085406.
[54] R. Fei, W. Kang, L. Yang, Ferroelectricity and phase transitions in monolayer group-IV monochalcogenides, Phys. Rev. Lett. 117 (9) (2016) 097601.
[55] H. Wang, X. Qian, Two-dimensional multiferroics in monolayer group IV monochalcogenides, 2D Mater. 4 (1) (2017) 015042.
[56] A. Rodin, et al., Valley physics in tin (II) sulfide, Phys. Rev. B 93 (4) (2016) 045431.
[57] M. Wu, X.C. Zeng, Intrinsic ferroelasticity and/or multiferroicity in two-dimensional phosphorene and phosphorene analogues, Nano Lett. 16 (5) (2016) 3236–3241.
[58] K. Chang, et al., Discovery of robust in-plane ferroelectricity in atomic-thick SnTe, Science 353 (6296) (2016) 274–278.
[59] V.L. Deringer, R.P. Stoffel, R. Dronskowski, Vibrational and thermodynamic properties of GeSe in the quasiharmonic approximation, Phys. Rev. B 89 (9) (2014) 094303.
[60] Y. Hu, et al., GeSe monolayer semiconductor with tunable direct band gap and small carrier effective mass, Appl. Phys. Lett. 107 (12) (2015) 122107.
[61] A. Shafique, Y.-H. Shin, Thermoelectric and phonon transport properties of two-dimensional IV–VI compounds, Sci. Rep. 7 (1) (2017) 506.

CHAPTER 4

Ferromagnetism in two-dimensional materials via doping and defect engineering

Yiren Wang[1] and Jiabao Yi[2]

[1]*School of Materials Science and Engineering, Central South University, Changsha, P.R. China,*
[2]*Global Innovative Centre for Advanced Nanomaterials, School of Engineering, The University of Newcastle, Callaghan, NSW, Australia*

Chapter Outline

4.1 Introduction

Electronic devices have been playing important roles in today's informationsociety. The key components of daily used electronic products, such as mobile phones, computers, and electronic appliances are electronic devices. Conventional electronic devices, which are made from semiconductors, such as field-effect transistor, computer processor, and random-access memories, are all based on the manipulation of charge transport. The charge flow on/off controlled by the gate voltage represents the digital states 1/0 of the devices. Different functions are realized via the integrated circuit composed of transistors, capacitors, resistors, and diodes. These semiconducting electronics have been developed rapidly since the 1970s, which are popularly considered as following Moore's law. The Moore's law has successfully predicted that the power of these semiconductor devices doubles in approximately 18 months for several decades. However, now it is almost reaching the limits due to the continuous shrink of the chip size and limitations of the current microelectronic technique. Spintronics device is proposed to be a good candidate to face the challenge by substituting the

Spintronic 2D Materials.
DOI: https://doi.org/10.1016/B978-0-08-102154-5.00008-4

established semiconductor devices. Spin, instead of charge, is used as the logical unit for spintronics devices. For one electron spin, it has two states, spin up and spin down. In ferromagnetic (FM) materials, the spins can be aligned in one direction and a small magnetic field can manipulate the spin state from up to down. The two states of up and down can be used as the logic "on" and "off" as that of a semiconductor device. Combined both charge and spin information of electrons, spintronics devices are expected to have broader properties and applications, since the spin-devices will have many advantages over conventional semiconductor devices, such as low power consumption, high data processing speed, high density, and capacity and can flow a spin current without dissipation.

The success of spin devices will strongly depend on the development of new materials. For the application of spin-related devices, high spin polarization is required, preferably, 100%, which is for the high efficiency of spin injection and spin detection. In addition, the spin alone cannot provide an amplifying function. Hence, spin-related devices, such as spin light-emitting diode and spin-field-effect transistor (spin-FET) require that the materials should possess both semiconductor and FM behavior. Most importantly, the spin of the materials must be able to be manipulated by the electric field, which is for the realization of device functionality. The structure of the proposed spin-FET is shown in Fig. 4.1 [1]. When the gate voltage is 0, the source-polarized spins can flow without flipping to the drain and detected by a diluted magnetic semiconductor (DMS), one kind of promising spintronics materials. If a gate voltage is applied, the polarized spins at source will be randomized by the applied gate voltage. Hence, polarized spins cannot be detected by the drain. The flow of polarized spins can be as "on" and "off" for the two states.

To achieve polarized spin for spin-field effect transistors, several ways have been used. One is spin injection, to inject polarized spin into the semiconductor by the contact of magnetic metal and semiconductors. However, the spin injection efficiency is generally very low. Hence, people have used tunneling junction effect by depositing a very thin layer of oxide (i.e., MgO and Al_2O_3) between the magnetic metal films and semiconductors. Therefore the spin injection efficiency has been greatly improved. Another method is to dope magnetic

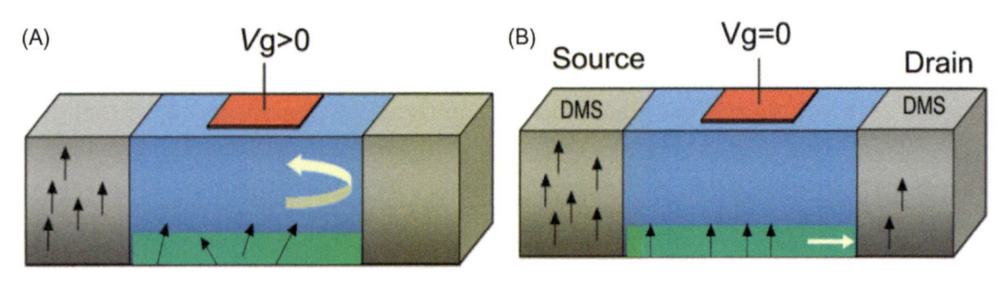

Figure 4.1
Conception of spin-field effect transistors.

elements into the semiconductor, which is also called DMS. The magnetic moment is to induce spin polarization and semiconductor host provides semiconductivity functions. The materials for producing DMS have been extensively investigated both experimentally and theoretically. Mn-doped GaAs is one of the model examples, which is very promising for spintronics devices. However, its low Curie temperature nature limits its practical applications. Subsequently, oxide-based DMSs have been extensively investigated and room-temperature ferromagnetism has been reported. However, clusters and secondary phases have been considered as the origins of ferromagnetism [2]. There is no conclusion for the origins of ferromagnetism yet and there is still a long way to achieve qualified DMS materials for spintronics devices.

Nanomaterials research focuses on the materials with structures at the nanometer scale, which can exhibit unique structural, optical, and electronic properties. However, at this scale, the properties can be greatly influenced by the number of atoms or molecules in the materials. Unique size-dependent chemical, thermal, magnetic, and other properties are widely reported in nanomaterials. For example, graphite is a layered structure, whereas, the monolayer of graphite—graphene has shown many promising properties, such as high mobility, high thermal conductivity, and excellent mechanical properties, which is totally different from graphite. Not only the physical properties have changed significantly when materials are in nanoscale, but also chemical properties such as reactivity and catalysis in nanomaterials may change evidently as a function of the size due to the large surface area and surface energy.

In order to better describe the nanosized materials, they are defined as three-dimensional (3D), two-dimensional (2D), one-dimensional (1D) materials, and zero-dimensional (0D) materials. Dimensionality is one of the fundamental material parameters, which has attracted extensive interest due to the unique properties in different dimensional materials. 0D material refers to quantum dots, usually with all dimensions limited in a few nanometers; 1D materials, such as nanoribbons, nanowires, and nanotubes are restricted with one dimension; 2D materials refer to monolayer thick materials with free in two dimensions; and 3D materials refer to nanoballs and nanocones, which are free at all dimensions.

Low-dimensional materials have attracted extensive attention due to the fact that they provide the possibility to follow Moore's Law continuously since many low-dimensional materials have shown promising properties for next-generation electronic devices, that is, high speed, high capacity, and low power consumption. The success in isolating graphene sheet has stimulated tremendous interests in 2D materials in the last decade due to their unique properties and great potential in a variety of applications. Currently, there are more than a dozen 2D materials have joined the family of 2D materials, including the graphene-like 2D materials, transition-metal dichalcogenides (TMDs), and layered metal oxides. The planar structures of 2D materials have enabled the mobile charge carriers to access directly in layers, high carrier mobility and high thermal conductivity at low or room temperature are expected.

In addition, high spin injection efficiency and long spin diffusion length in 2D materials have attracted the eyeball of a physicist to investigate the physical properties of 2D materials for the possibility of applications in spintronic devices. Therefore ferromagnetism in 2D materials has been investigated extensively. In the family of 2D materials, some materials are magnetic. Most of the other materials are not magnetic. How to make the 2D materials magnetic is one of the most interesting research focuses on the research of 2D materials.

At the beginning of the research of the ferromagnetism in TMD materials, most of the work focuses on theoretical calculations. First principles calculations based on density-functional theory (DFT) are widely used for the investigations. First principles calculation or ab initio calculation is an algorithmic method which is based on the interaction between atomic nuclei, electrons, and their motion patterns. It starts to calculate physical properties from established laws of physics without additional assumptions, such as empirical or fitted parameters. Derived from the molecular orbital theory [3], there are only five exclusive constants included in ab initio method, which are electron mass m_0, electron charge e, Planck constant h, velocity of light c, and Boltzmann constant K_b. However, in order to get satisfactory results, some less responsible factors will be reasonably ignored. Typical first principles calculations can deal with the atomic structure of materials by applying a series of approximations, such as systematic energy (entropy, enthalpy, and Gibbs free energy) to solve the Schrödinger's equation instead of fitting experimental data. This computational approach provides a direct prediction towards physical, chemical, and mechanical properties. Moreover, it does not require any input parameters, or a minimal set and the results are mainly self-converged by energy or force on an atom. Consequently, first principles calculations can reach a high precision in physical state. For instance, lattice constants and bulk modulus obtained from computations have been shown in good agreement with experiments. The relative error is only a few percentages. With the great enhancement in computing capacity, first principles calculation is regarded as a fundamental and critical technique in computational materials.

In 1964, Hohenberg and Kohn proposed the electron density function theory (DFT) [4]. Subsequently, Kohn and Sham developed its primary fulfillment in 1965 [5]. Using the electron density as the fundamental variable instead of the electron wave function, DFT has become a convincing quantum mechanical modeling method in exploring the electronic structure of many body systems. In the research of ferromagnetism of 2D materials, first principles calculations have become the major approach for identifying whether the magnetic element doping can induce magnetism/ferromagnetism in 2D materials, providing guidance for experimental investigations.

In this book chapter, we will introduce the recent development of the research of ferromagnetism in 2D materials. Both theoretical and experimental results will be introduced and the mechanism for the ferromagnetism is discussed.

4.2 Ferromagnetism in graphene

Graphene, an isolated single layer of graphite, as shown in Fig. 4.2, is the first successfully fabricated 2D material in nature. It has a zero bandgap which has shown many extraordinary properties such as room-temperature Hall effect [6], magnetotransport [7], and high mobility and long spin diffusion length. These extraordinary properties of graphene have potentials in the applications of energy storage, catalyst, and spintronics devices [8].

The magnetism in carbon-based nanomaterials is expected to have new technological applications since the conventional magnetic materials are generally based on d and f elements. However, the pristine graphite is nonmagnetic while defects are crucial in inducing magnetism in nanographites [9–11]. First principles calculations were then performed to investigate the defects-induced magnetism in graphene by Yazyev and Helm [12]. The results indicated that stable FM behavior of 1 μ_B per hydrogen could be achieved in a system with H-adsorption and concentration-dependent magnetism ranging from 1.12 to 1.53 μ_B per vacancy predicted in graphene. The vacancies in different positions can induce either ferromagnetism or antiferromagnetism. Later Boukhavalov et al., predicted that the hydrogen-induced magnetism could be obtained only at low-hydrogen concentrations and hydrogen pairs were found to be nonmagnetic [13].

Besides hydrogen addition, theoretical approaches towards 3d transition metal (TM) atoms embedding in graphene systems were conducted as well [14–16]. The study on TM like Cr-, Ti-, Co-, Fe-, and Mn-absorbed graphene nanoribbons with zigzag- and armchair-edge shapes show that the TM atoms prefer to be adsorbed on hollow site above the center of the hexagon structure of carbons [14]. FM or antiferromagnetic (AFM) spin alignment can be obtained in TM-adsorbed systems. However, it is found that magnetic properties may vary with different adsorbed TM atom species, adsorbate concentrations, and ribbon width.

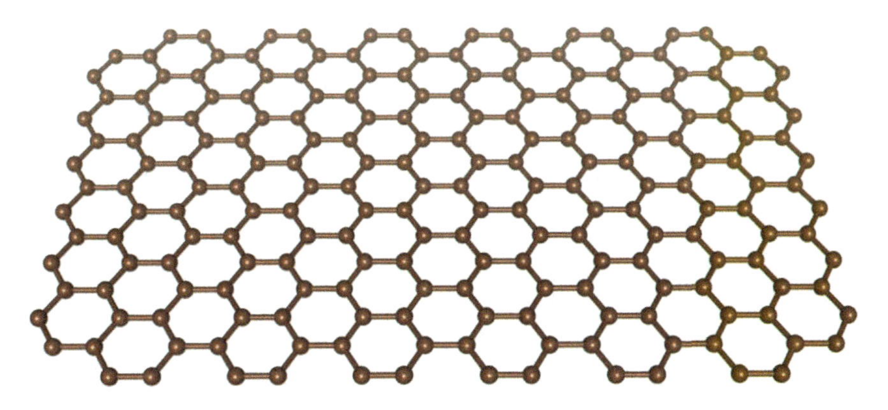

Figure 4.2
Atomic structure of two-dimensional graphene.

The adsorption of 12 different metal adatoms from IA, IIA, IIIA, and TMs groups on graphene was later reported using first principles DFT [15]. It was found that metallic element like Na and K from the IA group could induce very weak magnetism except for Li and the magnetism is slightly lower than an isolated atom due to partial charge transfer. The asymmetry of the 4s states of Ca can result in the magnetic moment and therefore graphene sheet with Ca adsorption can be magnetic. IIIA group such as Al, Ga, and In cannot induce magnetism when they are embedded in graphene and similar partial density of states (PDOS) pattern can be observed. Their calculations have also shown that TMs can alter the electronic properties significantly due to strong p–d hybridization between TMs and graphene sheet. However, no magnetism is found in Pd-adsorbed system though Pd atom has a filled 4d shell. The favored adsorption site and magnetic properties of different elements on graphene are shown in Table 4.1.

Adsorption of a series of TMs and precious metals including Sc–Zn, Pt, and Au on single and double vacancies in graphene was studied [16]. Though magnetism can be obtained in graphene with some TM adatoms, their calculations on the migration barriers suggested that the adatoms might move at room temperature which means it is difficult to have practical applications due to the need of low temperatures to stop mobile adatoms. Interesting magnetic behaviors can be observed in graphene with metal–vacancy complexes. In this study, the substitutional TM was regarded as a TM embedded with a single vacancy, which is called M@SV complex. Almost all TM except for those with almost full d-shells can

Table 4.1: Electronic properties for the favored adsorption site for the 12 adatoms considered in this work.

Atom	Site	p (D)	Φ (eV)	μ_{AG} (μ_B)	μ_A (μ_B)	IP (eV)
Li	H	3.46	2.72	0.00	1.00	5.39
Na	H	2.90	2.21	0.27	1.00	5.14
K	H	4.48	1.49	0.17	1.00	4.34
Ca	H	0.85	3.18	1.04	0.00	6.11
Al	H	0.93	3.08	0.00	1.00	5.99
Ga	H	1.83	2.66	0.00	1.00	6.00
In	H	2.57	2.34	0.00	1.00	5.79
Sn	T	0.19	3.81	1.81	2.00	7.34
Ti	H	1.39	3.16	3.41	4.00	6.83
Fe	H	1.84	3.24	2.03	4.00	7.90
Pd	B	1.23	3.61	0.00	0.00	8.34
Au	T	− 1.29	4.88	0.96	1.00	9.23

The properties listed are the electric-dipole moment per adatom (p), work function (φ), the magnetic moment per adatom of the adatom–graphene system (μ_{AG}), and magnetic moment of the isolated atom (μ_A). The calculated work function for isolated graphene is 4.26 eV. The electric-dipole moment of the unit of Debye (D) and the magnetic unit is given in Bohr magneton (μ_B). *IP*, ionization potential.

Also included the experimental IP of the isolated atom from W.C. Martin, W.L. Wiese, Atomic Molecular, and Optical Physics Handbook, In: G.W.F. Drake (Ed.), American Institute of Physics, New York, 1996, pp. 135–153 [17].

bind strongly with the vacancy site since binding energies of this complex can be very low. M@SV complex is magnetic for TM with single-filled d states, such as V, Cr, and Mn. It was also found that TMs with an odd number of electrons are magnetic while these with even number of electrons are nonmagnetic. The magnetization in Cu and Au@SV was found to originate from the s and p states from the TM and carbons. For TM absorbed with two carbon vacancies, namely M@DV, magnetism can be observed in V, Cr, Mn, Fe, and Co@DV. It was found that the value of the magnetic moments could be correlated to the electron and structural configurations of M@DV. For V, Cr, and Mn, the value of magnetic moments happens to be equal to the difference from the numbers of valence electrons and the surrounding carbon atoms. For Ti and Sc, the bonding states are fully filled, and therefore nonmagnetic. The configurations of Cu, Zn, and Au are quite flat in the graphene sheet, and no magnetism is observed either.

Substitutional metals in graphene systems were investigated as well [18–21]. Similar to the adsorption systems, the magnetic moments are associated with the numbers of the valence electrons and the structures [18]. The substitutional Ti and Sc are nonmagnetic since all the bands are occupied while V, Cr, and Mn show large spin moments due to the d states. Strong contribution from the carbon vacancy level is observed in the doping by magnetic elements, that is, Fe, Co, and Ni. Interestingly, substitutional Fe is nonmagnetic using generalized gradient approximation (GGA) method [19] while spin moments of 1.00 μ_B can be obtained with the local density approximation (LDA) + U method [18]. Substitutional Co is calculated to have a spin moment of 1 μ_B [20]. However, no magnetism was observed in flat graphene with Ni substitution [21]. Noble metal substitutions in graphene can stabilize the structure and induce magnetism from p states of nearby carbons. Substitutional Zn is nonmagnetic after Jahn–Teller distortion while Zn impurity in symmetric C_{3v} configuration has a magnetic moment of 2.00 μ_B [18].

Based on these studies, the magnetization of IA, IIA, IIA TMs, and noble metals are correlated with covalent bonding and electronic structures. The magnetism originates from the electronic states generated by the interaction of adatom and graphene. For ionically bonded adatoms, the charge transfer is calculated quantitatively using two methods: one is based on the DOS and the other is based on the real-space charge density. Variation in dipole moments and work–function shifts across the different adatoms are observed. In particular, the work–function shift shows a general correlation with the induced interfacial dipole of the adatom–graphene system and the ionization potential of the isolated atom.

4.3 Ferromagnetism in boron nitride

Boron nitride (BN) has three different structures: hexagonal BN (h-BN), cubic BN (c-BN), and wurtzite BN (w-BN). h-BN which has a graphite-like layered structure as shown in Fig. 4.3, is the most stable crystalline structure at room temperature. The layers are bonded

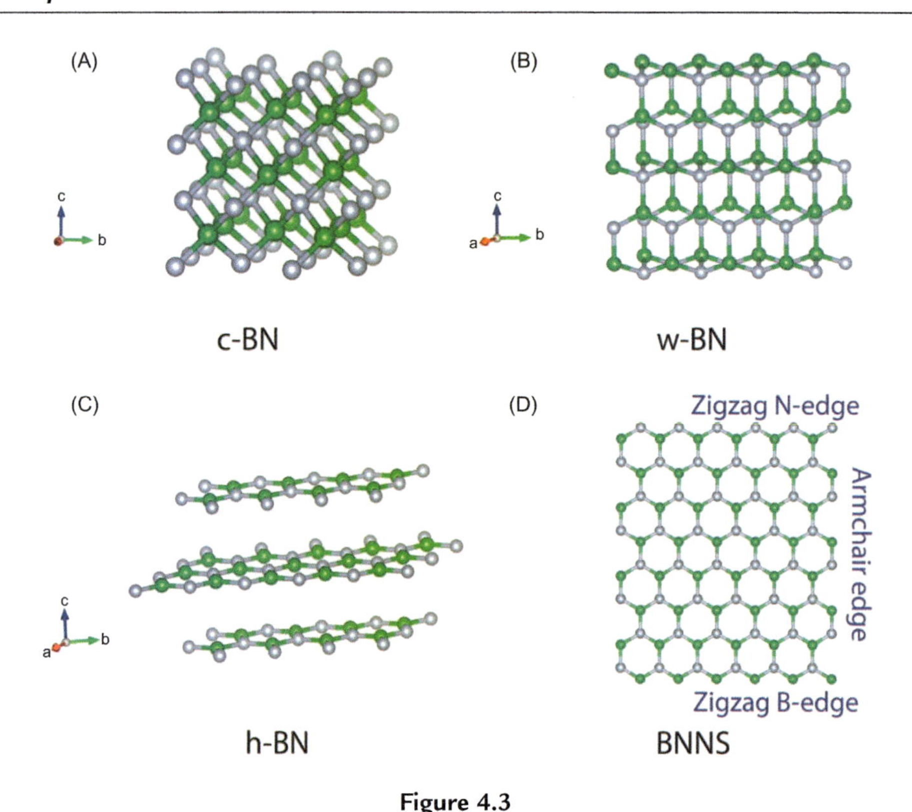

Figure 4.3
Atomic structures of some important phases of BN: (A) c-BN, (B) w-BN, (C) h-BN, and (D) BN nanosheet. *BN*, boron nitride; *c-BN*, cubic BN; *w-BN*, wurtzite BN; *h-BN*, hexagonal BN.

alternatingly with van der Waals forces while B atoms share the same total numbers of electrons with the nearby N atoms within the same atomic layer via sp^2 hybridization. Many kinds of nanostructures of BN have been successfully fabricated, such as nanosheets, nanocones, nanotubes, nanorods, and nanowires structures.

BN nanosheets, which can be an individual or a few layers of h-BN, as shown in Fig. 4.3D, have been successfully fabricated. BN nanosheet can be divided into two categories according to the edges of the nanosheet: zigzag (B- or N-edged) or armchair (BN pair-edged). Different from the AB stacking of graphene layers where the carbon atoms are located on the hollow site of the adjacent layer, the layers of h-BN prefer an AA stacking in forming few layers of BN nanosheet where the B atom sits on top of the N atom in the neighboring layer. This stacking order can induce strong lip–lip interaction between the interlayers of BN nanosheets, and therefore increase the difficulty in exfoliation. It is generally known that electronic properties can vary from different structures. The bandgap of bulk h-BN is 5.69 eV while the bandgap of BN nanosheet from chemical vapor deposition (CVD) is about 6.07 eV [22]. The calculations of bandgap suggest that it can be tuned by hydrogenation. BN nanosheet is an insulator with a bandgap of around 4.6 eV [23,24]. The insulating nature can attribute to the electronegativity changes among the B and N atoms which contribute to the localization of the electrons around the N atomic centers.

Previous studies have also shown a strong correlation between vacancy and doping defects and the electronic and magnetic properties of the graphene systems. Moreover, magnetic properties can be different in various nanostructures according to Lieb's theorem [25] and previous studies in graphene. Therefore it is crucial to understand the properties of different BN nanostructures with diverse defects and dopants.

Electronic and magnetic properties of boron vacancies (V_B) and nitride vacancies (V_N) were investigated using DFT calculations with a monolayer BN model [26–28]. Both defects were calculated to be spin-polarized, and V_N has a magnetic moment of 1.0 μ_B, while V_B has found to be half-metallic with a moment of 3.0 μ_B. The magnetism in vacancies comes from the 2p orbitals of the neighboring B atoms around the vacancy and the surrounding boron atoms via Ruderman–Kittel–Kasuya–Yosida (RKKY) interactions.

Nonmagnetic impurities, such as Be, B, C, N, O, Al, and Si were introduced into monolayer BN to replace either B or N for investigating the possible magnetism [26]. Different from the previous results, V_N was found to show stable magnetic moment of 1.0 μ_B/vacancy with different vacancy concentrations. However, increasing the concentration of V_B can lead to an increase of magnetic moment from 1.0 μ_B/vacancy to about 1.5 μ_B/vacancy. The stable magnetic moment of 1.0 μ_B/defect was obtained in C, Si, and O substitution at either B or N site, as well as the Be substitution on N site. The magnetization of Be_B decreases from 1.0 μ_B/defect to 0.5 μ_B/defect with increasing the doping concentrations. Al, B, and N substituting were found to be nonmagnetic. Later, a study on carbon substitution showed spontaneous magnetization when carbon substituted at boron site. Moreover, obvious reductions of bandgap energy and formation energies were observed, which can be very instructive for experimental researches [28].

Electronic and magnetic properties of monolayer BN with boron and nitride adatoms were calculated as well [29]. Both adatoms preferred to stand on the bridge site in BN. Weak spin polarization of about 0.12 μ_B was found in B adatom, while N adatom showed a magnetic moment of 0.38 μ_B. Both B and N adatom defects were found to be semiconducting or pseudo-semiconducting. Hydrogenation and fluorination of BN nanosheet were investigated as well, and the electronic and magnetic properties can be controlled by the hydrogenation and fluorination ratio. The fully hydrogenated BN nanosheet was found to be nonmagnetic with a direct bandgap of 3.76 eV [30] or 3.33 eV [31] or 3.05 eV [32] by different researchers. The calculations of the binding energies suggest that hydrogen atoms prefer bonding with boron atoms and the semihydrogenated BN nanosheet becomes FM with a magnetic moment of 1.0 μ_B/unit

[30,33,34]. The FM state originates from the unpaired 2p electrons of N atoms and the 1s orbitals of H atoms.

F ions preferred to bind with B instead of N as well. Long-range FM coupling was found in 1/8 fluorinated BN nanosheet [35]. The ferromagnetism comes mainly from the electron spin-polarization of the F and the three N atoms nearest to the sp^3-hybridized B atom underlying the F atom. Antiferromagnetism (AFM) semiconducting behavior was observed in semifluorinated BN nanosheet [33,36,37]. External strains are applied to tune the magnetic properties as well. In semifluorinated BN nanosheet, the AFM state can be transformed to FM state when 6% compressive strain is applied [37]. The bandgap of BN nanosheet is reduced when F atoms are fully decorated and it can be modified by applying different strains [38].

Room-temperature ferromagnetism has been observed in carbon-doped BN nanosheet from both experimentally and theoretically results [39]. Substitutional C doping can induce spontaneous spin polarization and lead to long-rangeFerromagnetism (FM) ordering between the local magnetic moments.

DFT calculations on structural, electronic, and magnetic properties of TM-adsorbed BN sheet were carried out subsequently [31,40,41]. With 1/8 Sc, Ti, and V adsorptions, the systems showed AFM ground state and O adsorptions led to ferromagnetism (Fig. 4.4). However, Sc, Ti, and V turned to FM ground states when the adsorption coverage was reduced to 1/32. The magnetic moments of Sc, Ti, and V are 3.0 μ_B, 4.0 μ_B, and 5.0 μ_B, respectively [31]. However, Ni, Pd, and Pt adsorption, as well as Al and Be substitution, are nonmagnetic.

Adsorptions of Fe, Co, and Ni were investigated by Zhou et al. [40]. Fe-doped BN system was found to be half-metallic with a magnetic moment of 2.0 μ_B; Co-doped BN sheet is semiconducting with a magnetic moment of 1.0 μ_B; and Ni doping had no influence on the bandgap energy and magnetic properties. Moreover, all the TM dopants preferred to adsorb on the hollow site.

In the later research, both the local-spin-density approximation (LSDA) and GGA methods were performed on TM (from V to Ni) adsorption [41]. Compared to the free-standing states of the TM, the magnetic moment of V atom is enhanced and those of Cr, Mn, Fe, and Co atoms are reduced, whereas Ni atom shows nonmagnetic characteristics same as the previous studies. The different degree of the filled d shell can cause different electron migration rates of 4p orbitals, therefore the variation in the magnetic moment of different TMs adsorbed on BN sheet is observed.

Due to the similar crystal structure to graphene, BN nanosheets are regarded as an important member in 2D materials family. The studies of their electronic and magnetic properties via doping have shown great potentials in fabricating spintronics devices

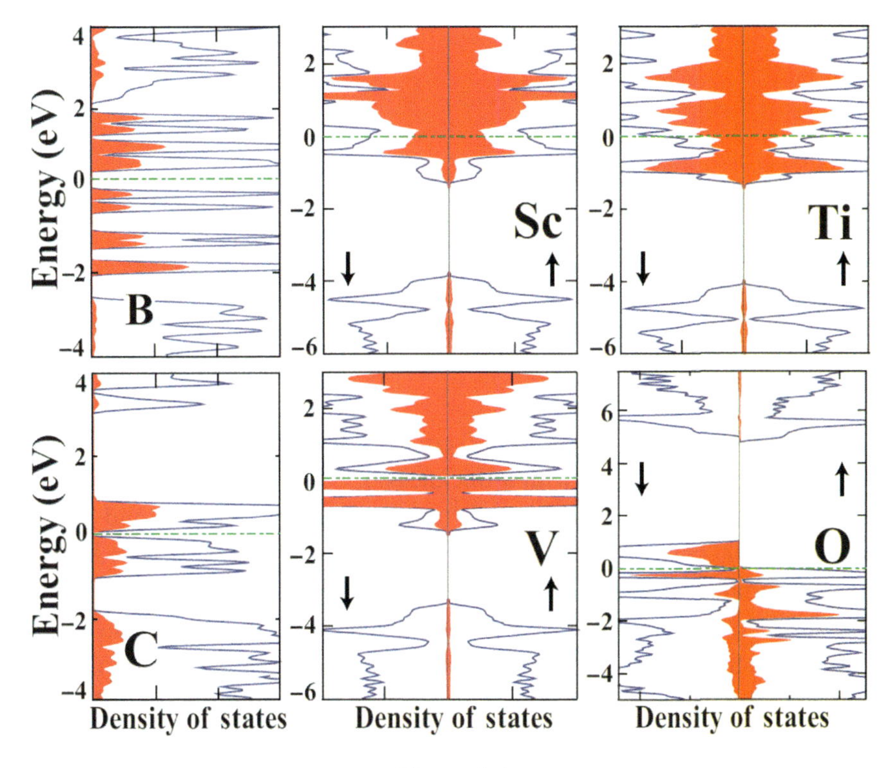

Figure 4.4
TDOS (*dark/blue line*) and PDOS (*shaded gray/red*) for B, C, O, Sc, Ti, and V adatoms at high coverage, $\theta = 1/8$. Up (down) arrow on the right (left) side of TDOS and PDOS indicates spin direction. Zero of the DOS energy is set to the Fermi energy, EF of the adatom + 2D BN system which is indicated with dashed-dotted gray/green line. All adatoms treated in (4×4) supercell to allow AFM interaction or reconstruction, whereby single adatom is adsorbed to each (2×2) supercells of (4×4) supercell amounting $\theta = 1/8$. Sc, Ti, and V have AFM ground state; O has FM state; and B and C are nonmagnetic. The state densities are given in arbitrary units but have the same scale for all adatoms. *EF*, Fermi energy; *TDOS*, total density of states; *PDOS*, partial density of states.

4.4 Ferromagnetism in phosphorene

Black phosphorus (BP) is the most stable allotrope of phosphorus at room temperature [42,43]. BP has a graphene-like layered structure as shown in Fig. 4.5. The individual P atomic layers in bulk BP are puckered together by van der Waals force and therefore can be easily exfoliated to a single layer with less lattice vacancies. Recently, monolayer black phosphorus (M-BP) or so-called phosphorene has been successfully exfoliated [44–48]. Different from graphene, phosphorene is not flat but has a buckled hexagonal structure as shown in the figure.

The bulk BP has an indirect bandgap of about 0.31–0.35 eV [49] while the M-BP is calculated to have a direct bandgap from 0.7 to 1.51 eV [49–51]. Moreover, high hole

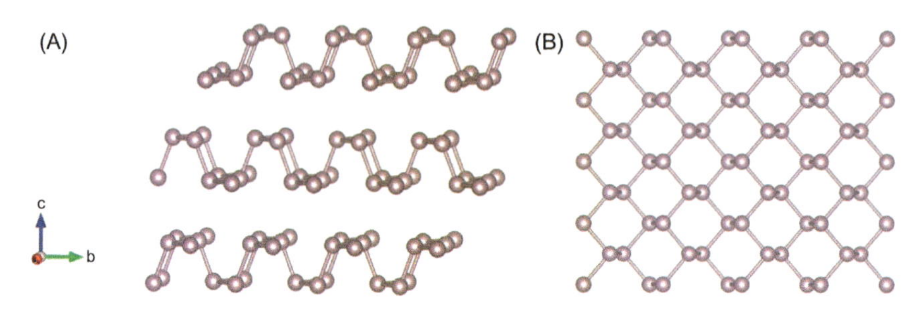

Figure 4.5
Atomic structures of (A) BP and (B) monolayer black phosphorus. *BP*, Black phosphorus.

mobility from 10,000 $cm^2 V^{-1} s^{-1}$ to 26,000 $cm^2 V^{-1} s^{-1}$ was predicted in phosphorene [51]. Experimentally, phosphorene with a thickness of 5 nm was found to have a relatively high field-effect mobility of 286 $cm^2 V^{-1} s^{-1}$ at room temperature [48] and the mobility can increase to ~1000 $cm^2 V^{-1} s^{-1}$ with a thickness of ~10 nm [47]. The high mobility of M-BP is very promising for high-performance FETs. In addition to the high-mobility property, the magnetic properties of M-BP have also been widely investigated. Defects, adsorption, and substitution of magnetic or nonmagnetic elements have been reported to show magnetic ordering [52–55]. Moreover, the Curie temperature of the FM ordering based on M-BP is expected to be higher than other 2D materials since phosphorene has a unique anisotropic structure made from the single elemental atoms, thus prevent the formation of antisite defects. Defects have been reported to be detrimental in achieving high Curie temperature in many DMS systems. Hence, the Curie temperatures of phosphorene-based DMSs are expected to be higher. These characteristics have made phosphorene a very promising material for semiconductor and spintronics devices. Literatures on the investigating magnetic properties of M-BP are growing rapidly.

Simulations on M-BP on various metal substrates including Ta (110), Nb (110), Cu (111), Zn (0001), and In (110) were performed to investigate the stability of the monolayer. [56] Stable monolayer BP is predicted to be able to achieve when fabricated on Cu, Zn, and In substrates. However, Ta and Nb surfaces are found to form strong bonds with P. Therefore M-BP is expected to grow on TaP (110) and NbP (110). Cu (111) was predicted to be the best choice as the substrate materials in fabricating monolayer BP electronic devices.

Defects, adsorption, and substitution of magnetic or nonmagnetic elements in BP have been reported to show magnetic ordering, making it possible to fabricate 2D type DMS for spintronics applications [52–55]. P vacancies were considered for tuning the magnetism since they are inevitable during the fabrication process. DFT study showed that pristine phosphorene is nonmagnetic while P vacancy could induce a magnetic moment of about 1.0 μ_B. However, the system with di-vacancies is nonmagnetic [57,58]. Further investigations suggested that adsorption with Ni, Fe, and Co in the neutral state could induce magnetism with a magnetic moment of 1.0, 2.0, and 1.0 μ_B, respectively [57]. They considered the adsorption of different elements together with the P vacancies as well. And magnetism was obtained in nonmetallic C, O, and N and TMs, such as Fe and Co with single P vacancy or di-vacancies using both PBE and PBE + U methods. Control of oxidation is regarded as one of the best methods to manipulate the properties of 2D materials. Surface oxidation in phosphorene can occur very easily during the exfoliation [59]. The stabilities of oxygen adsorbed on M-BP were evaluated by calculating the formation energies by Ziletti et al. [60]. Different values of the magnetizations are expected with the charged systems on these researches, therefore a way of tuning the electronic and magnetic properties via changing the oxidation state of the defective phosphorene is provided. From the above discussion, it has shown that defects have played an important role in the generation of ferromagnetism in BP. SV and DV have shown different properties.

Surface adatoms B, C, N, O, and F in M-BP were calculated by Wang et al. [61]. The preferable sites of the adatoms are determined by DFT calculations, and B and C prefer the interstitial site while N, O, and F atoms prefer the surface site of phosphorene. Adsorption of interstitial boron can induce a local magnetic moment of 1.0 μ_B, as well as N and F on the surface. The moment is contributed by the unpaired electron on the 2p orbitals of the surrounding P atoms.

The electronic and magnetic properties of nonmagnetic impurities, such as Al-, Si-, S-, and Cl-doped M-BP system were investigated as well [62]. The binding energies of the impurities decreased from Al to Cl. All adatoms exhibited metallic band structures except for Al which is semiconducting. Si-, S-, and Cl-doped systems were found to be spin-polarized. The magnetism in Si is mainly contributed by the impurity itself. However, the magnetism depends on the unpaired 2p orbitals of neighboring P atoms in S, and Cl-doped systems.

Subsequently, nonmetal elements including H, F, Cl, Br, I, B, C, Si, N, As, O, S, and Se are introduced to substitute the P atoms to manipulate the concerned properties [63]. Stable-doped systems can be obtained with all considered atoms. The calculated results also indicate C, Si, O, S, and Se substitution can induce magnetism with the same magnetic moment of 1.0 μ_B/atom for the doping and the magnetic moment is induced via the p–p interaction between the dopants and surrounding P atoms. Moreover, long-range AFM coupling was observed in monolayer with two Si, O, S, or Se atoms via electrons hybridizations. However, the substitutional doping of H, F, Cl, Br, I, B, N, As, C^-, Si^-, S^+, or Se^+ cannot induce magnetism in phosphorene monolayer.

Besides nonmagnetic doping above, other elements, such as alkaline element doping and TM doping have been introduced into BP monolayer to simulate the possible magnetic properties by theoretical calculations. Hu and Hong investigated nonmagnetic adsorptions including Li, Na, Mg, and Al, and TMs, such as Cr, Fe, Co, Ni, Mo, Pd, Pt, and Au on phosphorene. They found that all the adatoms prefer to attach on the hollow sites and 2D growth mode except for Mg, Cr, Mo, and Au which show 3D growth mode. Cr-, Fe-, Co-, and Au-doped phosphorene systems were found to be spin-polarized and Fe-doped M-BP is expected to be a potential candidate for DMS material.

Substitutional doping of a series of 3d TM atoms from Sc to Cu in phosphorene was calculated using first principles calculations [64]. For Sc, Co, and Cu doping in phosphorene, it is nonmagnetic. However, Ti-, Cr-, and Mn-doped systems are magnetic with magnetic moments of 1.0, 3.0, and 2.0 μ_B, respectively, from both GGA and GGA + U calculations. Further calculations indicated that V- and Fe-doped M-BP is half-metallic and the moments are 2.0 and 1.0 μ_B, respectively. FM ground state is obtained in Fe-doped phosphorene with GGA method and Ni-doped system with both GGA and GGA + U methods when two dopants are introduced. Cr- and Mn-doped systems are predicted to show finite bandgap over 0.5 eV, suggesting possible applications in spintronics. The calculated band structure of TM-doped phosphorene is shown in Fig. 4.6. Adsorptions of TMs from Sc to Zn on M-BP were studied as well [65]. According to the calculations, the

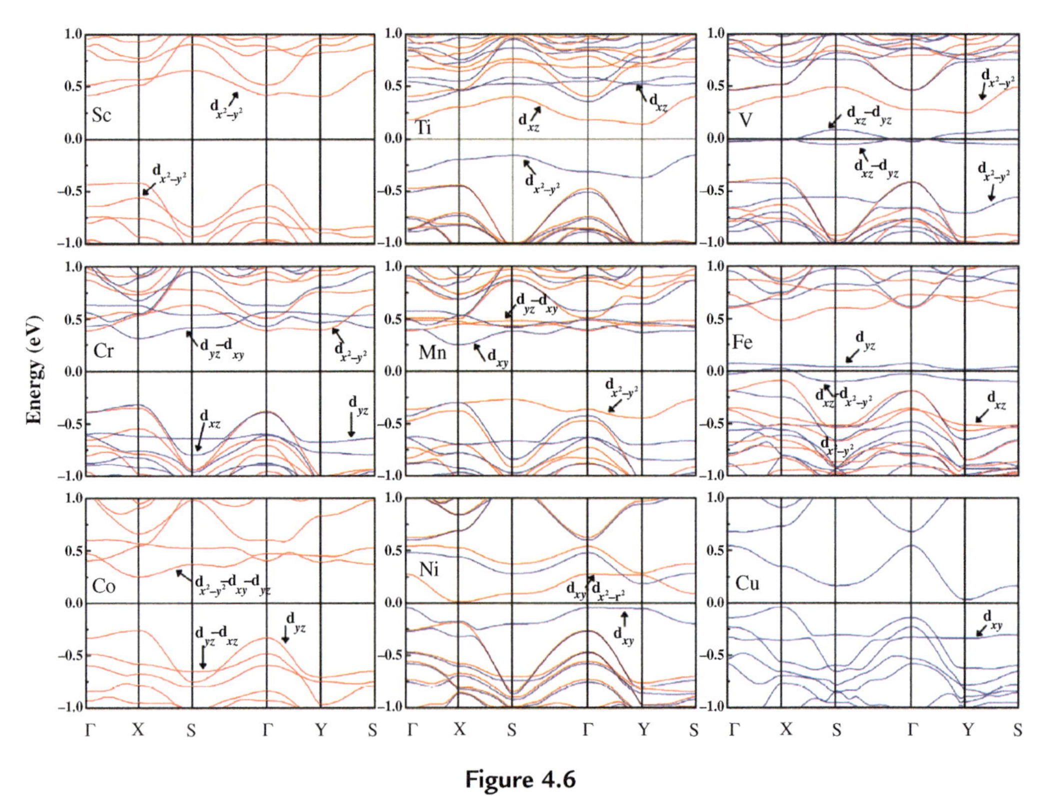

Figure 4.6

Calculated band structure of single TM-doped phosphorene systems. Arrow indicates the major orbital characteristics for a given band. Fermi energy is indicated by the solid black line. *TM*, transition metal.

adsorption of TMs, such as Sc, Ti, V, Cr, Mn, Fe, and Co all exhibits magnetic state due to the exchange split of the d orbitals of the TM. In addition, the magnetic moments can be tuned by applying biaxial strains. The spin states of Sc, V, and Mn adsorbed systems can change from low to high when a small strain is applied.

Co-doped phosphorene at either substitutional or interstitial site is investigated systematically [66]. Co substitutional is calculated to be nonmagnetic while Co interstitial presents a total magnetic moment of 1.0 μ_B from the p–d hybridization between Co and P atoms. Room-temperature ferromagnetism is expected in Co-doped phosphorene with Curie temperature around 466K when doping concentration is 2.7 at%, which is ideal for practical applications.

From afore discussion, magnetic element doping mostly can result in magnetism in BP. It is known that defects can be easily produced during sample fabrication. Therefore defect in BP should be carefully considered when doped with magnetic elements. Wang et al. have calculated magnetic element V-, Cr-, Mn-, Fe-, Co-, and Ni-doped MP by considering defects or defect complex (Fig. 4.7A). The results indicated that pristine M-BP is nonmagnetic, as well as P vacancy (V_P) and ($V_P + V_P$) complex. Without considering the defect complex, the calculated magnetic moment and formation energy are shown in Fig. 4.7B. Doping with TMs of V, Cr, Mn, Fe, or Ni can induce magnetism except Co doping. The magnetism is originated from their d orbitals. V and Mn both have magnetic moments of 2 μ_B/atom. Fe and Ni have magnetic moments of 1 μ_B/atom. Cr doping induces a magnetic moment of 3 μ_B/atom. From Fig. 4.7B, it also indicates that the formation energy is relatively high, all above 0. However, if considering defect and defect complex, the formation of energy has been strongly reduced, as shown in

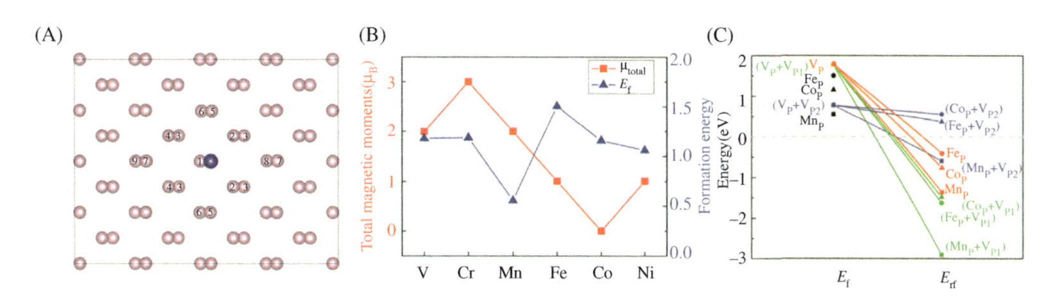

Figure 4.7
(A) The atomic structure of M-BP with TM_P and defects complex ($V_P + TM_P$). The blue ball stands for the substitutional TM. The various configurations of the complex can be obtained by removing the numbered P atoms. (B) The calculated formation energies and the total magnetic moments of M-BP doped with different TM elements. (C) The formation of energies and relative formation energies of TM-doped M-BP systems with different defects and defect complexes (TM = Mn, Fe, and Co). *M-BP*, monolayer black phosphorus.

Fig. 4.7C. These results have provided insights for the fabrication of DMSs based on 2D materials, considering defects and defects complexes.

Until now, there are few reports on the ferromagnetism observed in BP experimentally. Recently, Nakanishi et al. [67] reported that oxygen terminated zigzag pore edges of few-layer BP nanomeshes showed room-temperature ferromagnetism, which is due to FM spin coupling between edge P with O atoms, resulting in a strong spin localization at the edge valence band, and from uniform oxidation of full pore edges over a large area and interlayer spin interaction. Though the ferromagnetism of BP monolayer has not been experimentally explored, the large number of theoretical works has provided guidance for the fabrication of BP-based DMS, which may be one of the alternative candidates for the materials of spintronics devices.

4.5 Ferromagnetism of transition-metal dichalcogenide: MoS_2

TMDs monolayers, with the formula MX_2, (M stands for a TM atom, such as Mo, W, etc., and X represents a chalcogen atom including S, Se, or Te) are the most studied 2D materials in the last decade. TMD usually has a bulk structure containing several 2D layers vertically stacking with one layer of M atoms lying between two layers of X atoms via van der Waals force, as shown in Fig. 4.8 The TM and chalcogenides are strongly bonded with covalent bonds in a sandwich structure "X–M–X" within the layer. TMD monolayers, including MoS_2, WS_2, $MoSe_2$, $MoTe_2$, etc., have shown unique properties different from their bulk counterparts. Direct bandgaps ranging between 1 and 2 eV [68,69] are found in TMD monolayers, whereas, their bulk states are indirect bandgap semiconductors. TMD has strong spin–orbit coupling, high "on" and "off" ratio and relatively high mobility, which shows the possibility to be a new category of 2D materials for fabricating transistors, detectors, sensors, and spintronics devices.

Monolayer molybdenum disulfide (MoS_2) is a typical semiconductor from TMDs family which has attracted intensive research interests. The layers in MoS_2 bulk structures are

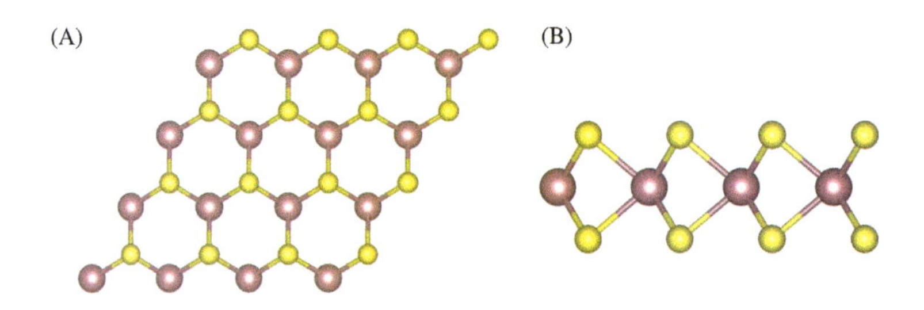

Figure 4.8
Atomic structure of monolayer MoS_2 from (A) top view and (B) side view.

bonded via van der Waals force, and the monolayer has been successfully exfoliated using micromechanical cleavage technique by Novoselov et al. [70].

Similar to graphene, monolayer MoS_2 (1H-MoS_2) exhibits unique properties from its bulk counterpart. For example, bulk MoS_2 is an indirect bandgap semiconductor with a bandgap of 1.29 eV while the monolayer has a direct bandgap of 1.8 eV [68]. 1H-MoS_2 can be fabricated via different methods like mechanical or liquid exfoliation, CVD and electron irradiation [71–73]. Different from graphene with zero bandgap and light carbon atom nature resulting in a very low spin–orbital coupling, MoS_2 has a relatively strong spin–orbital coupling, thus making the spin manipulation possible for the applications of spintronics devices [74]. These findings further confirm the possibility of fabricating monolayer MoS_2 spintronics devices. Theoretically, quantum Hall effect, spin Hall Effect, and spin manipulation by electric field have been predicted by first principles calculations [75–77]. Experimentally, high spin injection efficiency has been achieved using Fe as the electrode [78]. In particular, the carrier mobility of exfoliated TMD is relatively low comparing to graphene, therefore limits the applications of such kind of materials [79,80]. However, a record high electron mobility of fewer-layer MoS_2 is measured to be 34,000 $cm^2\ V^{-1}\ s^{-1}$ at low temperature using a heterostructure device platform [81]. These findings show the potential of monolayer MoS_2 for the applications of spintronics devices.

Abundant researches have been reported on the electrical, mechanical, optical properties, and magnetic properties of monolayer MoS_2 experimentally and theoretically [80,82–86]. In fact, the pristine bulk MoS_2 is nonmagnetic. However, recent theoretical and experimental results indicate that MoS_2 nanostructures are magnetic and the magnetic moment mainly comes from the zigzag edges or vacancies [87,88]. In addition, Togay et al. [87] discovered that the magnetic measurement was insensitive to the interlayer coupling and therefore proposed that the monolayer MoS_2 might share the same magnetic behaviors and can be possible to be a qualified DMS material.

Inspired by the experimental studies, theorists employed first principles calculations to investigate the magnetic behavior of MoS_2 monolayer. As a matter of fact, the calculation results show that similar to its bulk counterpart, the pristine MoS_2 monolayer is also nonmagnetic [89]. In this case, the impurity absorption on the surface of MoS_2 monolayer has been strategically used to modify the electronic and magnetic properties. First principles calculation results indicate that the absorption of B, C, and N on 1H-MoS_2 leads to the FM ordering while H and F absorptions induce weak AFM coupling [90]. On the other hand, the adsorption of TM atoms, such as Co, Cr, Fe, Ge, Mn, Mo, Sc, and V, result in a local magnetic moment [91]. The triple vacancy in MoS_2 was found to be magnetic due to the disturbances in charge transfer while other vacancy defects are nonmagnetic.

Besides the adsorption of impurity atoms, strain in the monolayer is also responsible for the magnetic properties. The calculations demonstrate that strain cannot induce any magnetism in the pristine monolayer MoS_2 [92,93]. However, the magnetic moment in MoS_2 monolayers can be generated by the particular native defects when tensile strain is applied [55,94]. Therefore the defects should be considered for the study of magnetic properties of MoS_2 monolayers.

In the research of oxide-based DMSs, doping is one of the most important techniques to realize the ferromagnetism at room temperature. Magnetic and nonmagnetic element dopants have both been successfully used to tune and tailor the semiconductor properties [54,95–99]. Monte Carlo calculations [100] have demonstrated that the Curie temperature can be above room temperature when the doping concentration of Mn is 10%–15% in MoS_2, which is very promising for achieving atomic layer DMS with room-temperature ferromagnetism. However, the reported results in this area are contradictive. For example, Mn-doped MoS_2 has shown a long-range FM ordering for the relatively high doping concentration [101]. The FM ordering is attributed to the exchange interaction between the localized Mn spins and the delocalized p spins of S atoms. Similarly, from the first principles calculations, the magnetic moment is produced when monolayer MoS_2 is doped with nonmetals (H, B, N, and F) or TMs (V, Cr, Mn, Fe, and Co) to substitute the S [102]. These nonmagnetic substitutions were found to have magnetic moments of 1.0 μ_B, while V-, Cr-, Mn-, Fe-, and Co-doped ones attain 1.0, 4.0, 3.0, 3.0, and 1.0 μ_B, respectively. The hybridization between 2p orbitals or 3d orbitals of the substitutional atoms and 4d orbitals of nearest Mo atoms together with a small contribution from the 3p of nearest S atoms results in the observed magnetism. The total density of states (TDOS) of the TM-doped MoS_2 is shown in Fig. 4.9. Theoretical calculations have also shown that the 6.25% doping with the TMs including Mn, Co, Fe, and Zn creates the magnetism in MoS_2 as well, with magnetic moments of 1.0, 2.0, 3.0, and 1.0 μ_B, respectively [103]. The FM ground state was found in Mn- and Zn-doped systems due to the asymmetry of 3d states of the doped TM elements.

As discussed in the TM-doped black phosphors, defects, and defect complex play a vital role in the generation of ferromagnetism and the formation energy. Wang et al. also calculated TM-doped MoS_2 by considering defects and defect complexes. From the results, pristine MoS_2 monolayer does not show spin-polarized state. In addition, Mo or S vacancy alone does not show spin-polarized state either. When TM dopants (Mn, Fe, Co, and Ni) are introduced to the system, it can bind easily with the Mo vacancy sites and form TM_{Mo}. And TM substitution in Mo site can produce magnetic moment. The magnetic moment originates from the d orbitals of the TM, nearest neighbor (NN) Mo atoms and the p orbitals of the NN S atoms via p–d hybridization. The magnetic moment varied significantly, dependent on the type of dopants and doping concentration. Using Co-doped system, for example, lower doping concentration (4 at% or 6.25 at%) has a stable magnetic moment of 3 μ_B, which is higher than that at higher doping level (8 at%, or 11.1 at%, or 12.5 at%). Co_{Mo} defects tend to cluster with higher doping concentration. Subsequently, the substitutional Co_{Mo} defects will not result in magnetic state. Further calculation results indicate that if there are Mo

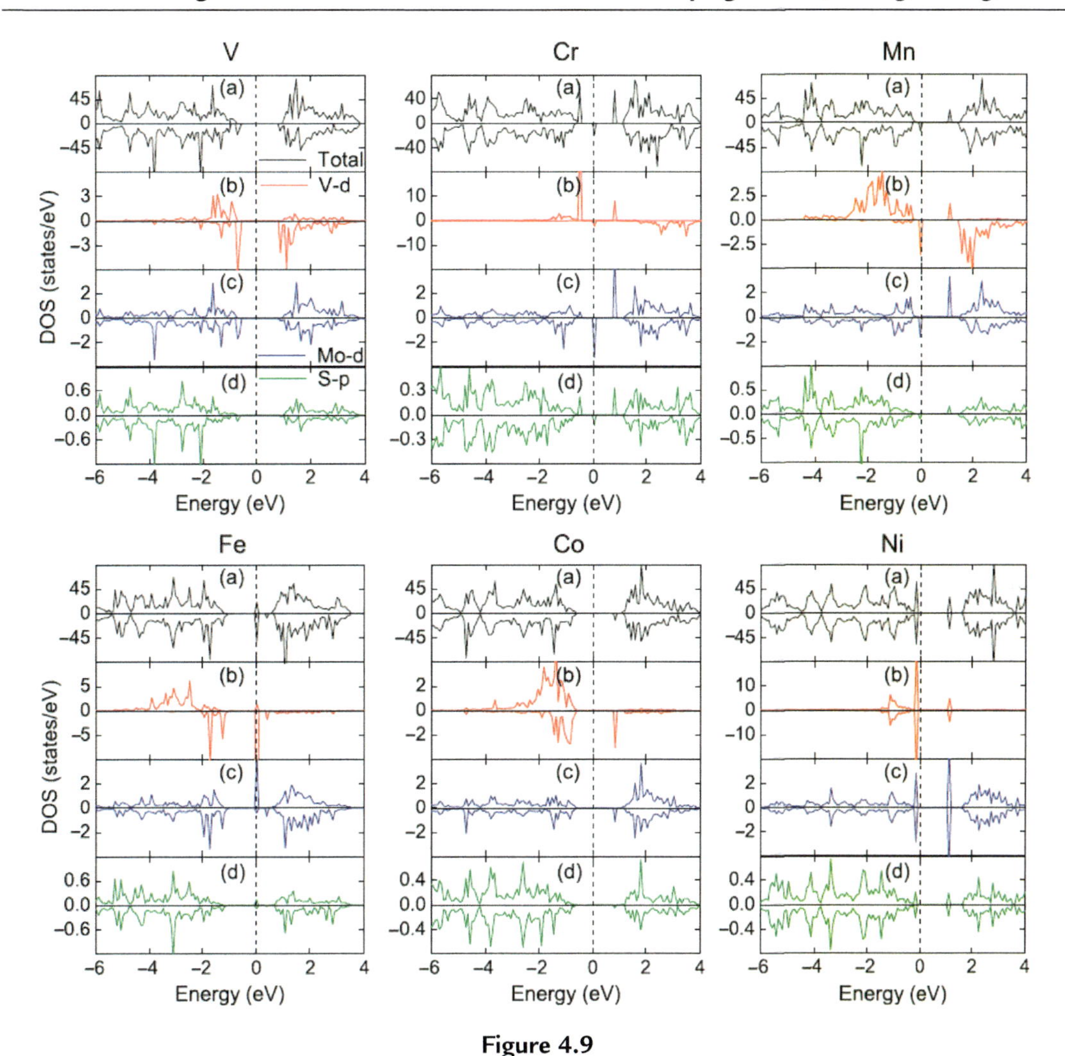

Figure 4.9
The TDOS and PDOS of V-, Cr-, Mn-, Fe-, Co-, and Ni-substituted 1H-MoS_2 at the cover of 1/48. The plane of (A) is the TDOS, and (B)–(D) are the d states of TM atom, d states of nearest Mo atom, and p states of nearest S atom, respectively. Fermi level is denoted by the vertical dashed line.

vacancies in the system, the dopants tend to separate from each other with a particular distance and the system shows a FM state. In addition, this tendency is energy favorable, which means that the existence of a defect can reduce the formation energy of doping.

Experimentally, Mn-doped MoS_2 samples with different concentrations have been prepared by ion implantation. Room-temperature ferromagnetism has been achieved, as evidenced by remnant magnetization and coercivity (Fig. 4.10A). In addition, relatively high magnetization has been achieved. 2% Mn-doped MoS_2 has achieved a saturation magnetization as high

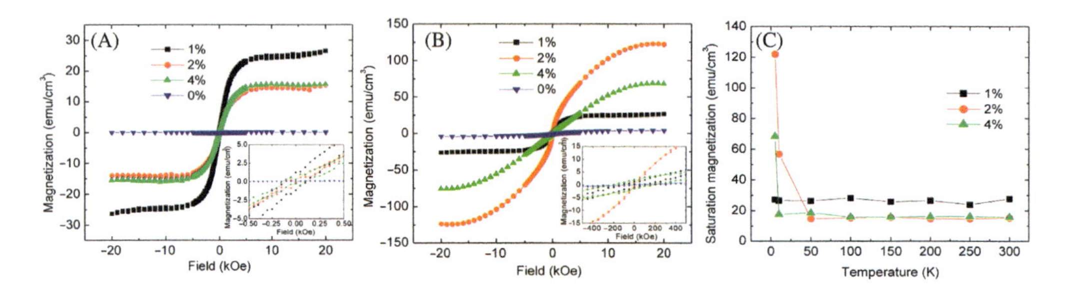

Figure 4.10

M−H loops of Mn-doped MoS_2 with different doping concentrations at (A) 300K. (B) 5K. (C) Temperature dependence of saturation magnetization.

as 120 emu/cm^3 (Fig. 4.10B). Temperature dependence of magnetization shows that a large volume of paramagnetic phases exists due to high magnetization at low temperature (Fig. 4.10C).

Similarly, Fe, Ni, and Co have also been used as a dopant to dope MoS_2 by implantation. All the doping has induced ferromagnetism at room temperature (Fig. 4.11). For Ni doping, there is a very high magnetization at low temperature but low saturation magnetization, suggesting that the interaction between magnetic moments is not strong enough to form FM ordering (Fig. 4.11B).

The different researches on chalcogenides suggest that monolayer MoS_2 can be a promising candidate for 2D dilute magnetic semiconductors. It should also be possible to obtain magnetic ordering in other 2D metal dichalcogenides and thus achieve 2D-based DMSs with different bandgaps and mobilities.

4.6 Ferromagnetism in two-dimensional metal oxide: SnO

Besides the well-studied TMDC materials, recent studies have gained insights into a new member of 2D materials family: metal oxide. The exfoliation of layered oxides has been achieved successfully in layered manganese oxide, titanium oxide, niobium oxide and cobalt oxide [72,104−106]. 2D metal oxide materials have shown various electronic properties: for instance, 2D TM oxide like TiO_2 is usually semiconducting while MnO_2 nanosheets are found to be redoxable or semimetallic [107], rendering wide applications in electronic devices with their unique electronic properties. High-performance supercapacitors have been fabricated using RuO_2 nanosheets [108]; FETs can be obtained with $Ti_{0.87}O_2$ and $Ti_{0.91}O_2$ nanosheets via layer-by-layer assembly [105]; Room-temperature ferromagnetism was observed in magnetic oxide nanosheets like $Ti_{1-x}Co_xO_2$ ($x \leq 0.2$) [109], $Ti_{1-x}Fe_xO_2$ ($x \leq 0.4$) [110], and $Ti_{1-x}Mn_xO_2$ ($x \leq 0.4$) [111]. Moreover, 2D $Ti_{1-x}Co_xO_2$ was calculated to be FM with spin−orbit coupling anisotropy [109,112]. 2D metal oxides are therefore

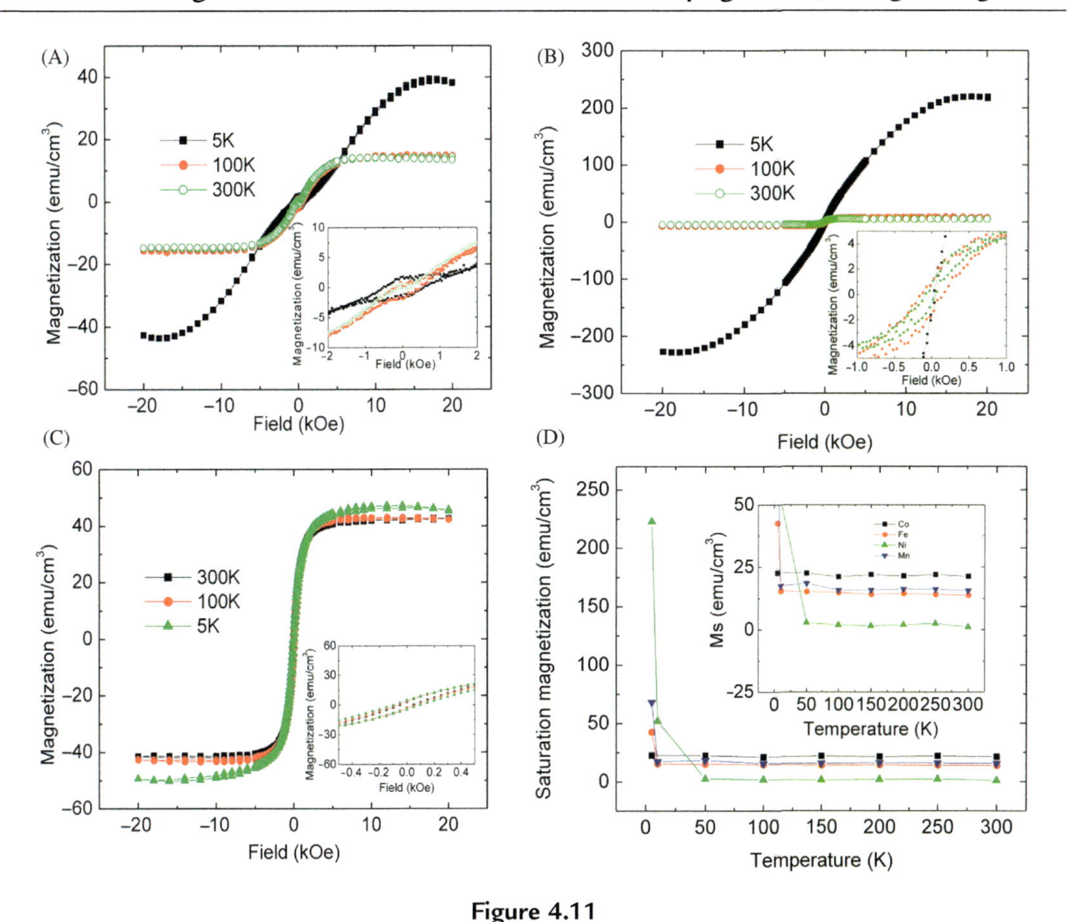

Figure 4.11

M–H loops of (A) 4% Fe-doped MoS_2. (B) 4% Ni-doped MoS_2. (C) 4% Co-doped MoS_2. (D) Temperature dependence of saturation magnetization 4% Fe-, Ni-, Co-, and Mn-doped MoS_2.

believed to have great potential in the applications of various electronic and spin-based electronic devices.

Tin monoxide (SnO) has a tetragonal litharge structure with four O atoms and one Sn atoms form a pyramid structure as shown in Fig. 4.12. Recently, 2D tin monoxide monolayer has been successfully fabricated experimentally and it possesses promising properties for the application of FETs, making it one of the most interesting candidates for the family of metal oxide 2D materials [113]. Tin monoxide (SnO) has a unique layered tetragonal structure with both Sn and O atoms in four coordinate pyramids. The bottom of the square pyramid is formed by the four oxygen atoms arranged in regular tetrahedral geometry while tin atoms sit alternatively on the top of pyramids. Electron densities are observed to be asymmetric along the vertical axis due to the long pairs of the 5s electrons of Sn. These electrons do not contribute to the bonding but dipole–dipole interaction between SnO layers.

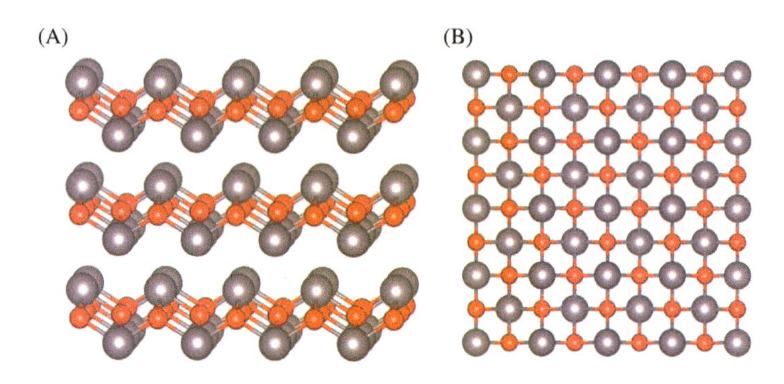

Figure 4.12
Atomic structures of (A) tetragonal SnO and (B) monolayer SnO.

The investigations of SnO mainly focus on thin films structures, which exhibit native p-type conductivity and are ideal for fabricating high-mobility p-channel thin-film electronic devices, such as gas sensors and thin-film transistors [114–120]. Several computational studies have been carried out to explore the structural, electronic, optical and magnetic properties of SnO to elucidate the defect chemistry of SnO [121–127]. Briefly, tin vacancy is the major native defect under oxygen-rich atmosphere and responsible for the observed p-type conductivity [121]. However, the existence of oxygen vacancy and tin interstitials can enhance the electron mobility under Sn-rich environment. It is important to clarify the defects chemistry since defects have shown a strong influence on the properties of oxide-based semiconductors [52–54,97,128–132]. Good examples are the effects of defects or defect complexes on the magnetic properties in the research of oxide-based magnetic semiconductors [52–54,132]. Cation vacancies or oxygen vacancies have been found to be the origin of the ferromagnetism in pure oxide semiconductors, such as ZnO and TiO_2 [97,131,133,134]. In addition, these defects play an important role in the ferromagnetism of TM-doped oxide-based DMSs and the ferromagnetism can be tuned by defect engineering [135–138].

In the research of DMSs, TMs have been widely used as dopants to realize ferromagnetism. Similarly, a theoretical study on SnO bulk systems doped with a series of TMs including Sc-, V-, Cr-, Mn-, Fe-, Co-, Ni-, and Cu-doped was also conducted [139]. Formation energies have been extensively calculated to predict the preferable doping sites and the calculation results indicate that the doping sites can strongly influence the magnetic properties. For example, the spin polarization induced by TM doping can be neutralized if two TM dopants are located closely. However, it is exceptional for V doping, which has strong FM behavior if the dopants are clustering [139].

Although intensive research efforts have been devoted to bulk SnO, the report related to 2D SnO is very limited. The research interest was triggered by the stability of 2D SnO as a semiconductor with a bandgap energy of ~3.95 eV predicated with HSE06 functionals. [140] Subsequently multiferroic was found from the monolayer SnO using HSE06 and PBE-vdW functionals, respectively [141]. It was believed that the ferromagnetism can be achieved in monolayer SnO with a hole density of $2-3 \times 10^{14}$ cm^{-2}. Such a phenomenon may originate from Mexican-hat band dispersion. Ferroelasticity and multiferroicity can be stabilized using a generalized Landau model, suggesting great potential of 2D SnO in achieving multifunctional electronic devices [141]. A theoretical study carried out by Ma et al. confirmed the stability of monolayer SnO and the semiconducting nature of monolayer SnO with an indirect bandgap of 2.62 and 3.33 eV using PBE and HSE06 functionals [142]. More recently, a few layers of SnO films were successfully fabricated on sapphire substrates using pulsed laser deposition [113]. The structural, optical and transport properties were measured, which showed that 2D SnO could be a potential candidate in fabricating room temperature 2D spintronics devices. However, there is almost no study on the magnetic properties of TM-doped 2D SnO. Wang et al. used first principles calculations to calculate Co-, Cr-, Mn-, Fe-, Ni-, and V-doped SnO monolayers, considering defect and defects complexes. The results are listed in Table 4.2. GGA and GGA + U calculations obtain same results. From the table, it can be seen that all dopants doping can induce magnetism, except Ni. The magnetic behavior is due to the interaction between the dopants and the surrounding Sn/O atoms. Further results indicate that substitution is preferred at oxygen-rich condition. From Table 4.2, For the substitution condition, Co shows a magnetic moment of 1 μ_B/atom; Mn and V show magnetic moments of 3 μ_B/atom; Cr and Fe show magnetic moments of 4 μ_B/atom. There is no much change for the dopant moment when complex defect TM_{sub} and V_O are formed expect that $Co_{sub} + V_O$ shows a magnetic moment of 5 μ_B. The trend of TM moment changes significantly when defect complex $Co_{sub} + V_{Sn}$ is formed. The moments of ($Co_{sub} + V_{Sn}$) and ($Ni_{sub} + V_{Sn}$) are enhanced compared to that of the substitutional alone. On the other hand, the magnetic moments of ($Fe_{sub} + V_{Sn}$) and ($Mn_{sub} + V_{Sn}$) maintain the same as that of substitutional dopants alone.

Table 4.2: The calculated total magnetic moments (μ_{total}, μ_B) for defects TM_{sub}, and defects complexes ($TM_{sub} + V_{Sn}$) and ($TM_{sub} + V_O$) using both GGA and GGA + U methods, and the local magnetic moments (μ_{local}, μ_B) for the TM in the relevant defects.

		Co		Cr		Fe		Mn		Ni		V	
		μ_{local}	μ_{total}	μ_{local}	μ_{total}	μ_{local}	μ_{total}	μ_{local}	μ_{total}	μ_{local}	μ_{total}	μ_{local}	μ_{total}
TM_{sub}	GGA	0.943	1	3.436	4	3.492	4	2.97	3	0	0	2.290	3
	GGA + U	0.992	1	3.441	4	3.494	4	4.545	5	0	0	2.549	3
TM_{sub}	GGA	1.343	1	3.431	4	3.477	4	3.00	3	0	0	2.332	3
$+V_O$	GGA + U	1.736	1	3.436	4	3.483	4	4.576	5	0	0	2.623	3
TM_{sub}	GGA	1.573	3	2.133	2	3.231	4	3.471	3	0.396	2	1.017	1
$+V_{Sn}$	GGA + U	1.498	3	2.17	2	3.260	4	3.909	5	0.411	2	1.241	1

TM, transition metal.

Whereas, ($Cr_{sub} + V_{Sn}$) and ($V_{sub} + V_{Sn}$) have lower magnetic moments than that of single TM_{sub}. The work has shown that most of the magnetic element doping can lead to magnetism except Ni doping. GGA and GG + U, as well as different sizes of supercells, have been used for the calculation. All the calculations have demonstrated consistency, suggesting the correction of the calculations. Though the magnetism of TM-doped SnO monolayer has been investigated in detail by first principles calculations, there is still no experimental report on the research of magnetism based on 2D SnO films. Similarly, the magnetism of other 2D oxides has not been investigated either. Theoretical work may provide guidance for the research of magnetism in 2D-based oxides.

4.7 Conclusion and prospective

Since the discovery of graphene, the magnetism in 2D materials has attracted extensive interest. Doping or adsorption is one of the approaches for inducing ferromagnetism in 2D materials. The idea is similar to that of DMS by doping magnetic elements into semiconductor host, which possess both high spin polarization and semiconductor behavior. Therefore magnetic 2D materials have the potential for future spintronics devices. However, most of the research is focusing on theoretical calculations. Experimental report is seldom presented. In addition, most of the magnetism/ferromagnetism attributes to the edges or defects. Doping and adsorption of magnetic elements are also of challenge. In 2017, the journal of Nature has published two papers in one issue to discover the ferromagnetism in monolayer 2D materials [143,144]. These two reports have further stimulated the research interest in searching ferromagnetism in different kinds of 2D structures. In the future research of magnetism in 2D materials, doping or adsorption of magnetic element into 2D materials maybe still the major approach to achieve ferromagnetism. Heterostructure to induce ferromagnetism may be another effective way to induce ferromagnetism, as reported recently [145]. Certainly, the research of magnetism in 2D is at the initial stage, advanced experimental techniques are needed to induce ferromagnetism in 2D materials, such as implantation, molecular doping, and surface adsorption, etc. To achieve high-quality FM-2D materials is a prerequisite for the investigation of physics phenomena such as spin injection and spin transport, leading to the ultimate goal to apply the materials for spintronics devices.

References

[1] S.D. Sarma, Am. Sci. 89 (2001) 516.
[2] H. Ohno, Making nonmagnetic semiconductors ferromagnetic, Science 281 (5379) (1998) 951–956.
[3] W.J. Hehre, Ab initio molecular orbital theory, Acc. Chem. Res. 9 (11) (1976) 399–406.
[4] P. Hohenberg, W. Kohn, Inhomogeneous electron gas, Phys. Rev. 136 (3B) (1964) B864.
[5] W. Kohn, L.J. Sham, Self-consistent equations including exchange and correlation effects, Phys. Rev. 140 (4A) (1965) A1133.

[6] K.S. Novoselov, Z. Jiang, Y. Zhang, S. Morozov, H. Stormer, U. Zeitler, et al., Room-temperature quantum Hall effect in graphene, Science 315 (5817) (2007). 1379.
[7] C. Berger, Z. Song, X. Li, X. Wu, N. Brown, C. Naud, et al., Electronic confinement and coherence in patterned epitaxial graphene, Science 312 (5777) (2006) 1191–1196.
[8] A.K. Geim, K.S. Novoselov, The rise of graphene, Nat. Mater. 6 (3) (2007) 183–191.
[9] J. Coey, M. Venkatesan, C. Fitzgerald, A. Douvalis, I. Sanders, Ferromagnetism of a graphite nodule from the Canyon Diablo meteorite, Nature 420 (6912) (2002) 156–159.
[10] P. Lehtinen, A.S. Foster, Y. Ma, A. Krasheninnikov, R.M. Nieminen, Irradiation-induced magnetism in graphite: a density functional study, Phys. Rev. Lett. 93 (18) (2004) 187202.
[11] K. Kusakabe, M. Maruyama, Magnetic nanographite, Phys. Rev. B 67 (9) (2003) 092406.
[12] O.V. Yazyev, L. Helm, Defect-induced magnetism in graphene, Phys. Rev. B 75 (12) (2007) 125408.
[13] D. Boukhvalov, M. Katsnelson, A. Lichtenstein, Hydrogen on graphene: electronic structure, total energy, structural distortions and magnetism from first-principles calculations, Phys. Rev. B 77 (3) (2008) 035427.
[14] H. Sevinçli, M. Topsakal, E. Durgun, S. Ciraci, Electronic and magnetic properties of 3d transition-metal atom adsorbed graphene and graphene nanoribbons, Phys. Rev. B 77 (19) (2008) 195434.
[15] K.T. Chan, J. Neaton, M.L. Cohen, First-principles study of metal adatom adsorption on graphene, Phys. Rev. B 77 (23) (2008) 235430.
[16] A. Krasheninnikov, P. Lehtinen, A.S. Foster, P. Pyykkö, R.M. Nieminen, Embedding transition-metal atoms in graphene: structure, bonding, and magnetism, Phys. Rev. Lett. 102 (12) (2009) 126807.
[17] W.C. Martin, W.L. Wiese, in: G.W.F. Drake (Ed.), Atomic Molecular, and Optical Physics Handbook, American Institute of Physics, New York, 1996, pp. 135–153.
[18] E.J. Santos, A. Ayuela, D. Sánchez-Portal, First-principles study of substitutional metal impurities in graphene: structural, electronic and magnetic properties, New J. Phys. 12 (5) (2010) 053012.
[19] P. Venezuela, R. Muniz, A. Costa, D. Edwards, S. Power, M. Ferreira, Emergence of local magnetic moments in doped graphene-related materials, Phys. Rev. B 80 (24) (2009) 241413.
[20] E.J. Santos, D. Sánchez-Portal, A. Ayuela, Magnetism of substitutional Co impurities in graphene: realization of single π vacancies, Phys. Rev. B 81 (12) (2010) 125433.
[21] E.J. Santos, A. Ayuela, S. Fagan, J. Mendes Filho, D. Azevedo, A. Souza Filho, et al., Switching on magnetism in Ni-doped graphene: density functional calculations, Phys. Rev. B 78 (19) (2008) 195420.
[22] K.K. Kim, A. Hsu, X. Jia, S.M. Kim, Y. Shi, M. Hofmann, et al., Synthesis of monolayer hexagonal boron nitride on Cu foil using chemical vapor deposition, Nano. Lett. 12 (1) (2011) 161–166.
[23] M. Topsakal, E. Aktürk, S. Ciraci, First-principles study of two-and one-dimensional honeycomb structures of boron nitride, Phys. Rev. B 79 (11) (2009) 115442.
[24] Y. Lin, T.V. Williams, J.W. Connell, Soluble, exfoliated hexagonal boron nitride nanosheets, J. Phys. Chem. Lett. 1 (1) (2009) 277–283.
[25] E.H. Lieb, Two theorems on the Hubbard model, Phys. Rev. Lett. 62 (10) (1989) 1201–1204.
[26] R.-F. Liu, C. Cheng, Ab initio studies of possible magnetism in a BN sheet by nonmagnetic impurities and vacancies, Phys. Rev. B 76 (1) (2007) 014405.
[27] M. Si, D. Xue, Magnetic properties of vacancies in a graphitic boron nitride sheet by first-principles pseudopotential calculations, Phys. Rev. B 75 (19) (2007) 193409.
[28] S. Azevedo, J. Kaschny, C. De Castilho, F. de Brito Mota, Electronic structure of defects in a boron nitride monolayer, Eur. Phys. J. B -Condens. Matter. Complex Syst. 67 (4) (2009) 507–512.
[29] J. Yang, D. Kim, J. Hong, X. Qian, Magnetism in boron nitride monolayer: adatom and vacancy defect, Surf. Sci. 604 (19) (2010) 1603–1607.
[30] S. Tang, Z. Cao, Structural and electronic properties of the fully hydrogenated boron nitride sheets and nanoribbons: insight from first-principles calculations, Chem. Phys. Lett. 488 (1) (2010) 67–72.
[31] C. Ataca, S. Ciraci, Functionalization of BN honeycomb structure by adsorption and substitution of foreign atoms, Phys. Rev. B 82 (16) (2010) 165402.
[32] W. Chen, Y. Li, G. Yu, C.-Z. Li, S.B. Zhang, Z. Zhou, et al., Hydrogenation: a simple approach to realize semiconductor – half-metal – metal transition in boron nitride nanoribbons, J. Am. Chem. Soc. 132 (5) (2010) 1699–1705.

[33] J. Zhou, Q. Wang, Q. Sun, P. Jena, Electronic and magnetic properties of a BN sheet decorated with hydrogen and fluorine, Phys. Rev. B 81 (8) (2010) 085442.
[34] Y. Wang, Electronic properties of two-dimensional hydrogenated and semihydrogenated hexagonal boron nitride sheets, Phys. Stat. Sol. (RRL) 4 (1-2) (2010) 34–36.
[35] M. Du, X. Li, A. Wang, Y. Wu, X. Hao, M. Zhao, One-step exfoliation and fluorination of boron nitride nanosheets and a study of their magnetic properties, Angew. Chem. 126 (14) (2014) 3719–3723.
[36] Y. Ma, Y. Dai, M. Guo, C. Niu, L. Yu, B. Huang, Magnetic properties of the semifluorinated and semihydrogenated 2D sheets of group-IV and III-V binary compounds, Appl. Surf. Sci. 257 (17) (2011) 7845–7850.
[37] Y. Ma, Y. Dai, M. Guo, C. Niu, L. Yu, B. Huang, Strain-induced magnetic transitions in half-fluorinated single layers of BN, GaN and graphene, Nanoscale 3 (5) (2011) 2301–2306.
[38] A. Bhattacharya, S. Bhattacharya, G. Das, Strain-induced band-gap deformation of H/F passivated graphene and h-BN sheet, Phys. Rev. B 84 (7) (2011) 075454.
[39] C. Zhao, Z. Xu, H. Wang, J. Wei, W. Wang, X. Bai, et al., Carbon-doped boron nitride nanosheets with ferromagnetism above room temperature, Adv. Funct. Mater. 24 (38) (2014) 5985–5992.
[40] Y. Zhou, J. Xiao-Dong, Z. Wang, H.Y. Xiao, F. Gao, X.T. Zu, Electronic and magnetic properties of metal-doped BN sheet: a first-principles study, Phys. Chem. Chem. Phys. 12 (27) (2010) 7588–7592.
[41] J. Li, M. Hu, Z. Yu, J. Zhong, L. Sun, Structural, electronic and magnetic properties of single transition-metal adsorbed BN sheet: a density functional study, Chem. Phys. Lett. 532 (2012) 40–46.
[42] D. Warschauer, Electrical and optical properties of crystalline black phosphorus, J. Appl. Phys. 34 (7) (1963) 1853–1860.
[43] T. Nishii, Y. Maruyama, T. Inabe, I. Shirotani, Synthesis and characterization of black phosphorus intercalation compounds, Synth. Met. 18 (1) (1987) 559–564.
[44] H.O. Churchill, P. Jarillo-Herrero, Two-dimensional crystals: phosphorus joins the family, Nat. Nanotechnol. 9 (5) (2014) 330–331.
[45] E.S. Reich, Phosphorene excites materials scientists, Nature 506 (7486) (2014) 19.
[46] W. Lu, H. Nan, J. Hong, Y. Chen, C. Zhu, Z. Liang, et al., Plasma-assisted fabrication of monolayer phosphorene and its Raman characterization, Nano. Res. 7 (6) (2014) 853–859.
[47] L. Li, Y. Yu, G.J. Ye, Q. Ge, X. Ou, H. Wu, et al., Black phosphorus field-effect transistors, Nat. Nanotechnol. 9 (5) (2014) 372–377.
[48] H. Liu, A.T. Neal, Z. Zhu, Z. Luo, X. Xu, D. Tománek, et al., Phosphorene: an unexplored 2D semiconductor with a high hole mobility, ACS Nano. 8 (4) (2014) 4033–4041.
[49] Y. Du, C. Ouyang, S. Shi, M. Lei, Ab initio studies on atomic and electronic structures of black phosphorus, J. Appl. Phys. 107 (9) (2010) 093718.
[50] A. Rodin, A. Carvalho, A.C. Neto, Strain-induced gap modification in black phosphorus, Phys. Rev. Lett. 112 (17) (2014) 176801.
[51] J. Qiao, X. Kong, Z.-X. Hu, F. Yang, W. Ji, High-mobility transport anisotropy and linear dichroism in few-layer black phosphorus, Nat. Commun. (2014) 5.
[52] J.B. Yi, C.C. Lim, G.Z. Xing, H.M. Fan, L.H. Van, S.L. Huang, et al., Ferromagnetism in dilute magnetic semiconductors through defect engineering: Li-doped ZnO, Phys. Rev. Lett. 104 (13) (2010) 137201.
[53] Y.R. Wang, J.Y. Piao, G.Z. Xing, Y.H. Lu, Z.M. Ao, N. Bao, et al., Zn vacancy induced ferromagnetism in K doped ZnO, J. Mater. Chem. C (2015).
[54] Y. Wang, X. Luo, L.-T. Tseng, Z. Ao, T. Li, G. Xing, et al., Ferromagnetism and crossover of positive magnetoresistance to negative magnetoresistance in Na-doped ZnO, Chem. Mater. 27 (4) (2015) 1285–1291.
[55] P. Tao, H.H. Guo, T. Yang, Z.D. Zhang, Strain-induced magnetism in MoS_2 monolayer with defects, J. Appl. Phys. 115 (5) (2014) 054305.
[56] K. Gong, L. Zhang, W. Ji, H. Guo, Electrical contacts to monolayer black phosphorus: a first-principles investigation, Phys. Rev. B 90 (12) (2014) 125441.

[57] P. Srivastava, K. Hembram, H. Mizuseki, K.-R. Lee, S.S. Han, S. Kim, Tuning the electronic and magnetic properties of phosphorene by vacancies and adatoms, J. Phys. Chem. C 119 (12) (2015) 6530–6538.
[58] W. Hu, J. Yang, Defects in phosphorene, J. Phys. Chem. C 119 (35) (2015) 20474–20480.
[59] S.P. Koenig, R.A. Doganov, H. Schmidt, A. Castro Neto, B. Özyilmaz, Electric field effect in ultrathin black phosphorus, Appl. Phys. Lett. 104 (10) (2014) 103106.
[60] A. Ziletti, A. Carvalho, D.K. Campbell, D.F. Coker, A.C. Neto, Oxygen defects in phosphorene, Phys. Rev. Lett. 114 (4) (2015) 046801.
[61] G. Wang, R. Pandey, S.P. Karna, Effects of extrinsic point defects in phosphorene: B, C, N, O, and F adatoms, Appl. Phys. Lett. 106 (17) (2015) 173104.
[62] I. Khan, J. Hong, Manipulation of magnetic state in phosphorene layer by non-magnetic impurity doping, New J. Phys. 17 (2) (2015) 023056.
[63] H. Zheng, J. Zhang, B. Yang, X. Du, Y. Yan, A first-principles study on the magnetic properties of nonmetal atom doped phosphorene monolayers, Phys. Chem. Chem. Phys. 17 (25) (2015) 16341–16350.
[64] A. Hashmi, J. Hong, Transition metal doped phosphorene: first-principles study, J. Phys. Chem. C 119 (17) (2015) 9198–9204.
[65] X. Sui, C. Si, B. Shao, X. Zou, J. Wu, B.-L. Gu, et al., Tunable magnetism in transition-metal-decorated phosphorene, J. Phys. Chem. C 119 (18) (2015) 10059–10063.
[66] L. Seixas, A. Carvalho, A.C. Neto, Atomically thin dilute magnetism in Co-doped phosphorene, Phys. Rev. B 91 (15) (2015) 155138.
[67] Y. Nakanishi, A. Ishi, C. Ohata, D. Soriano, R. Iwaki, K. Nomura, et al., Nano. Res 10 (2017) 718–728.
[68] K.F. Mak, C.G. Lee, J. Hone, J. Shan, T.F. Heinz, Atomically thin MoS_2: a new direct-gap semiconductor, Phys. Rev. Lett. 105 (13) (2010) 136805.
[69] Q.H. Wang, K. Kalantar-Zadeh, A. Kis, J.N. Coleman, M.S. Strano, Electronics and optoelectronics of two-dimensional transition metal dichalcogenides, Nat. Nanotechnol. 7 (11) (2012) 699–712.
[70] K.S. Novoselov, A.K. Geim, S.V. Morozov, D. Jiang, Y. Zhang, S.V. Dubonos, et al., Electric field effect in atomically thin carbon films, Science 306 (5696) (2004) 666–669.
[71] Y.H. Lee, X.Q. Zhang, W. Zhang, M.T. Chang, C.T. Lin, K.D. Chang, et al., Synthesis of large-area MoS_2 atomic layers with chemical vapor deposition, Adv. Mater. 24 (17) (2012) 2320–2325.
[72] J.N. Coleman, M. Lotya, A. O'Neill, S.D. Bergin, P.J. King, U. Khan, et al., Two-dimensional nanosheets produced by liquid exfoliation of layered materials, Science 331 (6017) (2011) 568–571.
[73] A. Castellanos-Gomez, M. Barkelid, A. Goossens, V.E. Calado, H.S. van der Zant, G.A. Steele, Laser-thinning of MoS_2: on demand generation of a single-layer semiconductor, Nano. Lett. 12 (6) (2012) 3187–3192.
[74] N. Zibouche, A. Kuc, J. Musfeldt, T. Heine, Transition-metal dichalcogenides for spintronic applications, Ann. Phys. 526 (9-10) (2014) 395–401.
[75] X. Qian, J. Liu, L. Fu, J. Li, Quantum spin Hall effect in two-dimensional transition metal dichalcogenides, Science 346 (6215) (2014) 1344–1347.
[76] Z. Zhu, Y. Cheng, U. Schwingenschlögl, Giant spin-orbit-induced spin splitting in two-dimensional transition-metal dichalcogenide semiconductors, Phys. Rev. B 84 (15) (2011) 153402.
[77] D. Xiao, G.-B. Liu, W. Feng, X. Xu, W. Yao, Coupled spin and valley physics in monolayers of MoS_2 and other group-VI dichalcogenides, Phys. Rev. Lett. 108 (19) (2012) 196802.
[78] K. Dolui, A. Narayan, I. Rungger, S. Sanvito, Efficient spin injection and giant magnetoresistance in $Fe/MoS_2/Fe$ junctions, Phys. Rev. B 90 (4) (2014) 041401.
[79] B. Radisavljevic, A. Kis, Mobility engineering and a metal–insulator transition in monolayer MoS_2, Nat. Mater. 12 (9) (2013) 815–820.
[80] B.W. Baugher, H.O. Churchill, Y. Yang, P. Jarillo-Herrero, Intrinsic electronic transport properties of high-quality monolayer and bilayer MoS_2, Nano. Lett. 13 (9) (2013) 4212–4216.
[81] X. Cui, G.-H. Lee, Y.D. Kim, G. Arefe, P.Y. Huang, C.-H. Lee, et al., Multi-terminal transport measurements of MoS_2 using a van der Waals heterostructure device platform, Nat. Nanotechnol. (2015).

[82] A. Castellanos-Gomez, M. Poot, G.A. Steele, H.S. van der Zant, N. Agraït, G. Rubio-Bollinger, Elastic properties of freely suspended MoS_2 nanosheets, Adv. Mater. 24 (6) (2012) 772–775.
[83] N. Singh, G. Jabbour, U. Schwingenschlögl, Optical and photocatalytic properties of two-dimensional MoS_2, Eur. Phys. J. B 85 (11) (2012) 1–4.
[84] Y. Yoon, K. Ganapathi, S. Salahuddin, How good can monolayer MoS_2 transistors be? Nano. Lett. 11 (9) (2011) 3768–3773.
[85] G. Eda, H. Yamaguchi, D. Voiry, T. Fujita, M. Chen, M. Chhowalla, Photoluminescence from chemically exfoliated MoS_2, Nano. Lett. 11 (12) (2011) 5111–5116.
[86] Q. Yue, J. Kang, Z. Shao, X. Zhang, S. Chang, G. Wang, et al., Mechanical and electronic properties of monolayer MoS_2 under elastic strain, Phys. Lett. A 376 (12) (2012) 1166–1170.
[87] S. Tongay, S.S. Varnoosfaderani, B.R. Appleton, J. Wu, A.F. Hebard, Magnetic properties of MoS_2: existence of ferromagnetism, Appl. Phys. Lett. 101 (12) (2012) 123105.
[88] L. Cai, J. He, Q. Liu, T. Yao, L. Chen, W. Yan, et al., Vacancy-induced ferromagnetism of MoS_2 nanosheets, J. Am. Chem. Soc. 137 (7) (2015) 2622–2627.
[89] J.D. Fuhr, A. Saúl, J.O. Sofo, Scanning tunneling microscopy chemical signature of point defects on the MoS_2 (0001) surface, Phys. Rev. Lett. 92 (2) (2004) 026802.
[90] J. He, K. Wu, R. Sa, Q. Li, Y. Wei, Magnetic properties of nonmetal atoms absorbed MoS_2 monolayers, Appl. Phys. Lett. 96 (8) (2010) 082504.
[91] C. Ataca, S. Ciraci, Functionalization of single-layer MoS_2 honeycomb structures, J. Phys. Chem. C 115 (27) (2011) 13303–13311.
[92] E. Scalise, M. Houssa, G. Pourtois, V. Afanas'ev, A. Stesmans, Strain-induced semiconductor to metal transition in the two-dimensional honeycomb structure of MoS_2, Nano. Res 5 (1) (2012) 43–48.
[93] P. Johari, V.B. Shenoy, Tuning the electronic properties of semiconducting transition metal dichalcogenides by applying mechanical strains, ACS Nano. 6 (6) (2012) 5449–5456.
[94] H.L. Zheng, B.S. Yang, D.D. Wang, R.L. Han, X.B. Du, Y. Yan, Tuning magnetism of monolayer MoS_2 by doping vacancy and applying strain, Appl. Phys. Lett. 104 (13) (2014) 132403.
[95] N. Bao, H. Fan, J. Ding, J. Yi, Room temperature ferromagnetism in N-doped rutile TiO_2 films, J. Appl. Phys. 109 (7) (2011) 07C302.
[96] Y. Ma, J. Yi, J. Ding, L. Van, H. Zhang, C. Ng, Inducing ferromagnetism in ZnO through doping of nonmagnetic elements, Appl. Phys. Lett. 93 (4) (2008) 042514.
[97] N.H. Hong, J. Sakai, N. Poirot, V. Brizé, Room-temperature ferromagnetism observed in undoped semiconducting and insulating oxide thin films, Phys. Rev. B 73 (13) (2006) 132404.
[98] Y. Li, R. Deng, Y. Tian, B. Yao, T. Wu, Role of donor-acceptor complexes and impurity band in stabilizing ferromagnetic order in Cu-doped SnO_2 thin films, Appl. Phys. Lett. 100 (17) (2012) 172402.
[99] X. Luo, W.-T. Lee, G. Xing, N. Bao, A. Yonis, D. Chu, et al., Ferromagnetic ordering in Mn-doped ZnO nanoparticles, Nanoscale Res. Lett. 9 (1) (2014) 1–8.
[100] A. Ramasubramaniam, D. Naveh, Mn-doped monolayer MoS_2: an atomically thin dilute magnetic semiconductor, Phys. Rev. B 87 (19) (2013) 195201.
[101] R. Mishra, W. Zhou, S.J. Pennycook, S.T. Pantelides, J.C. Idrobo, Long-range ferromagnetic ordering in manganese-doped two-dimensional dichalcogenides, Phys. Rev. B 88 (14) (2013) 144409.
[102] Q. Yue, S. Chang, S. Qin, J. Li, Functionalization of monolayer MoS_2 by substitutional doping: a first-principles study, Phys. Lett. A 377 (19) (2013) 1362–1367.
[103] Y. Cheng, Z. Guo, W. Mi, U. Schwingenschlögl, Z. Zhu, Prediction of two-dimensional diluted magnetic semiconductors: doped monolayer MoS_2 systems, Phys. Rev. B (2013).
[104] R. Ma, T. Sasaki, Nanosheets of oxides and hydroxides: ultimate 2D charge-bearing functional crystallites, Adv. Mater. 22 (45) (2010) 5082–5104.
[105] M. Osada, T. Sasaki, Exfoliated oxide nanosheets: new solution to nanoelectronics, J. Mater. Chem. 19 (17) (2009) 2503–2511.
[106] T.W. Kim, E.J. Oh, A.Y. Jee, S.T. Lim, D.H. Park, M. Lee, et al., Soft-chemical exfoliation route to layered cobalt oxide monolayers and its application for film deposition and nanoparticle synthesis, Chem: Eur. J. 15 (41) (2009) 10752–10761.

[107] Y. Omomo, T. Sasaki, Wang, M. Watanabe, Redoxable nanosheet crystallites of MnO_2 derived via delamination of a layered manganese oxide, J. Am. Chem. Soc. 125 (12) (2003) 3568–3575.
[108] K. Naoi, P. Simon, New materials and new configurations for advanced electrochemical capacitors, J. Electrochem. Soc. 17 (1) (2008) 34–37.
[109] M. Osada, Y. Ebina, K. Fukuda, K. Ono, K. Takada, K. Yamaura, et al., Ferromagnetism in two-dimensional $Ti_{0.8}Co_{0.2}O_2$ nanosheets, Phys. Rev. B 73 (15) (2006) 153301.
[110] M. Osada, Y. Ebina, K. Takada, T. Sasaki, Gigantic magneto-optical effects in multilayer assemblies of two-dimensional titania nanosheets, Adv. Mater. 18 (3) (2006) 295–299.
[111] M. Osada, M. Itose, Y. Ebina, K. Ono, S. Ueda, K. Kobayashi, et al., Gigantic magneto-optical effects induced by (Fe/Co)-cosubstitution in titania nanosheets, Appl. Phys. Lett. 92 (25) (2008) 253110.
[112] Y. Kotani, T. Taniuchi, M. Osada, T. Sasaki, M. Kotsugi, F.Z. Guo, et al., X-ray nanospectroscopic characterization of a molecularly thin ferromagnetic $Ti_{1-x}Co_xO_2$ nanosheet, Appl. Phys. Lett. 93 (9) (2008) 093112.
[113] K.J. Saji, K. Tian, M. Snure, A. Tiwari, 2D tin monoxide—an unexplored p-type van der Waals semiconductor: material characteristics and field effect transistors, Adv. Electron. Mater. (2016).
[114] X. Pan, L. Fu, Tin oxide thin films grown on the (1012) sapphire substrate, J. Electroceram. 7 (1) (2001) 35–46.
[115] Y. Ogo, H. Hiramatsu, K. Nomura, H. Yanagi, T. Kamiya, M. Hirano, et al., p-channel thin-film transistor using p-type oxide semiconductor, SnO, Appl. Phys. Lett. 93 (3) (2008) 2113.
[116] J.A. Caraveo-Frescas, P.K. Nayak, H.A. Al-Jawhari, D.B. Granato, U. Schwingenschlögl, H.N. Alshareef, Record mobility in transparent p-type tin monoxide films and devices by phase engineering, ACS Nano. 7 (6) (2013) 5160–5167.
[117] L.Y. Liang, H.T. Cao, X.B. Chen, Z.M. Liu, F. Zhuge, H. Luo, et al., Ambipolar inverters using SnO thin-film transistors with balanced electron and hole mobilities, Appl. Phys. Lett. 100 (26) (2012) 263502.
[118] K. Nomura, T. Kamiya, H. Hosono, Ambipolar oxide thin-film transistor, Adv. Mater. 23 (30) (2011) 3431–3434.
[119] P.C. Chen, Y.C. Chiu, Z.W. Zheng, C.H. Cheng, Y.H. Wu, P-type tin-oxide thin film transistors for blue-light detection application, Phys. Stat. Sol. (RRL) (2016).
[120] Y. Ogo, H. Hiramatsu, K. Nomura, H. Yanagi, T. Kamiya, M. Kimura, et al., Tin monoxide as an s-orbital-based p-type oxide semiconductor: electronic structures and TFT application, Phys. Stat. Sol. (A) 206 (9) (2009) 2187–2191.
[121] A. Togo, F. Oba, I. Tanaka, K. Tatsumi, First-principles calculations of native defects in tin monoxide, Phys. Rev. B 74 (19) (2006) 195128.
[122] J.P. Allen, D.O. Scanlon, S.C. Parker, G.W. Watson, Tin monoxide: structural prediction from first principles calculations with van der Waals corrections, J. Phys. Chem. C 115 (40) (2011) 19916–19924.
[123] D. Granato, J. Caraveo-Frescas, H. Alshareef, U. Schwingenschlögl, Enhancement of p-type mobility in tin monoxide by native defects, Appl. Phys. Lett. 102 (21) (2013) 212105.
[124] J.P. Allen, D.O. Scanlon, L.F. Piper, G.W. Watson, Understanding the defect chemistry of tin monoxide, J. Mater. Chem. C 1 (48) (2013) 8194–8208.
[125] K. Govaerts, R. Saniz, B. Partoens, D. Lamoen, van der Waals bonding and the quasiparticle band structure of SnO from first principles, Phys. Rev. B 87 (23) (2013) 235210.
[126] D.B. Granato, A. Albar, U. Schwingenschlögl, Ab initio study of native defects in SnO under strain, EPL 106 (1) (2014) 16001.
[127] M. Meyer, G. Onida, M. Palummo, L. Reining, Ab initio pseudopotential calculation of the equilibrium structure of tin monoxide, Phys. Rev. B 64 (4) (2001) 045119.
[128] A. Kohan, G. Ceder, D. Morgan, C.G. Van de Walle, First-principles study of native point defects in ZnO, Phys. Rev. B 61 (22) (2000) 15019.
[129] A. Janotti, C.G. Van de Walle, Native point defects in ZnO, Phys. Rev. B 76 (16) (2007) 165202.
[130] S. Na-Phattalung, M.F. Smith, K. Kim, M.-H. Du, S.-H. Wei, S. Zhang, et al., First-principles study of native defects in anatase TiO_2, Phys. Rev. B 73 (12) (2006) 125205.

[131] Q. Wang, Q. Sun, G. Chen, Y. Kawazoe, P. Jena, Vacancy-induced magnetism in ZnO thin films and nanowires, Phys. Rev. B 77 (20) (2008) 205411.

[132] H. Pan, J. Yi, L. Shen, R. Wu, J. Yang, J. Lin, et al., Room-temperature ferromagnetism in carbon-doped ZnO, Phys. Rev. Lett. 99 (12) (2007) 127201.

[133] H. Peng, J. Li, S.-S. Li, J.-B. Xia, Possible origin of ferromagnetism in undoped anatase TiO_2, Phys. Rev. B 79 (9) (2009) 092411.

[134] G. Xing, D. Wang, J. Yi, L. Yang, M. Gao, M. He, et al., Correlated d0 ferromagnetism and photoluminescence in undoped ZnO nanowires, Appl. Phys. Lett. 96 (11) (2010) 112511.

[135] H. Hsu, J.-C.A. Huang, Y. Huang, Y. Liao, M. Lin, C. Lee, et al., Evidence of oxygen vacancy enhanced room-temperature ferromagnetism in Co-doped ZnO, Appl. Phys. Lett. 88 (24) (2006) 24250–24257.

[136] N. Khare, M.J. Kappers, M. Wei, M.G. Blamire, J.L. MacManus-Driscoll, Defect-induced ferromagnetism in Co-doped ZnO, Adv. Mat. 18 (11) (2006) 1449–1452.

[137] W. Yan, Z. Sun, Q. Liu, Z. Li, Z. Pan, J. Wang, et al., Zn vacancy induced room-temperature ferromagnetism in Mn-doped ZnO, Appl. Phys. Lett. 91 (6) (2007) 2113.

[138] K.A. Griffin, A. Pakhomov, C.M. Wang, S.M. Heald, K.M. Krishnan, Intrinsic ferromagnetism in insulating cobalt doped anatase TiO_2, Phys. Rev. Lett. 94 (15) (2005) 157204.

[139] A. Albar, U. Schwingenschlögl, Magnetism in 3d transition metal doped SnO, J. Mat. Chem. C 4 (38) (2016) 8947–8952.

[140] A.K. Singh, R.G. Hennig, Computational prediction of two-dimensional group-IV mono-chalcogenides, Appl. Phys. Lett. 105 (4) (2014) 042103.

[141] L. Seixas, A. Rodin, A. Carvalho, A.C. Neto, Multiferroic two-dimensional materials, Phys. Rev. Lett. 116 (20) (2016) 206803.

[142] Z. Ma, B. Wang, L. Ou, Y. Zhang, X. Zhang, Z. Zhou, Structure and properties of phosphorene-like IV-VI 2D materials, Nanotechnology. 27 (41) (2016) 415203.

[143] C. Gong, L. Li, Z. Li, H. Ji, A. Stern, Y. Xia, et al., Discovery of intrinsic ferromagnetism in two-dimensional van der Waals crystals, Nature 546 (2017) 265–269.

[144] B. Huang, G. Clark, E. Navarro-Moratalla, D.R. Klein, R. Cheng, K.L. Seyler, et al., Layer-dependent ferromagnetism in a van der Waals crystal down to the monolayer limit, Nature (2017) 546273.

[145] M. Bonilla, S. Kolekar, Y. Ma, H. Coy Diaz, V. Kalappattil, R. Das, et al., Strong room-temperature ferromagnetism in VSe_2 monolayers on van der Waals substrates, Nat. Nanotechnol. 13 (2018) 289–293.

CHAPTER 5

Charge-spin conversion in 2D systems

Yong Pu

Nanjing University of Posts and Telecommunications, Nanjing, Jiangsu, China

Chapter Outline

5.1 Overview

Two-dimensional (2D) material is one of the most important scientific discoveries in the past decade since the discovery of graphene in 2004. Generally, 2D materials can be categorized as a family of materials and structures with thickness of single or few atomic layers, including graphene, transition metal dichalcogenides (TMDs), and many other layered materials and compounds (so-called van der Waals materials). Besides the isolated 2D materials with atomic thickness, some other systems such as surfaces and interfaces also have 2D characteristics for certain aspect that can be viewed as 2D as well. The 2D systems can have various festinating physical properties, such as high mobility, dissipationless transport, quantum Hall effect, which make them very attractive for both fundamental research and applications of future electronic devices. For spintronics the 2D systems can serve as ideal material platforms to study the mechanisms for basic operations on spin and

Spintronic 2D Materials.
DOI: https://doi.org/10.1016/B978-0-08-102154-5.00004-7

generate high-speed, high-density, and low-cost spintronic devices. In fact "2D spintronics" has been becoming one of the most emerging research fields in spintronics, due to the long spin lifetime, gate-control, long spin diffusion length, and many other exotic spin properties in 2D systems.

At the same time, charge-spin conversion (CSC), which is a process to convert electrical signal to spin signal and vice versa, has attracted great attention as a novel technique to generate new functionalities of electronic devices. In the aspect of CSC, 2D systems are particularly interesting and have shown promising properties quite different with bulk materials, including ultrahigh efficiency, spin-momentum locking, strong spin–orbit coupling (SOC), high spin-accumulation, and more is about to be discovered. In 2D systems the spin-relevant particles (such as electron spins) and quasiparticles (such as magnons) are confined in 2D planes and, on the other hand, fewer spins to operate if compared with bulk, which make 2D system a clean environment for spin operations and enables efficient control on spins by means of electrical method through CSC process. Thus CSC in 2D systems has attracted a lot attention for both fundamental research and applications in spintronics in recent years.

In this chapter, basic concepts of CSC are introduced and most recent developments of CSC in 2D systems are reviewed. Section 5.2 gives a general introduction, Section 5.3 is about the spin generation, Section 5.4 is about the spin detection, Section 5.5 is the summary and outlook.

5.2 Introduction

Electron spin has two degenerate states, namely "up" and "down," in the contrary with the electron charge that has a fixed value. In information technology the two states of electron spin can represent "0" and "1" states for data storage, and the transition between the two states can be used for logic operation. Down to atomic level ~ 1 nm, the spin coupling energy between nearby electrons could be orders of magnitude smaller than the coulomb interaction. Compared with charge-based devices, spin-based devices could in principle have higher density, lower energy cost, and faster speed.

The main goal of spintronics is to utilize spin as basic unit in logic to build electronic devices for information storage and processing. To achieve that, four fundamental operations on spins are generally required: spin generation, spin detection, spin manipulation, and spin transport, as sketched in Fig. 5.1. Among the four operations the spin generation is usually the first step since there is always required to generate enough spins for further operation, from the device point of view a spintronic device needs to be driven by a spin source or spin current; spin detection is used to determine the orientation of spins for information read-out; spin manipulation and spin transport are usually used for information processing and communication, respectively [1].

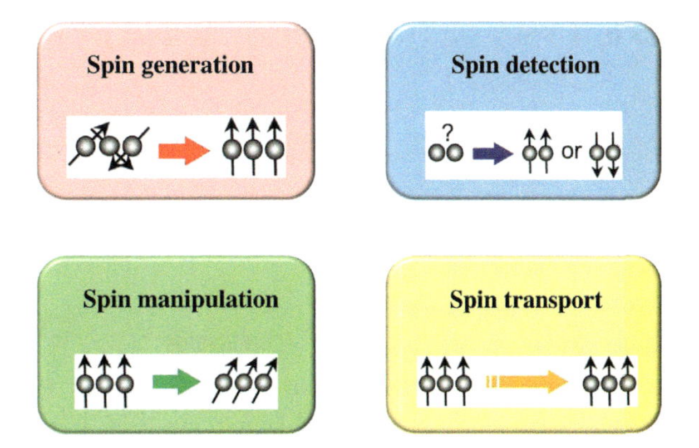

Figure 5.1
Basic spin operations: spin generation, spin detection, spin manipulation, and spin transport.

These four operations on spins have been realized in practice by different methods, including electrical, optical, mechanical, and thermal. Although in principle a spintronic device can be operated by either one or combination of these methods, the electrical control that has been well developed in modern electronics is the dominant method for spintronic devices, which is also the main focus of this chapter. From the material point of view the four spin operations have been achieved in conventional bulk materials. However, the fast development of modern spintronics demands new devices with much better performance, which become more and more challenging for bulk materials. More and more attentions have been attracted on the spin operations in 2D systems to design new devices with higher density, lower energy cost, and faster speed.

Although spin-based devices have advantages over charge-based devices for many aspects, the spin generation and spin detection are still the major challenges for the development of modern spintronics. In the contrary with charge, spin is not a conserved number. Given the fact that the spin lifetime is generally very short (picosecond or less in metals and nanosecond in semiconductors), spin information can be easily lost due to spin-flip or spin-dephasing. Unlike various electrical sources and detections for charge-based devices, to date suitable spin sources and sensitive spin detection are still lacking in most of cases.

Due to the development of spintronics and the discoveries of new materials including 2D materials, CSC has been becoming an emerging approach for efficient spin generation and spin detection. Since electron has both charge and spin, through the interplay between these two degrees of freedoms of electron, CSC can be used to generate spin currents by electrical source or convert spin signal to electrical signal for sensitive detection.

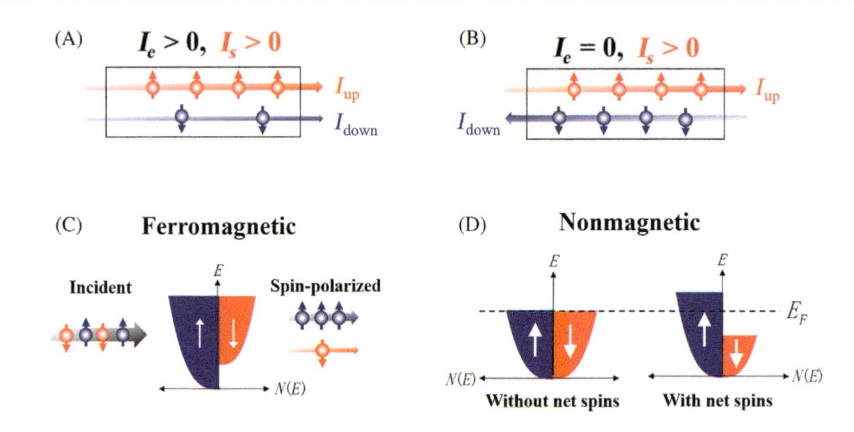

Figure 5.2
(A) Spin-polarized charge current; (B) pure spin current; (C) incident un-polarized charge current got spin-polarized in FM; (D) band-diagram of NM without net spins, or with net spins by spin injection. *NM*, Nonmagnetic materials.

5.3 Spin generation

Like the role of electrical current in electronics, spin current is crucial in spintronics. A main goal of spin generation is to produce sufficient spin current to drive spintronic device. As definition, spin current is a flow of spin angular momentum. Spin current could be with or without accompanying a charge flow in collinear direction; former is called spin-polarized charge current and latter is usually named as pure spin current. As sketched in Fig. 5.2A and B, spin-polarized charge current is the case that different amount of spin-up and spin-down electrons move in same direction, therefore both charge and spin currents present; pure spin current can be understood in concept that same amount of spin-up and spin-down electrons move in opposite directions, thus only spin current presents since no net charge flow. Both types of spin current can be generated by electrical current through the CSC process, and the efficiency of CSC is the key parameter that usually strongly depends on many factors including material, device structure, temperature.

5.3.1 Spin-polarized charge current

Spin-polarized charge current can be obtained by applying charge current in a ferromagnetic conductor, in which the majority and minority spin subbands have different density of states $N(E)$ due to spin splitting in the ferromagnet (FM), as sketched in Fig. 5.2C. In a simple two-channel model, electrons in FM move in majority and minority spin channels separately, with corresponding currents I_{maj} and I_{min}. The total charge current

$I_{charge} = I_{maj} + I_{min}$, and the net spin current in collinear direction $I_{spin} = I_{maj} - I_{min}$. The efficiency of the CSC process, namely spin polarization is defined as $P = I_{charge}/I_{spin}$, which is a material-dependent parameter. The typical value of spin polarization of conventional metals, such as, Fe, Ni, and CoFeB, is at level of 30%–50%.

Spin-polarized charge current is widely used to drive various spintronic devices, such as read head of hard disk, magnetoresistance (MR) random access memory (MRAM), and spin-transfer-torque-MRAM (STT-MRAM), which all have been commercialized [2]. However, the relatively low efficiency of the CSC of conventional metals is still the major limiting factor for further development: one issue is the Joule heating and relevant problems induced by the charge current, and another is the on/off ratio for logic that requires operation nearly 100% spin polarization for practice.

The recent discovery of the 2D ferromagnetic materials may provide promising solution. In 2017 intrinsic ferromagnetic order was demonstrated for the first time in single-layer CrI_3 [3] and few-layer $Cr_2Ge_2Te_6$ [4]. Theoretically, CrI_3 could be fully spin-polarized due to strong spin splitting and unique band structure. Experimentally, the devices consisting of four layers CrI_3 were demonstrated to have the spin polarization P higher than 95%, and the on/off ratio exceeds 19,000% [5]. The high spin polarization of the 2D ferromagnetic materials can be used in low power consumption devices, and the high on/off ratio can benefit the new devices for logic operation. In addition the atomic thickness of the 2D ferromagnetic materials shed the new light on the ultrahigh-density and low-power operation on the spintronic devices.

5.3.2 Spin injection into nonmagnetic materials

Although remarkable success has been achieved in ferromagnetic materials (FM) for data storage, data processing remains challenging because in FM the spin lifetime is extremely short $\sim 10^{-14}$ seconds due to strong SOC, which is considered the dominant mechanism for spin relaxation. Alternative has been proposed to inject spins into nonmagnetic materials (NM) with low SOC, such as Al, Si, GaAs, and more recently 2D systems (graphene, 2DEG, etc.), in which the spin lifetime could reach 10^{-6} second level.

The initial proposal was to apply electrical current directly across an FM/NM interface to inject spin-polarized charge current into the NM, but the polarization of the injected electrons was found extremely low due to the famous impedance mismatch problem, which is induced by the mismatch of the product of resistivity and spin diffusion length between FM and NM. In general, both resistivity and spin diffusion length of NM can be orders of magnitude higher than FM; therefore the injected spins tend to flow back to FM instead of stay in NM. The problem can be solved by inserting an insulating tunneling layer between FM and NM to form an FM/I/NM junction that can significantly suppress the back flow of the injected spins.

By this strategy the spin polarization of the injection can reach the same level of the injector to approximately tens of percent; significant spin imbalance can be induced in NM, and sufficient net spins can be used for further operations, as sketched in Fig. 5.2D.

Though the electrical spin injections through the FM/I/NM structures have been demonstrated in conventional NM, such as Al, Cu, Si, and GaAs, some critical issues still need to be solved: first is the high resistance of the tunnel junction that brings the problem of heating and other unexpected effects; second is the pinholes in the tunneling layer that would introduce significant leakage, to avoid that relatively thick tunneling layer is usually required, but the heating problem would be worse since the resistance of the tunnel junction increases with the thickness exponentially; in addition, the relatively large volume and carrier number of bulk material can dilute the spin signal. Thus the desired FM/I/NM structure for spin injection should consist of an FM with high spin polarization, a thin and flat tunneling layer, and a thin NM with low SOC and low carrier density.

The promising candidates can be found in 2D systems: graphene is single/few layer of carbon with extremely low SOC among known materials; 2D hexagonal boron nitride (hBN) is atomic flat and insulating down to single layer, which can serve as a good tunneling layer. Using graphene as NM [6] and hBN/graphene/hBN heterostructure [7], hundreds of ohms of nonlocal spin signal and 30 μm spin diffusion length have been demonstrated at room temperature, which both holds the records for electrical spin injection. Notable works have been extended into other 2D materials such as MoS_2, WS_2, black phosphorus, and silicene. Taking into account the newly discovered 2D ferromagnetic materials, all 2D devices or heterostructures are possible, which can potentially improve the device performance and introduce new functionalities.

5.3.3 Pure spin current

In the contrary with spin-polarized charge current, spin angular momentum can flow without accompanying an electrical current in collinear direction, which is usually named as pure spin current. Pure spin current itself generally has very low energy consumption since there is no Joule heating induced by charge current. And in case of pure spin current, conduction of electrons is not anymore a necessary, so the material choice is much more flexible, from metal, semiconductor, to even insulator. These advantages in fundamental make pure spin current a promising alternative of spin-polarized charge current and emerging research direction for the next generation of spintronic devices.

Pure spin current can be obtained by CSC via spin Hall effect (SHE), in which charge current generates pure spin current in transverse direction [8]. The efficiency of CSC (the ratio between charge and spin currents) is sometimes called spin Hall angle (SHA), which is a material-dependent parameter and can be viewed as the counterpart of spin polarization in the case of spin-polarized charge current. SHE was first found in GaAs [9], a nonmagnetic semiconductor, and detected via optical method. Fundamentally SHE originates from SOC, that is, proportional to Z^4, where Z is the atomic number; thus heavy elements, such as Pt, W, Ta, generally have large SHA at level of 10%–50%, that is comparable with the spin polarization of conventional ferromagnetic metals [8].

In 2012 thin layer of Ta was used to generate pure spin current that was strong enough to switch the magnetization of adjacent FM, which is particularly interesting in data storage [10]. Compared with strategies of magnetic switching used in commercialized magnetic memory devices, such as magnetic field used in MRAM that is difficult to confine in nanoscale, and spin-polarized charge current used in STT-MRAM that has to pass through highly resistive tunnel junction, the magnetic switching via SHE is basically driven by charge current that can be easily scaled down to nm, and importantly, charge current only travels in the highly conductive metal, so it can significantly benefit the high-density and low-energy consumption devices. Since the first discovery, lots of attentions have been attracted to explore new materials and architectures.

Just 2 years later, breakthroughs were made in 2D systems, the surface states of topological insulators (TIs). In NiFe/Bi_2Se_3 device the SHA was found to reach 3.5 [11]; In Cr-doped $(Bi_{0.5}Sb_{0.5})_2Te_3$ bilayer heterostructures the efficiency of CSC was found to exceed 425 [12], which is the highest of known materials and three orders of magnitude higher than the conventional magnetic and nonmagnetic bulk materials, such as Fe, Ni, Pt, and W. The giant efficiency of CSC in TIs is generally attributed to the strong SOC and the unique spin-momentum lock in TIs. Recently, high efficiency of CSC has been demonstrated in many other 2D systems including TMDs and other TIs [13]. These 2D systems are promising candidates to serve as efficient sources of pure spin currents for various spintronic devices. Though further investigations are still needed and debates are still remaining, the high efficiency of CSC of the 2D systems may originate from strong SOC of heavy elements, inversion symmetry broken that can enhance the SOC, and topological protection in certain cases.

5.4 Spin detection

By CSC, spin signal can be converted to electrical signal that can be detected easily using techniques well-established in electronics. Depending on the source of spin signal, such as spin current or spin accumulation, and strategy of detection, the electrical read-out can be in forms of resistance, current, or voltage. Apparently the efficiency of CSC is still a key parameter for sensitive spin detection. In many cases the spin detection and the spin generation share same physical origins and similar material preferences.

5.4.1 Spin accumulation voltage

When net spins accumulate in NM such as Si or graphene, the spin imbalance will introduce an energy difference (ΔE) between the two spin subbands. The energy difference is usually detected via an FM/I/NM tunneling junction, as sketched in Fig. 5.3A: the spin imbalance in the NM will introduce spin currents I_{up} and I_{down} tunnel into and from the FM, respectively. Due to the spin polarization of FM, I_{up}, and I_{down} are different in magnitudes, say $I_{up} > I_{down}$, so there will be a charge accumulation in FM until the difference is balanced, resulting in an electrical voltage V_s between FM and NM. Mathematically $V_s = \Delta E \times P/2q$, where P is the spin polarization of FM, and q is a unit charge. The spin accumulation can be detected locally that same FM contact serves as both spin injector and spin detector or nonlocally that the injector and the detector are separated in space; the latter is usually used for spin manipulation and logic operations, see Fig. 5.3B. In nonlocal geometry the spin signal decays exponentially with the distance between injector and detector; the decay constant is usually called spin diffusion length that is a material-dependent parameter.

The spin accumulation voltage is independent with the device size so can be compatible with high-density circuit. The spin accumulation voltage is apparently proportional to the spin splitting energy ΔE: for given spin imbalance (net spins), $\Delta E \sim \langle S \rangle / \mathrm{N}$, where $\langle S \rangle$ is the spin density, and N is the density of states of carriers; therefore, systems with low carrier density are desired. For spin manipulation, spin transport and logic operation, long spin lifetime and long spin diffusion length are required.

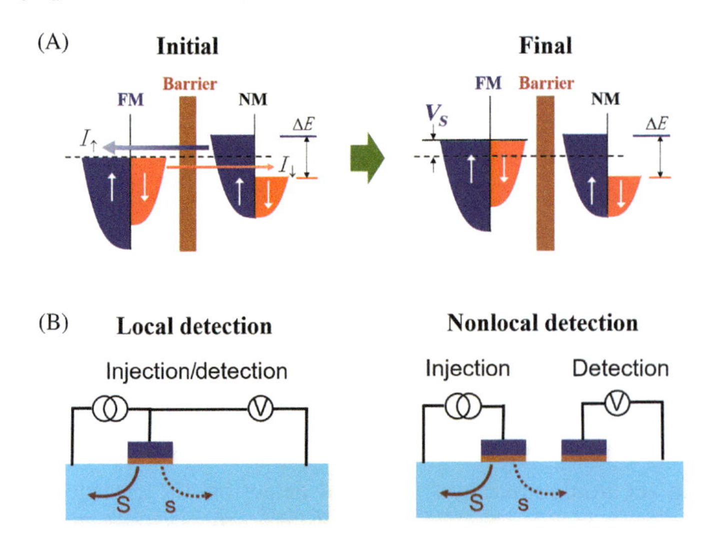

Figure 5.3
(A) Sketch of spin detection via spin accumulation voltage and (B) local and nonlocal spin detection.

2D systems such as 2DEG and graphene have both low carrier density and long spin lifetime/spin diffusion length, which enable sensitive spin detection, spin logic devices such as spin field-effect transistor and high-density spin circuit. In fact, nonlocal devices using graphene as spin channel have achieved hundreds of ohms of spin signal (spin accumulation voltage divided by injection current) and tens of μm of spin diffusion length at room temperature [6,7], both hold the records for electrical spin injection.

5.4.2 Inverse spin Hall effect

Electrical spin detection by the inverse SHE (ISHE) is basically a reciprocal effect of electrical spin generation by SHE, that is, a pure spin current I_s generating a charge current I_e in transverse direction and $I_e = I_s \times \theta_{SH}$, where θ_{SH} is the SHA. In open circuit a voltage will build up that can be detected electrically, given by $V_{ISHE} = J_e \times \rho L$, where J is the density of charge current, ρ is the resistivity, and L is the sample length. Thus sensitive detection usually demands materials with strong SOC and high resistivity, and large sample size as well. The early measurements of ISHE were mostly qualitative to demonstrate the pure spin injection, and more quantitative experiments have been performed later to accurately determine the SHAs of heavy metals [8]. FM/NM bilayer is the typical sample geometry for ISHE detection, where FM is the injector, and the ISHE voltage is measured in NM. Pure spin current can be generated by techniques, including spin-pumping driven by microwave and spin Seebeck effect driven by heat flow [8].

Many 2D systems have both large SHA and high resistivity that are desired for ISHE detection, for example, TMDs and surface states of TIs. Another key advantage of 2D systems is the flexibility to tune the properties of materials via gate, magnetic field, thickness, etc. In 2016 MoS_2 was found to have a large SHA of 2.3−12.7 [14]; in 2017 it was demonstrated that the SHA of Bi_2Se_3 increased with layer number decreasing and reached 1.5 at five layers [15]. In 2016 the SHA of Cr-(Bi,Sb)Te_3 was enhanced by factor of 4 by gate [16]; in 2016 the SHA of (Bi,Sb)Te_3 was increased by one order of magnitude by Fermi-level tuning [17]. The 2D systems mentioned here also have resistivity orders of magnitudes higher than conventional heavy metals than can significantly enhance the sensitivity. All these exciting developments suggest potential applications of 2D systems in spin detection with better performance and new functionalities. Although the final read-out of ISHE detection is generally an electrical voltage, the original source of signal is still charge current, so large sample size ($\sim$mm) is typically required for the ISHE detection, which somehow limits the material selection and the application in high-density devices.

5.4.3 Magnetoresistance

MR in general refers to the effect that the electrical resistance of material or device is sensitive to the orientation of magnetization, such as giant MR (GMR) effect, anisotropic

MR (AMR) effect, and tunneling MR (TMR) effect. Spin current, either pure spin current or spin-polarized charge current, can apply torque to magnetization by transfer spin angular momentum to local magnetic momentum, enabling spin detection by measuring the MR of detector that depends on the orientation of magnetization. Because the read-out is basically an electrical resistance, this type of spin detection is widely used in nanoscale devices and can be easily assembled in the high-density electronic circuits.

Spin-valve is one of the most important phenomena for the application of spintronics. The basic device structure to realize spin-valve consists of two magnetic layers separated by a nonmagnetic metallic layer or a tunneling layer, which are usually called GMR device and TMR device, respectively. When the magnetizations of two magnetic layers are in parallel or antiparallel configurations, the resistance of the device will be in low and high states, R_{low} and R_{high}, respectively, and the ratio of $(R_{\text{high}} - R_{\text{low}})/R_{\text{low}}$ is called the MR ratio of device. The GMR effect was first discovered in 1988 by Albert Fert and Peter Grünberger [18,19], who were awarded the Nobel Prize in Physics in 2007 for the discovery. The discovery of GMR effect is also viewed as the birth of modern spintronics. Later on TMR effect in magnetic tunnel junction (MTJ) was discovered, which significantly enhanced the MR ratio, enabling sensitive spin detection for commercial high-density hard disk and MRAM. Up to date more than 70% of data is saved in the magnetic media using MTJ to read out.

AMR is the effect that the sample resistance depends on the angle between the incident charge current and the magnetization. Although the MR ratio of AMR is typically at level of 0.1%–1%, much smaller than that of GMR (~10%) and TMR (up to 100% or even bigger), spin detection via AMR only requires a single layer of magnetic metal, so the sample preparation of experiment can be much easier than GMR and TMR so suitable for various materials.

The measurement is generally performed by applying an AC spin current $I_s(\omega)$ injected into a magnetic detector, due to angular momentum transfer and AMR effect, the resistance of detector will change correspondingly with the same frequency ω. If apply an additional charge current $I_e(\omega)$ across the device, the voltage $V = I_e(\omega) \times R(\omega) = V_0 + V_2(2\omega)$ can provide the information of incident spin current. In the equation, V_0 is the DC term due to rectification, and the second term has a frequency of 2ω. The recent developed Spin-torque FMR technique and second harmonic technique basically measure the first and the second term, respectively. In 2D systems, a major application of AMR detection is to determine the CSC efficiency or SHA. The typical sample structure is a 2D/FM bilayer. By applying a charge current through the bilayer the charge current in 2D system will generate magnetic field and spin current, both apply torques on FM detector. Analyzing the relative strength of the two torques can give the ratio of charge current and spin current, which is the CSC efficiency of the 2D system.

5.5 Outlook

After 30 years of development of CSC in spintronics, and a decade of exploring new materials and exotic properties in 2D systems, remarkable progress has been achieved on the CSC in 2D systems, including fundamental principles and proof-of-concept devices. However, challenges still need to be addressed to further understand the relevant mechanisms in 2D spintronics and deliver practical 2D spintronic devices.

5.5.1 New materials and heterostructures

The discovery of 2D ferromagnetic materials in 2017 is a milestone in 2D spintronics. Although various 2D ferromagnetic materials have been demonstrated or proposed, in most of them the Curie temperature is far lower than 300K, and wafer-scale sample growth is still lack in most cases. 2D ferromagnetic materials with Curie temperature higher than room temperature and large sample size are required for applications. 2D systems with strong SOC, such as TMDs and TIs, have shown high CSC efficiency, making them promising candidates for spin generation. However, most of these 2D systems are highly resistive that brings big issue of energy consumption when charge current passing through. Materials with high conductivity are yet to be discovered.

5.5.2 New techniques and characterizations

In bulk materials the operations of spin generation, spin detection, spin manipulation, and spin transport have been well developed, and fundamental parameters, including spin polarization, SHA, spin lifetime, magnetic damping parameters, can be measured accurately. Due to the small sample size and lack of suitable techniques, performing basic spin operations and measuring fundamental spin parameters in 2D systems are still challenging. The measurement results are sometimes inconsistent from different research groups even on same material. Standard procedure, new techniques, and even commercial equipment still need to be developed specifically for 2D systems.

5.5.3 New device architectures and functionalities

2D systems have shown exotic properties superior than bulk materials, as predicted by theories and demonstrated in proof-of-concept devices adopted from bulk spintronic devices. Further development can open up the new functionalities and fully take the advantages of 2D systems. For example, 2D ferromagnetic materials have high spin polarization that can replace the magnetic metals such as CoFeB in memory devices; on the other hand the resistance or even Curie temperature can be tuned by electrical gating. Integration of the two key properties may potentially achieve logic and memory on same chip, enabling fast and low energy consumption spintronic devices.

References

[1] I. Zutic, J. Fabian, S. Das Sarma, Rev. Mod. Phys. 76 (2004) 323.
[2] G. Prenat, K. Jabeur, G.D. Pendina, et al., Spintronics-Based Computing, vol. 145, Springer, 2015.
[3] B. Huang, G. Clark, E. Navarro-Moratalla, D. Klein, R. Cheng, K. Seyler, et al., Nature 546 (2017) 270.
[4] C. Gong, L. Li, Z. Li, H. Ji, A. Stern, Y. Xia, et al., Nature 546 (2017) 265.
[5] T. Song, X. Cai, M. Tu, X. Zhang, B. Huang, N. Wilson, et al., Science 360 (2018) 1214.
[6] W. Han, et al., Nat. Nanotechnol. 9 (2014) 794.
[7] M. Drögeler, et al., Nano Lett. 16 (2016) 3533.
[8] J. Sinova, S.O. Valenzuela, J. Wunderlich, C.H. Back, T. Jungwirth, Rev. Mod. Phys. 87 (2015) 1213.
[9] Y. Kato, R. Myers, A. Gossard, D. Awschalom, Science 306 (2004) 1910.
[10] L. Liu, C.-F. Pai, Y. Li, H.W. Tseng, D.C. Ralph, R.A. Buhrman, Science 336 (2012) 555.
[11] A.R. Mellnik, J.S. Lee, A. Richardella, J.L. Grab, P.J. Mintun, M.H. Fischer, et al., Nature 511 (2014) 449.
[12] Y. Fan, P. Upadhyaya, X. Kou, M. Lang, S. Takei, Z. Wang, et al., Nat. Mater. 13 (2014) 699.
[13] A. Soumyanarayanan, N. Reyren, A. Fert, C. Panagopoulos, Nature 539 (2016) 509.
[14] Q. Shao, G. Yu, Y.W. Lan, Y. Shi, M.Y. Li, C. Zheng, et al., Nano Lett. 16 (2016) 7514.
[15] J. Han, A. Richardella, S.A. Siddiqui, J. Finley, N. Samarth, L. Liu, Phys. Rev. Lett. 119 (2017) 077702.
[16] Y. Wang, D. Zhu, Y. Wu, Y. Yang, J. Yu, R. Ramaswamy, et al., Nat. Commun. 8 (2017) 1364.
[17] K. Kondou, R. Yoshimi, A. Tsukazaki, Y. Fukuma, J. Matsuno, K.S. Takahashi, et al., Nat. Phys. 12 (2016) 1027.
[18] M.N. Baibich, J.M. Broto, A. Fert, F. Nguyen Van Dau, F. Petroff, P. Etienne, et al., Phys. Rev. Lett. 61 (1988) 2472.
[19] G. Binasch, P. Grünberg, F. Saurenbach, W. Zinn, Phys. Rev. B 39 (1989) 4828.

CHAPTER 6

Magnetic properties of graphene

Nujiang Tang[1], Tao Tang[1,2], Hongzhe Pan[1,3], Yuanyuan Sun[1,3], Jie Chen[1] and Youwei Du[1]

[1]*Physics Department, National Laboratory of Solid State Microstructures, Jiangsu Provincial Key Laboratory for Nanotechnology, Nanjing University, Nanjing, P.R. China,* [2]*College of Science, Guilin University of Technology, Guilin, P.R. China,* [3]*School of Physics and Electronic Engineering, Linyi University, Linyi, P.R. China*

Chapter Outline

6.1 Significance of magnetic graphene

Magnetic materials are essential for modern industry. Almost all the presently used magnetic materials involve the elements of 3d- or 4f-transition metals such as Fe, Co, and Ni, and they are generally ferromagnets at room temperature. Magnetic ordering comes from the partially filled d- or f-electron bands. Development of transition-metal noninvolved magnets is a longtime interest because controlling the spin of s- or p-electrons will greatly extend the research boundary of magnetism. Particularly, carbon-based materials are of extreme interest since their structures are generally stable, simple, versatile and easy to be modified, which results in easier theoretical magnetism prediction and more likely spin induction [1–7]. More importantly, the coexistence of π- and σ-bonding in graphene is so unique that it provides the possibility to simultaneously generate the localized spins and couple these spins when modifying these bonds in certain modes, that is, magnetic ordering appears [6]. In fact, the earlier magnetism studies of graphite [8],

Spintronic 2D Materials.
DOI: https://doi.org/10.1016/B978-0-08-102154-5.00005-9

carbon nanotube [9], and fullerene [10] can be assorted with graphene since it is the basic block of all the graphitic forms [11]. In brief, graphene-based material is deemed as a promising s- or p-electron-based magnet.

In recent years, the demand of the lightweight magnets to open up new ways to design wearable, adaptable, and flexible information storage systems further highlights the importance of magnetic graphene. Moreover, the greatly potential application of graphene-based magnets in spintronics is promising, since graphene has extraordinary carrier mobility and may provide an easy way to integrate spin and molecular electronics. The long spin diffusion lengths and coherent times arising from the weak spin–orbit and hyperfine interactions in graphene can provide ideal conditions for coherent spin manipulation which can act as the next generation spintronic devices [12]. However ideal graphene is intrinsically nonmagnetic and lacks localized magnetic moments due to a delocalized π-bonding network, which limits its applications in spintronic devices. Therefore it is urgent and of great significance to develop effective methods for synthesizing magnetic graphene.

6.2 Primary theory of magnetism of graphene—Lieb's theorem

As graphene of honeycomb is a bipartite lattice, which can be partitioned into two sublattices A and B (see Fig. 6.1) [13]. Carbon atoms of sublattice A only connect to the atoms of sublattice B and *vice versa*. The Lieb's theorem gives a counted rule for the total spin of a bipartite lattice system [14]. According to Lieb's theorem [14], the number of nonbonding states is given by: $N = N_A - N_B$, where N_A and N_B respectively are the numbers of sites in sublattices A and B. The electrons in these nonbonding states degenerate at the Fermi level with spins obey the Hund rule and aligns parallelly. Then, the total spin of a bipartite system

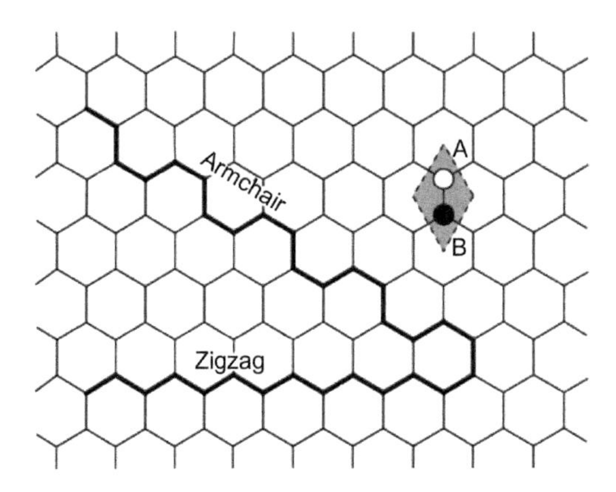

Figure 6.1
Two-dimensional crystalline lattice of graphene. The shaded area denotes the unit cell of graphene containing two carbon atoms which belong to the two sublattices of graphene, A (*empty circle*), and B (*filled circle*) [13].

can be characterized by Lieb's theorem as: $S = (1/2)(N_A - N_B)$. Accordingly, the spin magnetism of graphene is crucially dependent on its microstructure, which can be counted by Lieb's theorem [14].

6.3 General methods for inducing localized magnetic moments in graphene

As known, carbon allotropes generally exhibit diamagnetic susceptibility (χ_d) in the range of -10^{-5} to 10^{-7} emu/gOe. Despite that ideal graphene is intrinsic nonmagnetic, experimentally, many groups reported the unusual positive magnetic signal (paramagnetic and/or ferromagnetic) in graphene in recent years. Theoretically, point defects, such as vacancies [1,15–18], zigzag edges [19–21], and chemical doping of foreign atoms [3,4] can induce localized magnetic moments in graphene, which is the preliminary of the existence of ferromagnetic ordering. Experimentally, creating robust magnetic moments in graphene remains very difficult. Approximately, the approaches can be roughly divided into three categories: (1) the vacancy approach (creation of the magnetic moments on the basal-plane sites by vacancy via ion irradiation), (2) the edge approach (creation of the edge magnetic moments at the edge sites by edge-type defects), and (3) the sp^3 approach.

6.3.1 Vacancy approach

Atomic vacancies, including single, double, and multiple vacancies, have a strong impact on the mechanical, electronic, and magnetic properties of graphene. These vacancies may appear during growth or processing of graphene, or can also be artificially generated by irradiation with energetic particles [22–25] or chemical treatment [26,27]. The simplest vacancy defect is the single vacancy which means missing one carbon atom in graphene sheet. As seen in Fig. 6.2A and B, the single vacancy undergoes a Jahn–Teller distortion, which leads to two saturation bonds and one dangling bond towards the missing atom, and forms a five-membered and a nine-membered ring. Due to the increase in the local density of states (DOS) at the Fermi energy which is spatially localized on the dangling bonds, the single vacancy appears as a protrusion in STM images (Fig. 6.2C). Thus according to the Lieb's theorem [14], elimination of one atom from sublattice A introduces a magnetic moment of $|(N_A - 1) + N_B| = 1\ \mu_B$ per supercell, that is, the presence of a single vacancy defect can induce ferromagnetic ordering. This result has been widely confirmed using both first-principles [17,28,29] and mean-field Hubbard-model calculations [16].

Fig. 6.3A shows the spin-resolved DOS plots for the single vacancy defects in graphene obtained using first-principles calculations [17]. In addition to the quasi-localized state, there is also a localized nonbonding state due to the presence of a σ-symmetry dangling bond in this defect (Fig. 6.3B). Thus the peak is fully split by exchange, and the system is

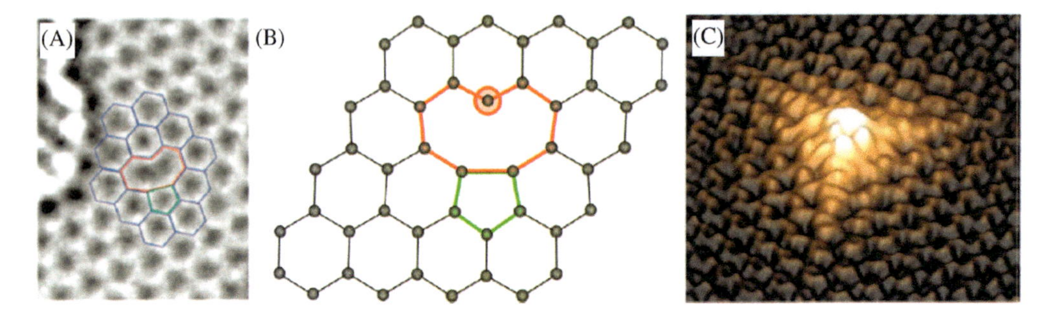

Figure 6.2
Single vacancy (A) as seen in an experimental TEM image; (B) its atomic structure obtained from the DFT calculations; and (C) an experimental STM image of a single vacancy, appearing as a protrusion due to an increase in the local DOS at the dangling bond [22,23,26]. *DOS*, density of states; *STM*, scanning tunneling microscopy; *DFT*, discrete Fourier transform.

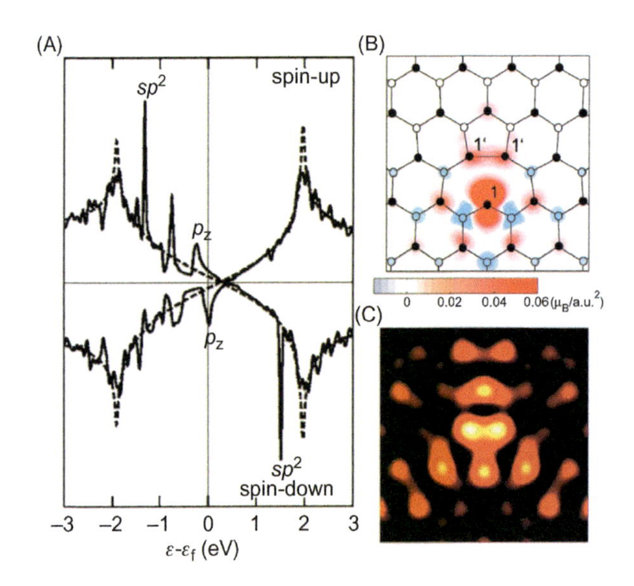

Figure 6.3
(A) Spin-resolved DOS for the single vacancy in graphene calculated from first principles. Both majority and minority spin channels are shown. Dashed curve shows the reference DOS of the ideal graphene. Zero energy corresponds to the Fermi level. Labels indicate the character of the defect states. (B) Spin-density projection (in μ_B/a.u^2) on the graphene plane around the vacancy defect in the α-sublattice. Carbon atoms corresponding to the α-sublattice (○) and to the β-sublattice (●) are distinguished. (C) Simulated STM images of the single vacancy in graphene [17]. *DOS*, density of states; *STM*, scanning tunneling microscopy.

characterized by a magnetic moment of 1.12–1.53 μ_B per single vacancy depending on the concentration of single vacancy ranged from 20% to 0.5%. One can find in Fig. 6.3B that the distribution of spin density around the defective site clearly shows the spin delocalization states. However it is worth noting that the magnetic ordering induced by single vacancies in graphene is only a rough physical model and it is improbable to be achieved experimentally at high temperatures for two reasons. One is that the formation energy of such a defect ($E_f \approx 7.5$ eV) is very high due to the presence of an under-coordinated carbon atom [30]. Thus, it is impossible to ensure all single vacancies are located in the same sublattice of graphene because of the high formation energy. The other is that the calculated migration barrier for a single vacancy in graphene is about 1.3 eV [30]. This low migration barrier allows the single vacancy migrate freely in graphene at the temperature slightly above room temperature (100°C–200°C).

When two single vacancies coalesce with each other or two neighboring atoms are removed, double vacancies are created in graphene [31]. As shown in Fig. 6.4A, no dangling bond is present in a fully reconstructed double vacancy, thus, two pentagons and one octagon [V_2(5-8-5) defect] appear instead of four hexagons in graphene. In addition, there are some other energetically favored double vacancies are shown in Figs. 6.4B and C. The atomic network of graphene remains coherent with minor perturbations in the bond lengths around these double vacancies. Simulations indicate that the formation energy E_f of

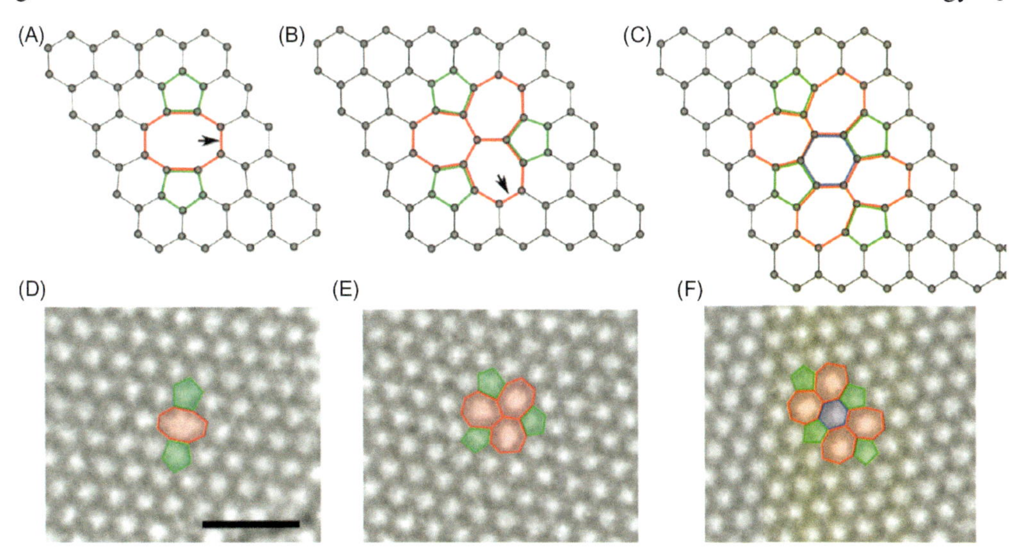

Figure 6.4
(A–C) Atomic structures of reconstructed double vacancy defects in graphene as obtained from the DFT calculations; and (D–F) experimental TEM images of the same structures. (A and D) Double vacancy V_2(5-8-5); (B and E) V_2(555-777) transformed from the V_2(5-8-5) defect by rotating a bond (marked in panel A); and (C and F) V_2(5555-6-7777) defect formed from V_2(555-777) by another bond rotation (bond marked in panel B) [26,32]. *DFT*, discrete Fourier transform.

a double vacancy is of the same order as that of a single vacancy (about 8 eV) [30]. As two atoms are now missing, the energy per missing atom (4 eV per atom) is much lower than that for a single vacancy. Hence, double vacancies are thermodynamically favored over single vacancies. Furthermore, the removal of more than two atoms may be expected to result in larger and more complex vacancy configurations, such as multiple vacancies or large hole in graphene [26]. Unfortunately, except single vacancy defects, most of the double and multiple vacancies cannot induce ferromagnetism in graphene [33].

The theoretical possibility of long-range ferromagnetic ordering has been predicted for randomly distributed point defects leaves little doubt that magnetism in graphene-based systems can in principle exist [33]. Although some previous experiments reported room-temperature ferromagnetism in graphene [34,35], no solid evidence can prove that this ferromagnetism is induced by vacancies. Is it possible to experimentally realize ferromagnetism in graphene only inducing by vacancies? To clarify this controversial issue, some targeted experiments were carried out [1,36]. In these experiments, no sign of ferromagnetism was detected at any temperature. For instance, Nair et al. [1] produced vacancies in graphene by the irradiation with high energy protons and carbon (C^{4+}) ions. The results of magnetization measurements for vacancies are plotted in Fig. 6.5. For all the concentrations, it is clear that irradiation defects only can give rise to the similar paramagnetism, independent of the ions used (H^+ or C^{4+}), and there is no sign of ferromagnetic ordering.

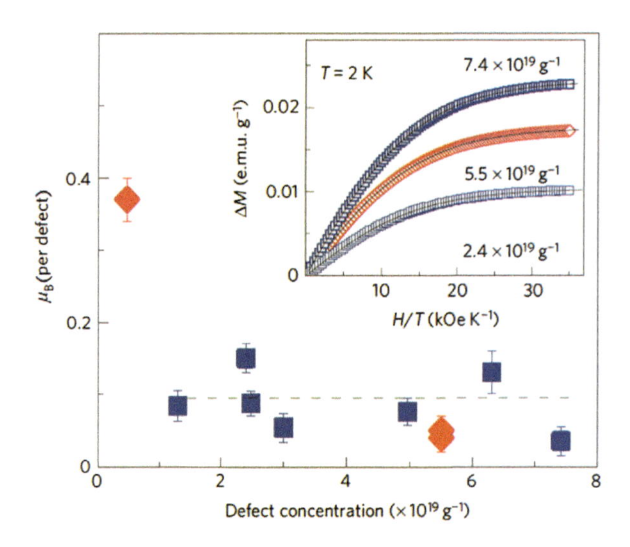

Figure 6.5

Magnetic properties of graphene laminates irradiated with protons (*blue squares*), and C^{4+} ions (*red diamonds*). Main panel: magnetic moment M normalized by the concentration of vacancies. Error bars indicate the accuracy of determination of the number of spins per vacancy. Inset: magnetic moment M due to vacancies as a function of parallel field H. The labels give the defect density; solid curves are Brillouin function fits for $J = 1/2$ [1].

6.3.2 Edge approach

6.3.2.1 Graphene quantum dots

Graphene quantum dots (GQDs) of nanometers with spin-polarized edge states has aroused continual and tremendous interest [37,38]. Due to the high edge-to-area ratio, GQDs has possible substantial spin-polarized edge states at the zigzag segments. Accordingly, the magnetism of GQDs was theoretically predicted to be especially intriguing [39–42]. The geometric shape of GQDs plays an important role in its magnetic properties [39–41,43]. Fig. 6.6 shows the geometric configuration and DOS of trigonal zigzag nanodisks calculated by tight-binding model [39]. The trigonal zigzag nanodisks exhibit metallic ferromagnetism due to their half-filled degenerate zero-energy states. Moreover, the degeneracy can be controllable arbitrarily by changing the size of the nanodisks. The emergence of magnetism in graphene nanoislands with triangular and hexagonal shapes terminated by zigzag edges have also been studied using both mean-field Hubbard model and discrete Fourier transform (DFT) calculations [40]. The results show that triangular graphene nanoislands have a finite S for all sizes whereas hexagons have $S = 0$ and develop local moments above a critical size of ≈ 1.5 nm. The correlation between sublattices and sign of the exchange interaction are also seen in this report that moments in the same sublattice couple ferromagnetically whereas moments in different sublattices couple antiferromagnetically.

Fig. 6.7 shows the magnetic susceptibility of GQDs with different geometrical shapes and characteristic sizes calculated by tight-binding approximation [41]. There are two types of edge states: (1) the zero-energy states (ZES) located exactly at the zero-energy Dirac point provides the temperature-dependent spin Curie paramagnetism, and (2) the dispersed edge states (DES) with the energy close, but not exactly equal to zero are responsible for the temperature-independent diamagnetic response. The hexagonal, circular, and randomly

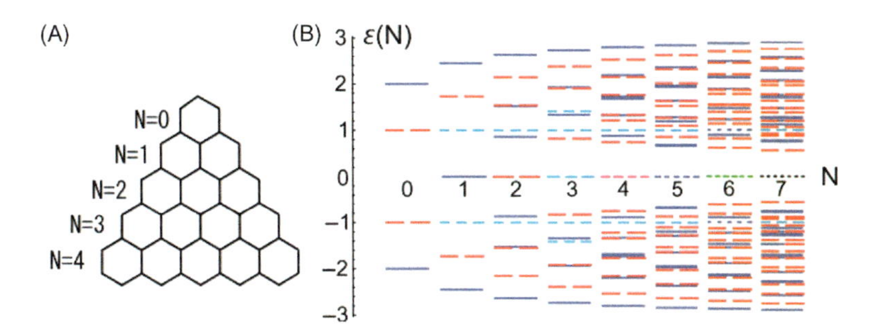

Figure 6.6
(A) Geometric configuration of trigonal zigzag nanodisks. (B) DOS of the N-trigonal nanodisk for $N = 0, 1, 2, \ldots, 7$. Dots on colored bar indicate the degeneracy of energy levels [39]. *DOS*, density of states.

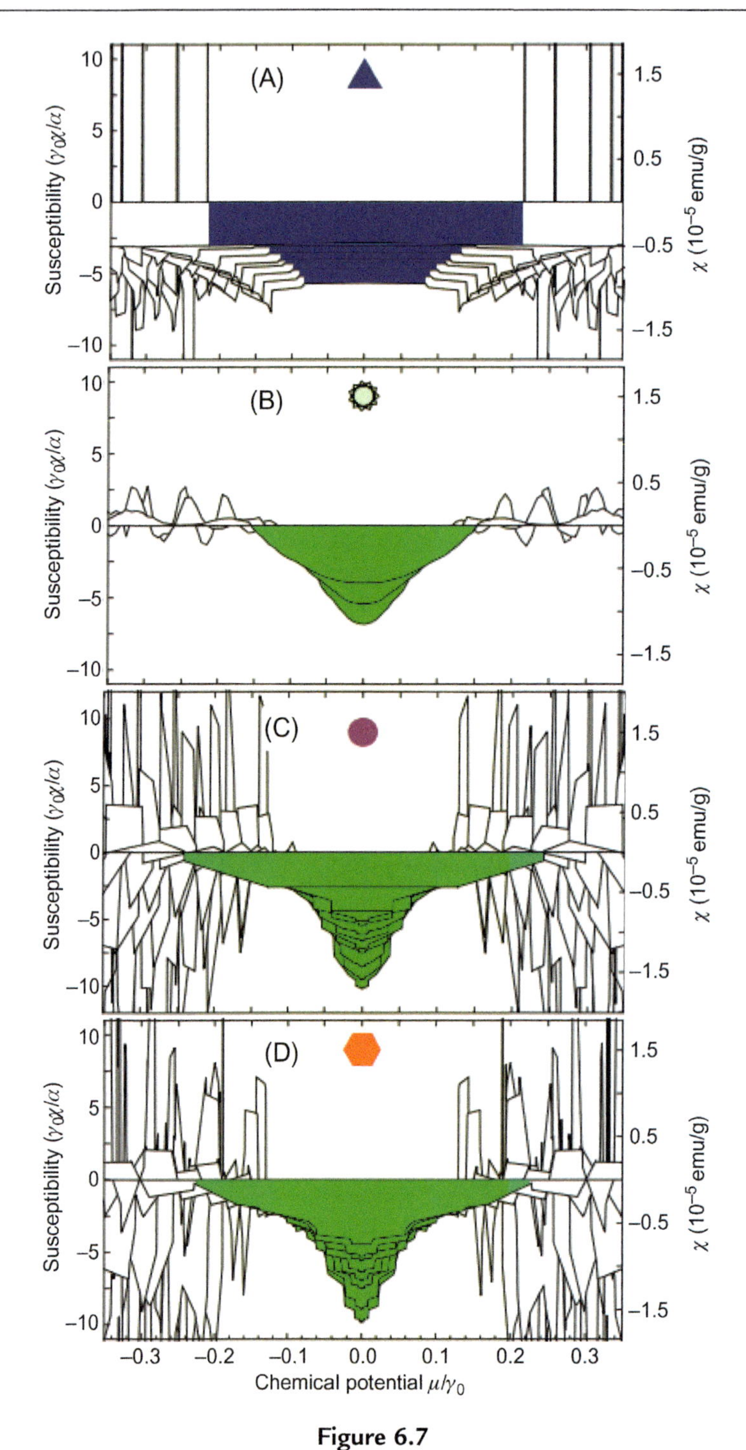

Figure 6.7

Susceptibility for triangular (A), random (B), circular (C), and hexagonal (D) GQDs of sizes 2–7 nm. The shaded region indicates the (pseudo)gap [41]. *GQDs*, graphene quantum dots.

shaped GQDs contain mainly DES, which is diamagnetic. The triangular GQDs have ZES. The small triangular GQDs would show spin paramagnetism at the low temperatures, while the large triangular GQDs exhibit orbital diamagnetism at the high temperatures.

Theoretically, the intrinsic edge magnetism of a hexagonal graphene nanoflake with zigzag edges can be manipulated via carrier doping, chemical modification at the edge, and temperature [42]. Fig. 6.8 shows the magnetic phase diagram for graphene nanoflakes with varied on-site Coulomb repulsion and carrier density. The local Coulomb interaction induces antiferromagnetic coupling, while carrier doping favors ferromagnetic coupling. Chemical modification of the edge atoms would give rise to a richer phase diagram consisting of antiferromagnetic, ferromagnetic, mixed, and nonmagnetic phases. Temperature can also alter the magnetic state of graphene nanoflakes.

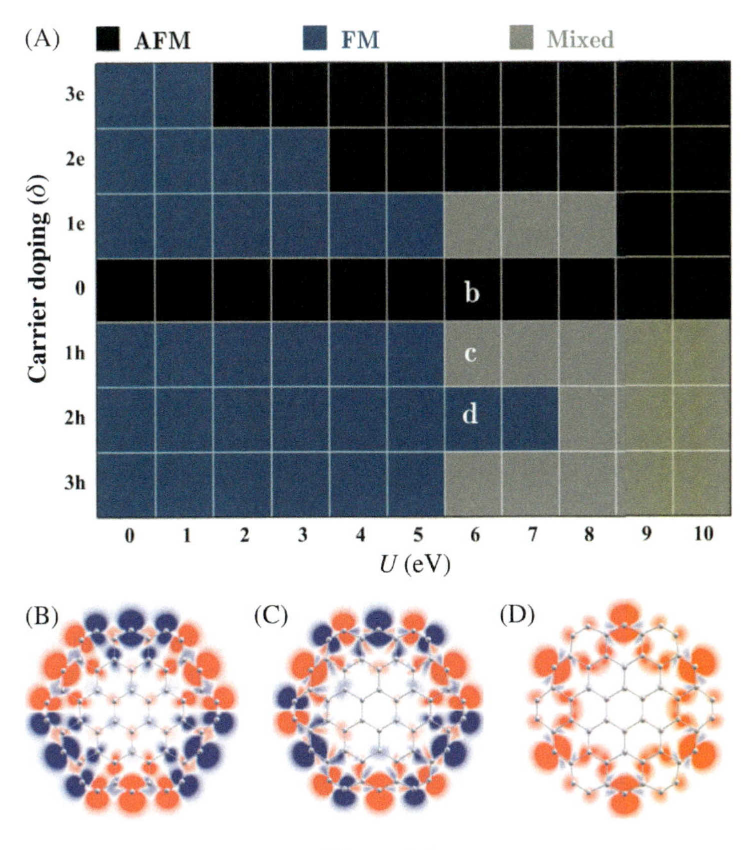

Figure 6.8

(A) Magnetic phase diagram for bare $N = 3$ nanoflakes with varied on-site Coulomb repulsion and carrier density, as obtained within DFT + U calculations. The antiferromagnetic, ferromagnetic, and the mixed state solutions are indicated with different color (*gray*) shades. Representative spin densities for (B) AFM, (C) mixed, and (D) FM configurations are shown for $U = 6$ eV [42]. *DFT*, discrete Fourier transform.

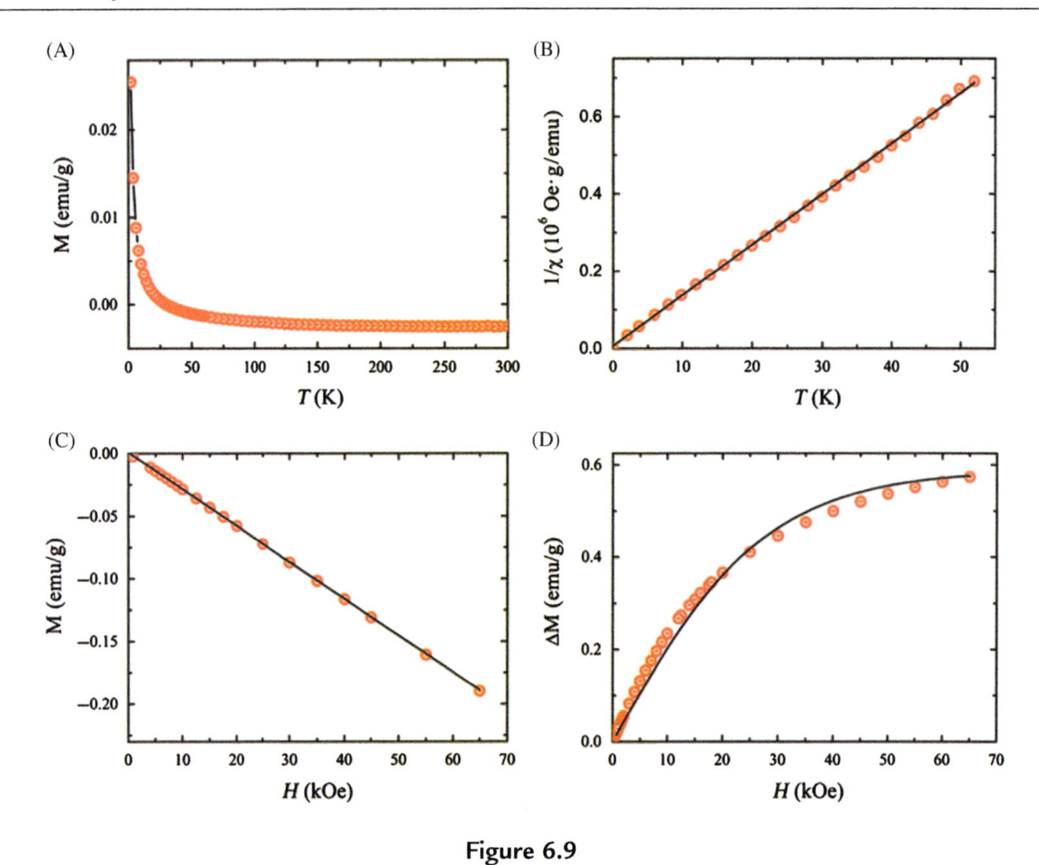

Figure 6.9

Magnetic properties of GQDs. (A) M–T and (B) $1/x$–T under the applied field $H = 1$ kOe. The round symbols are the measurements and the solid lines are fitted by the Curie law. (C) M–H measured at 300K. (D) ΔM–H measured at 2K [44]. *GQDs*, Graphene quantum dots.

Despite theoretical researches have predicted the intriguing magnetism of GQDs, experimental reports on the magnetism of GQDs are scarce. Research on the GQDs obtained by annealing graphene oxide quantum dots only shows Curie-like paramagnetism (Fig. 6.9) [44]. It is considered that since the GQDs obtained by thermal annealing enables both detection and reconstruction at edge switch off most of the spin polarization at the edge states [45–47].

6.3.2.2 Graphene nanoribbon

For the edge approach, it mainly includes introduction of line vacancies by formation of graphene nanomesh or synthesis of graphene nanoribbons (GNRs), or adsorption of heteroatoms such as nitrogen at the edge sites of GNRs, etc., GNRs can be classified into two basic categories, AGNRs (GNRs terminated by armchair edges), and ZGNRs (GNRs terminated by zigzag edges), as shown in Figs. 6.10A and B [19]. The electronic structure of GNR is determined by the edge termination. Theoretical and experimental studies

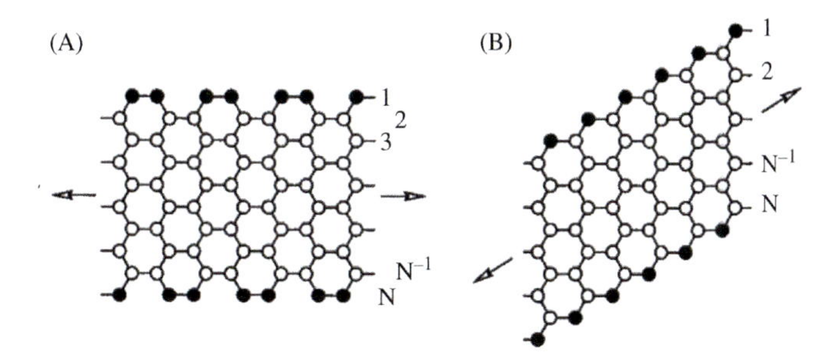

Figure 6.10
The network skeleton of GNR with armchair (A), and zigzag (B) edges. The edge sites are indicated by solid circles on each side. Periodic boundary conditions are assumed for the edges. The arrows indicate the translational directions of the GNRs [19]. *GNR*, graphene nanoribbon.

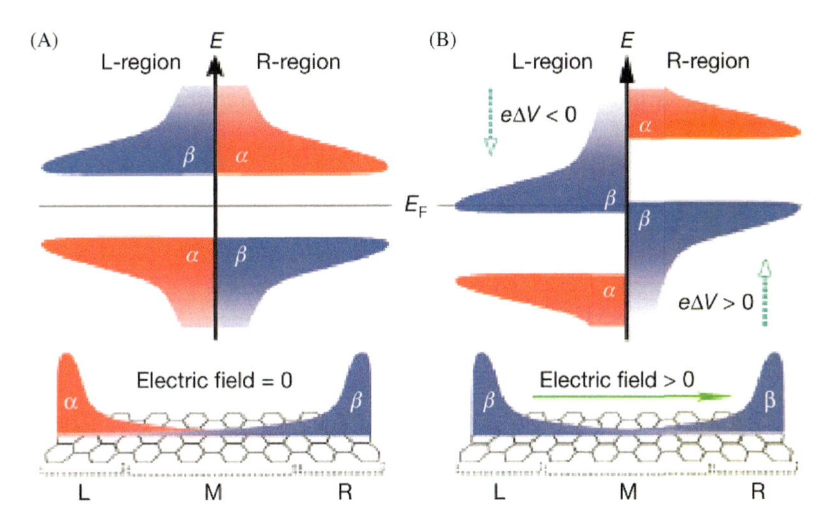

Figure 6.11
Schematic DOS diagram of the electronic states of a ZGNR (A) in the absence of an applied electric field and (B) under a transverse electric field [20]. *DOS*, density of states; *ZGNR*, GNRs terminated by zigzag edge.

showed the presence of localized electronic states at the zigzag edges (edge states), whereas the armchair edge does not exhibit this feature [19,48]. The edge states of ZGNRs extend along the edge direction and decay exponentially into the center of the ribbon. As a result of the electron–electron interactions, magnetism appears in ZGNRs by ordering the spins along the two ribbon edges with ferromagnetic coupling in the same edge, and antiferromagnetic coupling between opposite edges [20,49].

Fig. 6.11 shows the schematic DOS diagram of the electronic states of a ZGNR and a ZGNR under a transverse electric field, respectively [20]. In the absence of an applied electric field, the oppositely oriented spin states are located at the opposite sides of the ZGNR. Under a transverse electric field, the occupied and unoccupied β-spin states move closer in energy, while the occupied and unoccupied α-spin states move apart (Fig. 6.11B). Correspondingly, there is only one spin orientation at the Fermi level. Thus, the magnetic properties of ZGNR can be controlled by external electric fields.

Subsequently, the magnetic edge states in ZGNR have been experimentally observed. The near edge X-ray absorption fine structure and electron spin resonance (ESR) jointly confirm the existence of a magnetic edge state in ZGNR originated from zigzag edges [50]. Meanwhile, the edge states of ZGNR were spatially resolved by scanning tunneling microscopy (STM) and spectroscopy in 2011 [51]. There is a characteristic splitting in the dI/dV spectra, an unambiguous indication of magnetic ordering. The magnetic properties of potassium-split GNRs, and oxidative unzipped and chemically converted GNRs (CCGNRs) have been investigated by combination of magnetization and ESR measurements [50]. The $M-H$ of GNRs and CCGNRs are shown in Fig. 6.12, which indicates the occurrence of ferromagnetic features at low temperature for GNRs and CCGNRs.

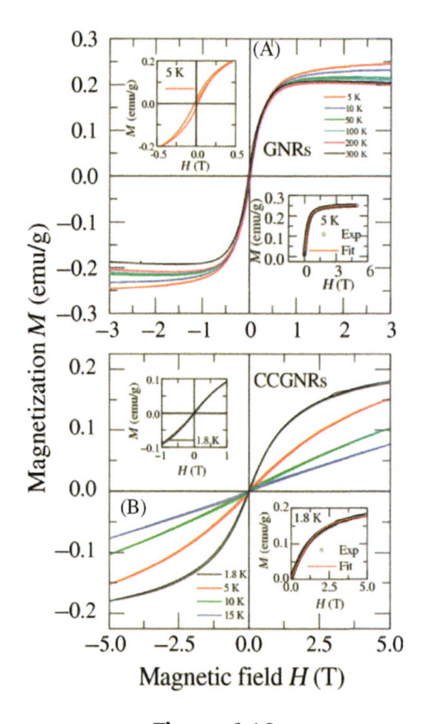

Figure 6.12

(A) Isothermal magnetization ($M-H$) observed on GNRs collected at the temperatures of 5K, 10K, 50K, 100K, 200K, and 300K. (B) Isothermal magnetization ($M-H$) observed on CCGNRs at various temperatures of 1.8K, 5K, 10K, and 15K [50]. *GNR*, graphene nanoribbon.

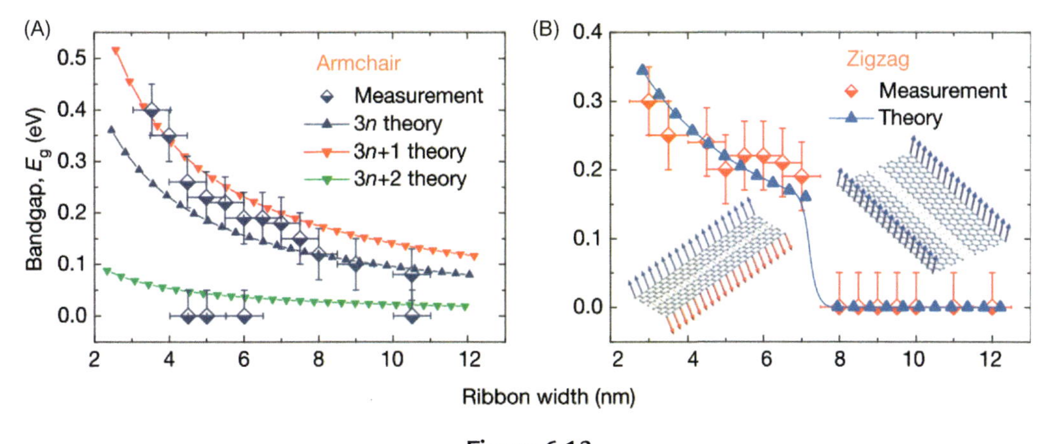

Figure 6.13
Edge-specific electronic and magnetic properties of graphene nanoribbons. The bandgap measured by tunneling spectroscopy as a function of ribbon width in armchair (A) and zigzag (B) ribbons [2].

It is found that the magnetic order on graphene edges of controlled zigzag orientation can be stable even at RT, raising hopes of graphene-based spintronic devices operating under ambient conditions [2]. As shown in Fig. 6.13A, ZGNRs that are narrower than 7 nm exhibit an electronic bandgap of about 0.2–0.3 eV, while a semiconductor-to-metal transition is revealed on increasing the ribbon width. This indicates a switching of the magnetic coupling between opposite ribbon edges from the antiferromagnetic to the ferromagnetic configuration. Thus, the width of ZGNRs is critical to its magnetic properties. The results provided a solid evidence that zigzag edge can induce magnetic moments, and the width-dependence magnetic ordering on the of ZGNRs by theoretical prediction.

6.3.2.3 Edge passivation

The shortage of both the vacancy and edge approach is that it can only induce limited magnetism at the vacancy or edge sites of the graphene sheet. For example although a single isolated vacancy can theoretically induce magnetic moment $\mu \sim 1\ \mu_B$ [13], the total vacancy density is restricted to a limited value to maintain the integrity of the graphene sheet, so the density of magnetic moments is very low [1]. And what is worse is that the vacancy and edge magnetic moments are unstable because they are fragile to be passivated by external surroundings [15]. Moreover, high -density vacancies can make graphene fragile and lose its structural stability, in contrast, chemical doping can keep its structural stability and can introduce more point defects in graphene. Theoretical study confirmed that N-doping is an effective method to introduce magnetic moments into graphene, and different types of N atoms contribute to different localized magnetic moments [52]. An N-5 atom which only can be absorbed at the vacancy of graphene sheet can introduce a net magnetic moment of 0.95 μ_B at either the vacancy or edge because of the contribution of the π-bonds [52,53]. By contrast, N-Q atom contributes to no magnetic moment, however it may act as a stable attractor, forming a magnetic defect complex N-Q + C-ad, which can contribute magnetic moment of 0.98 μ_B [54].

The experimental evidence was first obtained in N-doping of reduced graphene oxide (rGO) [4]. The magnetization was 0.312 emu/g for rGO, and 0.482 emu/g for NG-400, 0.514 emu/g for NG-500, 0.379 emu/g for NG-600, 0.246 emu/g for NG-700, and 0.173 emu/g for NG-900 (400, 500, 600 and 700 refer to the synthetic temperature in °C). The magnetic properties are attributed to the change in the proportion of pyrrolic nitrogen relative to graphitic nitrogen. The absence of ferromagnetism may attribute to the low doping level of N, and thus the long range between magnetic moments which is larger than the magnetic coupling length. Clearly, increasing the N-doping level may be the way to obtain ferromagnetic graphene. This work experimentally confirmed that N-doping can induce localized moments in graphene, which also demonstrate that N-doping can "repair" the vacancy or edge-induced magnetism canceled by oxidation.

The edge magnetic moments generated by N adatoms are stable because of the high stability of N adatoms. However because N adatoms are usually chemisorbed on the vacancy or edge sites, the density of N-induced magnetic moments is again low [55]. Fortunately, by the superdoping method, the N-doping level can reach high up to 29.82at.%, and that N-6 and N-5 dominate the N-doping in NG [27]. Interestingly, NG shows clear ferromagnetism with an obvious coercive field of 190 Oe and a remnant magnetization of 0.028 emu g^{-1} (bottom right inset of Fig. 6.14A). Moreover, the magnetization of NG is

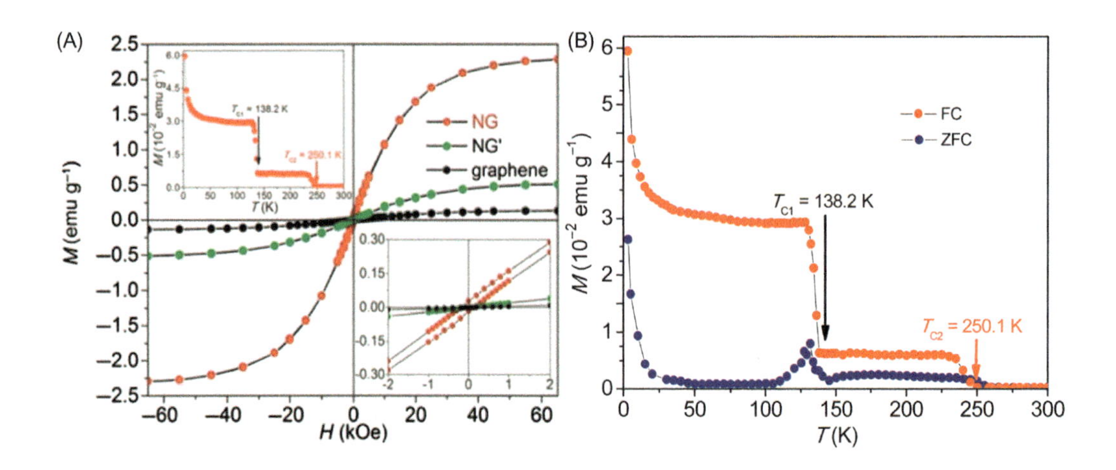

Figure 6.14
(A) Typical *M*–*H* curves of graphene, NG′, and NG measured at 2K. Insets are the *M*–*T* curve of NG measured from 2K to 300K under an applied field *H* = 500 Oe (top left) and part of the magnetization curves (bottom right). (B) Field cooling (FC) and zero field cooling *M*–*T* curves of NG under an applied field *H* = 500 Oe [27].

high, up to 2.3 emu g^{-1} (Fig. 6.14A), which is the highest recorded intrinsic magnetization for graphene derivatives. The $M-T$ measurement showed two clear T_C at 138.2 and 250.1K (top left inset of Fig. 6.14A). This can also be confirmed by the zero-field cooling $M-T$ curve (Fig. 6.14B). Apparently, N-superdoping results in a significant increase in the magnetization of graphene and the generation of near *RT* ferromagnetism. Clearly, the findings can offer the easy realization of ferromagnetic graphene with high magnetization, therefore push the way for potential applications in spintronic devices.

The magnetic coupling between the localized magnetic moments is the preliminary mode of the existence of ferromagnetic ordering. Theoretically, the coupling between the magnetic moments will appear via Ruderman–Kittel–Kasuya–Yosida interactions by delocalized π-electrons, which is expected to decay as D^{-r}, where D is the distance between the magnetic moments and r is the decay exponent [56,57]. As to the decay exponent r, it has been proposed that with the enhancing of electron–electron interaction, r may decrease and the power-law decay becomes more long ranged. Moreover, N-doping can make the Fermi level shifted upward due to the extra π-electrons, and make graphene electron-rich [58]. Thus N-doping can enhance the magnetic coupling between the magnetic moments because of the decrease of distance and decay exponent.

It should be noted that according to Li et al. [52] both pyridinic and pyrrolic N-doping can have significant effects on the spin distributions. For pyridinic nitrogen, the unpaired spins are mainly concentrated on edges and are localized on N atoms. Hence, it has less influence on the spin polarization of the edge states than pyrrolic nitrogen where the spin polarization on the doped edge was nearly removed. In the case of graphitic nitrogen, the opposite edges get localized and are completely canceled. Thus, increasing the pyrrolic nitrogens can increase the magnetization of GNRs.

6.3.3 The sp^3-type approach

To some extent, the effect of sp^3-type defects in covalently functionalized graphene to make graphene magnetic universally. When a weakly polar single covalent bond is established with the layer, a local spin moment of 1 μ_B always appears in graphene [59]. Fig. 6.15 shows a schematic of an isolated sp^3-type adatom or group chemisorbed on graphene. While when the sp^3-type bonding is carbon linked to other atoms, the gained magnetic moment is dependent on the electronegativity of the adatoms. With the electronegativity increasing, the ionic degree of C–X bonding increases and the covalent degree decreases, and thus the gained net spins decreases. Among them the weakest polar covalent bond is C–H, the gained spin is 1 μ_B, totally as the same as C–C bond.

Quite differently from the vacancy or edge-type approach to create defects only at vacancy or edge sites, the basal-plane approach is to introduce sp^3-type defects on the basal plane of

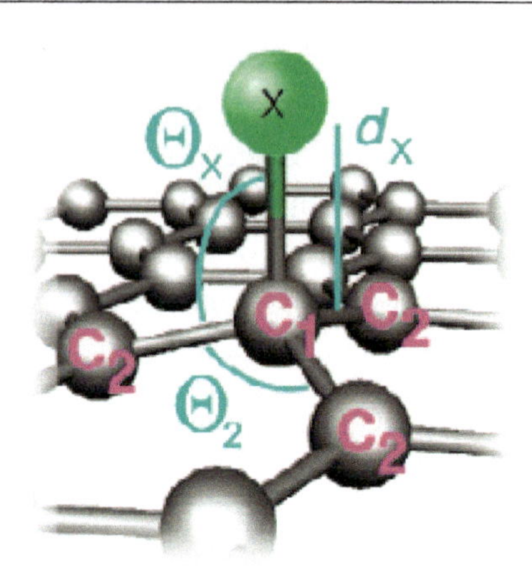

Figure 6.15
The schematic of a single sp^3 defect. X is the sp^3 adatom or functionalized group covalently bonded to graphene carbon atom (C_1). C_1 and C_2 are neighboring carbon atoms. Θ_x and Θ_2 are the bonding angulars, and d_x is the bonding length [59].

the graphene sheet and, thus more defects can be introduced. Therefore higher spin density is expected by this approach, and which is considered as hitherto the most promising approach to induce robust magnetic moments on the graphene sheet. The chemical stability of graphene makes precise sp^3-functionalization difficult to control, which is the obstacle for tuning its electronic or magnetic properties.

6.3.3.1 Hydrogen-doping approach

The experiment reported by Gonzalez—Herrero et al. presented the direct evidence that hydrogen atoms indeed yield a magnetic moment and that the magnetic moments can order ferromagnetically over relatively large distances [6]. Through the STM tip, a single H atom could be deposited on or removed of the graphene lattice which grown on SiC substrate, changing the sp^2 hybridization to sp^3 or vice versa, and thus removing or restoring the corresponding p_z orbital. According to Lieb's theory [14], it is easy to know that a single H atom chemisorbed on graphene could induce 1 μ_B magnetic moment. The differential conductance spectra (dI/dV) of STM have completely proven this conclusion. As seen in Figs. 6.16A and B, when a carbon atom bonded to a hydrogen atom, the symmetry of π-graphene system is broken and threefold $\sqrt{3} \times \sqrt{3}$ pattern spatial distribution of spin-polarized electronic state of the unpaired electron could be visualized. The total magnetic moment 1 μ_B distributes nearby the H atom (Fig. 6.16C) and fades with the distance to extend more than 3.5 nm (Fig. 6.16A). The spin-down magnetic moments sit on the sublattice which H atom chemisorbed and the spin-up ones sit on the opposite (Fig. 6.16D). The long extension of electronic state indicates that two unpaired electrons induced by two

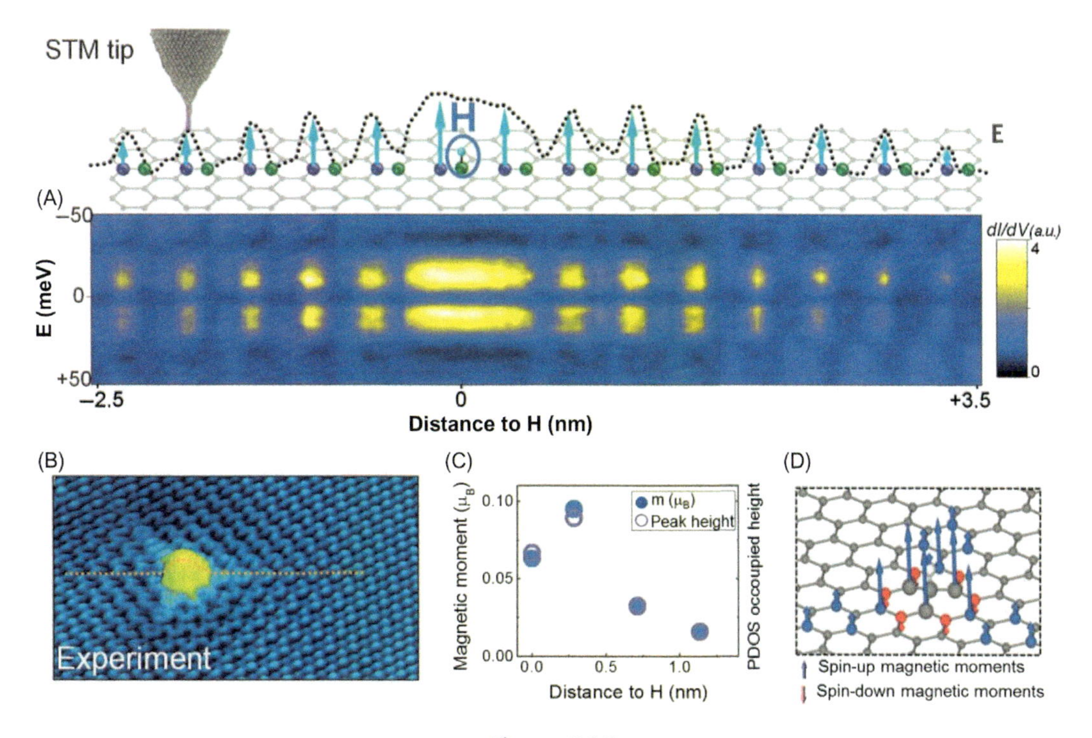

Figure 6.16

Spatial extension of the spin-polarized electronic state induced by H atoms in graphene. (A) Conductance map (*dI/dV*) of STM along the dashed line in (B). (B) STM topography of a single H atom on graphene. (C) the local magnetic moment distribution. (D) Calculated magnetic moments induced by H chemisorption (the lengths of the arrows signify the relative magnitudes of the magnetic moments) [6]. *STM*, scanning tunneling microscopy.

H atoms chemisorbed on graphene could interact with each other within a distance more than 3.5 nm. Also according to Lieb's theory [14], if the two H atoms sit on the same sublattice (AA dimer), the spins would couple ferromagnetically, and thus, the total magnetic moment is 2 μ_B; if the two H atoms sit on the different sublattice (AB dimer), the spins would couple nonmagnetically and the total magnetic moment is zero.

The DFT calculations and STM characterizations interpret the results very well. By DFT calculations, when a pair of H atoms is put on the different places of graphene lattice, the system had different energies. And the total energy would be lower when the two H atoms are closer (Fig. 6.17A). The lower energy indicates a stable configuration. Obviously, the AA dimer has a distinct spin-polarized electronic state distribution and the AB dimer has not (Fig. 6.17B–D). The total magnetic moment of 2 μ_B (0 μ_B) has nothing to do with the distance between the two H atoms in AA (AB) dimer, only if they are far enough not to interact with each other. However STM could only characterize the distribution of

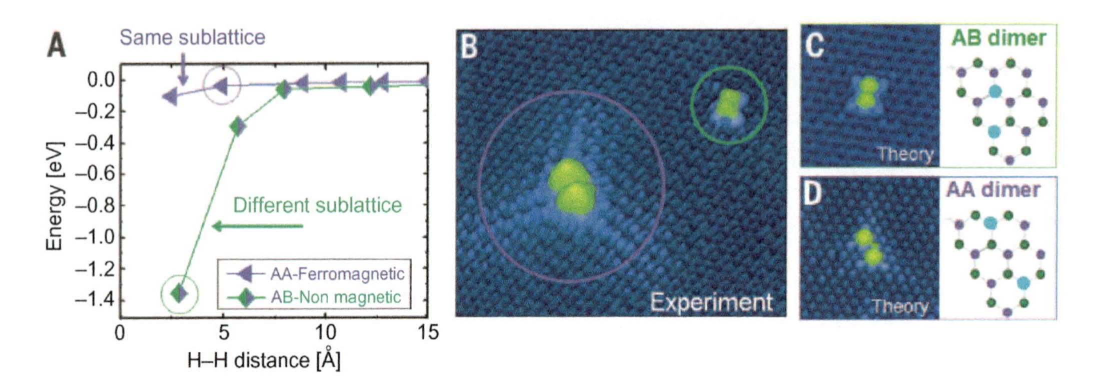

Figure 6.17
The magnetic coupling between neighboring H atoms on graphene. (A) Calculated total energy as a function of the H–H distance, relative to twice the adsorption energy of a single H atom, and magnetic state of a pair of H atoms adsorbed on the same (AA dimer) and different (AB dimer) sublattices. (B) STM image showing two different pairs of H atoms, with one pair in an AA (*purple circle*), and the other pair in an AB (*green circle*) configuration. (C) Calculated STM image of the AB dimer, and (D) the AA dimer, with the corresponding diagrams for H atoms (*blue balls*) on graphene (*purple and green balls*) [6]. *STM*, scanning tunneling microscopy.

spin-polarized electronic density within a very small scale, and the total magnetic moment gained by chemisorbed H atoms could only be calculated instead of visualized. Nevertheless, the ferromagnetically coupled spins in AA dimer enlightened the feasibility of preparing magnetic H-doping graphene.

In fact, the ineffective contribution of magnetic moment of H-doping is revealed in Fig. 6.17A. Since the nearest AB H-dimer has the lowest energy, a large of H atoms chemisorbed on graphene would tend to form such kind of nonmagnetic structures, and only very small fraction of H atoms could be arranged on the same sublattice without their counterpart locus on the opposite to form long-range ferromagnetic ordering. That is to say, within a nanoscale range on the graphene basal plane, if the number of H atoms chemisorbed on both sublattice is equal to form a "balanced" structure (all AB dimers), it would be nonmagnetic; while the number is not equal, the imbalance of the H-doping structure would contribute magnetic moments.

6.3.3.2 Fluorine-doping approach

Just like H-doping on graphene basal plane, F-doping will produce sp^3 defects on graphene to make it magnetic. As shown in Fig. 6.18, Nair et al. [1] fluorinated graphene via exfoliating graphite and measured their magnetic property, and found that: (1) The fluorinated graphene are all $J = S = 1/2$ paramagnetic in spite of the fluorination degree by fitting with Brillouin function; and (2) the magnetic moments are localized, and the

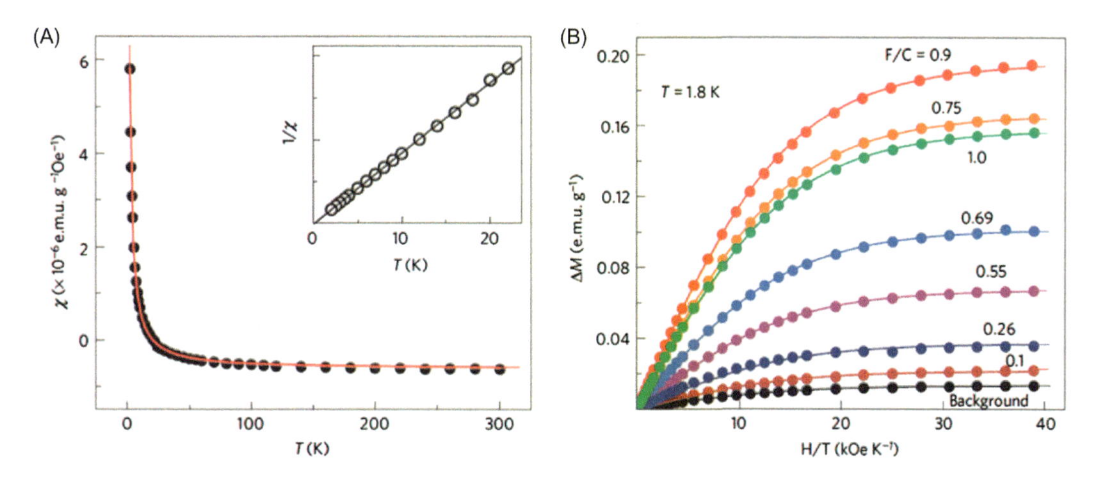

Figure 6.18

Spin-half paramagnetism of F-doped graphene. (A) Magnetic moment M as a function of parallel field H for different F/C ratios. Symbols are the measurements and solid curves are fits to the Brillouin function with $S = 1/2$ and assuming $g = 2$. (B) Dependence of susceptibility $\chi = M/H$ on T in parallel $H = 3$ kOe for $CF_{0.9}$; symbols are the measurements and the solid curve is the Curie law calculated self-consistently using the M/H dependence found in (A). Inset: Inverse susceptibility versus T demonstrating a linear, purely paramagnetic behavior [1].

susceptibility χ fits the Curie law $\chi = M/H = NJ(J+1)g^2\mu_B^2/3k_BT$ very well, where T is temperature. The uncoupled paramagnetic behavior of $J = S = 1/2$ indicates that all magnetic moments come from the contribution of independent single electrons. If we adopt the counting rule of Lieb [14], the fluorine atoms which contribute magnetic moments should all arranged as the configuration of $S_{N_A}^{N_B}$ where $N_B = N_A - 1$, along with some nonmagnetic configuration of $S_{N_A}^{N_B}$ where $N_B = N_A$. It is hard to know why the SQUID-characterization results have such a big difference from the DFT-calculated ones, however they completely assure that F-doping is an effective way to induce magnetic moments on graphene basal plane. Since the microstructure of experimental F-doped graphene remains unrevealed, it is meaningless to discuss about which kind of configuration of F-adatom indeed exists. The results showed that the maximal saturated magnetic moment of F-doped graphene is $\sim$0.2 emu/g. Apparently, the fact that about 1000C atoms could have the magnetic moment of 1 μ_B, is far less than the theoretical prediction. It is considered that such F adatoms tend to aggregate arising from the low migration barrier and, thus the inducing efficiency of magnetic moments is quite low.

Introducing vacancies into graphene is a feasible way to favor the formation of small F clusters [3]. GO is a common graphene derivative which heated in high temperature to obtain rGO with many vacancies after the oxygen groups are removed. By fluorinating rGO in XeF_2, F-doped rGO (F-RGO) with different F/C ratio is prepared via changing the mass ratio of XeF_2 to rGO. F adatoms tend to form F small clusters (cyan ball in Fig. 6.19A) nearby the vacancies and facilitate some solitary F adatoms (red ball in Fig. 6.19A) without their counterparts present in the neighboring sites. The spin density increases as the ratio of solitary F adatoms increases. The highest spin density of F-rGO with F/C = 0.46 is about 8 μ_B per thousand C atoms (Fig. 6.19B). However such kind of fluorinated vacancy graphene is still spin-half paramagnetic without any signals of spin coupling.

In fact, except for H-doping, it is very hard to deposit a single adatom or functionalized group on graphene to experimentally verify whether the spin exists or not, as well hard to compare the values of net spins. Anyway, the computational and experimental results convince us that sp^3 functionalization is a universal way to make graphene magnetic.

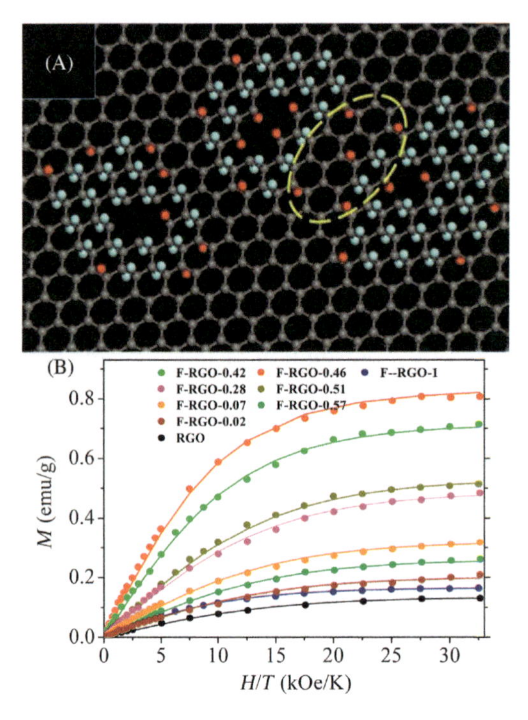

Figure 6.19
(A) The vacancy in graphene helps to form small F clusters to improve spin density. The cyan and red ball denote F adatoms with counterparts on the neighboring sites and those without, respectively. (B) $M(H/T)$ curves measured at 2K indicate that higher spin density could be obtained by fluorinating vacancy graphene (rGO). X in F-rGO-X denotes the fraction of F/C. Symbols are the measurements and the solid curves are fits to the Brillouin function with $J = S = 1/2$ [3].

6.3.3.3 Hydroxyl-doping approach

As known, both hydrogenated and fluorinated graphene are easy to lose their adatoms at moderate temperatures (hydrogenated graphene, ~200°C and the fluorinated one, ~400°C) [60,61]. Consequently, the magnetic graphene will convert back into the nonmagnetic one with the recovery of the aromatic structure. Therefore it is important to search another sp^3-type alternative that can effectively induce robust magnetic moments on the basal plane of the graphene sheet. Theoretical studies confirmed that the chemisorption of a single OH group on the basal plane of graphene can introduce magnetic moment ~1 μ_B [36,56,62]. But the great difference is, not like F or H adatoms clustering [1,3], OH groups on graphene sheet can be stable because of the suitable binding energy and relatively high migration barrier of C–OH bond [57]. More importantly, the stability can be strengthened by the emergence of ripples and surface topology [63,64] and/or the coincidence of symmetry of the clusters with the graphene lattice, suggesting the high feasibility of creating robust magnetic moments by OH groups.

Experimentally, highly oxidative debris (OD) was obtained by aqueous ammonia wash of as-prepared graphene oxide (aGO) which composed of OD and lightly oxidative GO sheets (awGO) [65]. The magnetic properties of OD and aGO were studied. The results showed that OD has a low magnetization of 0.16 emu/g, and the magnetization of aGO can be increased from 0.38 to 0.42 emu/g by discarding low magnetization OD. The M–H curves of the samples measured at 2K are well fitted using the Brillouin function. The Brillouin function provides the best fits by using $J = 2.18$, 2.65, and 2.15 for awGO, aGO, and OD, respectively. These J values are close to the quantum number 5/2 corresponding to 5 spins coupled together. Boukhvalov et al. [62] had explained the phenomena as 7 hydroxyl groups to form a cluster generating 4–5 μ_B with six of them sit at sublattice A, and the other at sublattice B of graphene. The results suggest that hydroxyl groups in GO contribute the magnetic moments (Fig. 6.20).

The above assumption was confirmed further by the following experiment [5]. Hydroxylated graphene (OHG) with a very high magnetization of 2.41 emu/g was prepared by annealing of graphene oxide (GO) to remove the unstable oxygen groups and leave the stable OH groups. The results demonstrate that OH groups can effectively induce robust magnetic moments on the basal plane of the graphene sheet. The results also highlight the two great superiorities of OH groups: high magnetic inducing efficiency (217 μ_B per 1000 OH groups) and high stability (surviving even at 900°C). Moreover, opening epoxy ring to form some newly arranged OH-group cluster can further enhance the spin density [66].

The results demonstrate that OH group is an effective candidate for inducing robust magnetic moments on the basal plane of the graphene sheet. These results have important implications for the synthesis of high-magnetization graphene for future applications of spintronics. In particular, the robust magnetic moments we observed in OHG provide the

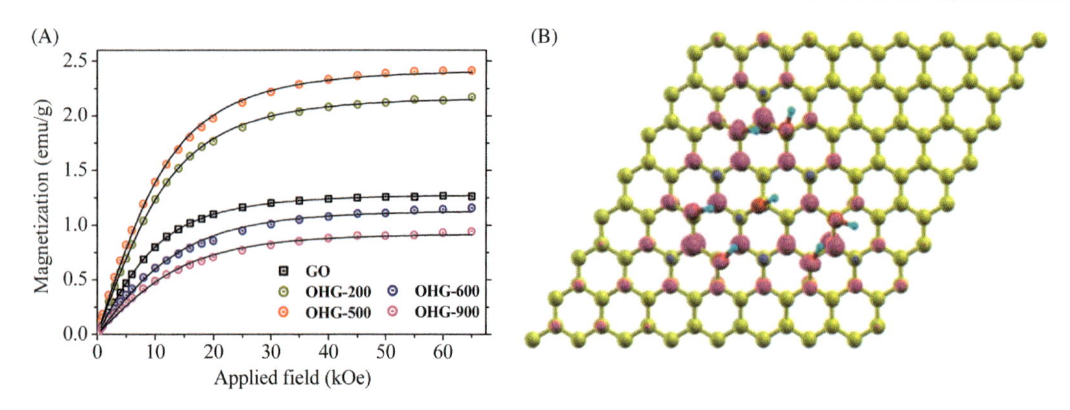

Figure 6.20
(A) Magnetic properties of GO and the samples obtained at different annealing temperatures measured at 2K by a superconducting quantum interference device (SQUID) magnetometer. X in OHG-X denotes the annealing temperature of GO. Dots are measurements and curves are fitted by Brillouin function with $J = S \approx 5/2$. (B) A probable threefold symmetric S_6^1 configuration composed of 7 OH groups to induce ~5 coupled spins [5]. *OHG*, hydroxylated graphene.

prerequisite to induce long-range magnetic ordering. Conversely, the ferromagnetic coupling between these basal-plane magnetic moments still needs to be addressed for the realization of ferromagnetic graphene with high magnetization. Most importantly, the findings highlight the importance of OH group as an effective sp^3-type candidate for inducing robust magnetic moments on the basal plane of the graphene sheet.

6.4 Conclusion and outlook

In this chapter, we highlighted the relevance of theoretical calculations and experimental results on the magnetism of graphene. The long spin diffusion length makes graphene very attractive for novel spintronic devices, and thus has triggered a quest for integrating the charge and spin degrees of freedom. Magnetic materials and nanostructures based on graphene offer unique opportunities for future technological applications such as spintronics. We also provided an overview of the challenges and opportunities for theory and experiments to promote the advancement of RT ferromagnetic graphene. We anticipate the significant future development of synthesis of ferromagnetic graphene with high magnetization, which is of great significance due to both fundamental and technological importance.

6.5 Acknowledgments

This work was financially supported by the State Key Program for Basic Research (Grant Nos. 2014CB921102 and 2017YFA0206304), and National Natural Science Foundation of China (Grant No. 51572122), P.R. China.

References

[1] R.R. Nair, M. Sepioni, I.L. Tsai, O. Lehtinen, J. Keinonen, A.V. Krasheninnikov, et al., Spin-half paramagnetism in graphene induced by point defects, Nat. Phys. 8 (2012) 199.
[2] G.Z. Magda, X. Jin, I. Hagymási, P. Vancsó, Z. Osváth, P. Nemes-Incze, et al., Room-temperature magnetic order on zigzag edges of narrow graphene nanoribbons, Nature 514 (2014) 608.
[3] Q. Feng, N. Tang, F. Liu, Q. Cao, W. Zheng, W. Ren, et al., Obtaining high localized spin magnetic moments by fluorination of reduced graphene oxide, ACS Nano 7 (8) (2013) 6729–6734.
[4] Y. Liu, Q. Feng, N. Tang, X. Wan, F. Liu, L. Lv, et al., Increased magnetization of reduced graphene oxide by nitrogen-doping, Carbon 60 (2013) 549–551.
[5] T. Tang, N. Tang, Y. Zheng, X. Wan, Y. Liu, F. Liu, et al., Robust magnetic moments on the basal plane of the graphene sheet effectively induced by OH groups, Sci. Rep. 5 (2015) 8448.
[6] H. González-Herrero, J.M. Gómez-Rodríguez, P. Mallet, M. Moaied, J.J. Palacios, C. Salgado, et al., Atomic-scale control of graphene magnetism by using hydrogen atoms, Science 352 (6284) (2016) 437–441.
[7] L.A. Gonzalez-Arraga, J.L. Lado, F. Guinea, P. San-Jose, Electrically controllable magnetism in twisted bilayer graphene, Phys. Rev. Lett. 119 (10) (2017) 107201.
[8] A. Ney, P. Papakonstantinou, A. Kumar, N.G. Shang, N.H. Peng, Irradiation enhanced paramagnetism on graphene nanoflakes, Appl. Phys. Lett. 99 (10) (2011) 3.
[9] J.P. Lu, Novel magnetic-properties of carbon nanotubes, Phys. Rev. Lett. 74 (7) (1995) 1123–1126.
[10] A. Lappas, K. Prassides, K. Vavekis, D. Arcon, R. Blinc, P. Cevc, et al., Spontaneous magnetic-ordering in the fullerene charge-transfer salt (TDAE) C-60, Science 267 (5205) (1995) 1799–1802.
[11] A.K. Geim, K.S. Novoselov, The rise of graphene, Nat. Mater. 6 (3) (2007) 183–191.
[12] N. Tombros, C. Jozsa, M. Popinciuc, H.T. Jonkman, B.J. van Wees, Electronic spin transport and spin precession in single graphene layers at room temperature, Nature 448 (7153) (2007) 571–U4.
[13] V.Y. Oleg, Emergence of magnetism in graphene materials and nanostructures, Rep. Prog. Phys. 73 (5) (2010) 056501.
[14] E.H. Lieb, Two theorems on the Hubbard model, Phys. Rev. Lett. 62 (10) (1989) 1201–1204.
[15] R.R. Nair, I.L. Tsai, M. Sepioni, O. Lehtinen, J. Keinonen, A.V. Krasheninnikov, et al., Dual origin of defect magnetism in graphene and its reversible switching by molecular doping, Nat. Commun. 4 (2013) 2010.
[16] J.J. Palacios, J. Fernández-Rossier, L. Brey, Vacancy-induced magnetism in graphene and graphene ribbons, Phys. Rev. B 77 (19) (2008) 195428.
[17] O.V. Yazyev, L. Helm, Defect-induced magnetism in graphene, Phys. Rev. B 75 (12) (2007) 125408.
[18] M.P. López-Sancho, F. de Juan, M.A.H. Vozmediano, Magnetic moments in the presence of topological defects in graphene, Phys. Rev. B 79 (7) (2009) 075413.
[19] K. Nakada, M. Fujita, G. Dresselhaus, M.S. Dresselhaus, Edge state in graphene ribbons: nanometer size effect and edge shape dependence, Phys. Rev. B 54 (24) (1996) 17954–17961.
[20] Y.-W. Son, M.L. Cohen, S.G. Louie, Half-metallic graphene nanoribbons, Nature 444 (2006) 347.
[21] H. Karimi, I. Affleck, Towards a rigorous proof of magnetism on the edges of graphene nanoribbons, Phys. Rev. B 86 (11) (2012) 115446.
[22] J.C. Meyer, C. Kisielowski, R. Erni, M.D. Rossell, M.F. Crommie, A. Zettl, Direct imaging of lattice atoms and topological defects in graphene membranes, Nano Lett. 8 (11) (2008) 3582–3586.
[23] M.M. Ugeda, I. Brihuega, F. Guinea, J.M. Gómez-Rodríguez, Missing atom as a source of carbon magnetism, Phys. Rev. Lett. 104 (9) (2010) 096804.
[24] A.V. Krasheninnikov, F. Banhart, Engineering of nanostructured carbon materials with electron or ion beams, Nat. Mater. 6 (2007) 723.
[25] H. Ohldag, T. Tyliszczak, R. Höhne, D. Spemann, P. Esquinazi, M. Ungureanu, et al., pi-Electron ferromagnetism in metal-free carbon probed by oft X-ray dichroism, Phys. Rev. Lett. 98 (18) (2007) 187204.

[26] F. Banhart, J. Kotakoski, A.V. Krasheninnikov, Structural defects in graphene, ACS Nano 5 (1) (2011) 26–41.
[27] Y. Liu, Y. Shen, L. Sun, J. Li, C. Liu, W. Ren, et al., Elemental superdoping of graphene and carbon nanotubes, Nat. Commun. (2016) 7.
[28] E.J. Duplock, M. Scheffler, P.J.D. Lindan, Hallmark of perfect graphene, Phys. Rev. Lett. 92 (22) (2004) 225502.
[29] P.O. Lehtinen, A.S. Foster, Y. Ma, A.V. Krasheninnikov, R.M. Nieminen, Irradiation-induced magnetism in graphite: a density functional study, Phys. Rev. Lett. 93 (18) (2004) 187202.
[30] A.A. El-Barbary, R.H. Telling, C.P. Ewels, M.I. Heggie, P.R. Briddon, Structure and energetics of the vacancy in graphite, Phys. Rev. B 68 (14) (2003) 144107.
[31] G.-D. Lee, C.Z. Wang, E. Yoon, N.-M. Hwang, D.-Y. Kim, K.M. Ho, Diffusion, coalescence, and reconstruction of vacancy defects in graphene layers, Phys. Rev. Lett. 95 (20) (2005) 205501.
[32] J. Kotakoski, A.V. Krasheninnikov, U. Kaiser, J.C. Meyer, From point defects in graphene to two-dimensional amorphous carbon, Phys. Rev. Lett. 106 (10) (2011) 105505.
[33] O.V. Yazyev, Magnetism in disordered graphene and irradiated graphite, Phys. Rev. Lett. 101 (3) (2008) 037203.
[34] Y. Wang, Y. Huang, Y. Song, X. Zhang, Y. Ma, J. Liang, et al., Room-temperature ferromagnetism of graphene, Nano Lett. 9 (1) (2009) 220–224.
[35] H.S.S.R. Matte, K.S. Subrahmanyam, C.N.R. Rao, Novel magnetic properties of graphene: presence of both ferromagnetic and antiferromagnetic features and other aspects, J. Phys. Chem. C 113 (23) (2009) 9982–9985.
[36] M. Sepioni, R.R. Nair, S. Rablen, J. Narayanan, F. Tuna, R. Winpenny, et al., Limits on intrinsic magnetism in graphene, Phys. Rev. Lett. 105 (20) (2010) 207205.
[37] W.L. Wang, O.V. Yazyev, S. Meng, E. Kaxiras, Topological frustration in graphene nanoflakes: magnetic order and spin logic devices, Phys. Rev. Lett. 102 (15) (2009) 157201.
[38] W.L. Wang, S. Meng, E. Kaxiras, Graphene nano flakes with large spin, Nano Lett. 8 (1) (2008) 241–245.
[39] M. Ezawa, Metallic graphene nanodisks: electronic and magnetic properties, Phys. Rev. B 76 (24) (2007) 245415.
[40] J. Fernández-Rossier, J.J. Palacios, Magnetism in graphene nanoislands, Phys. Rev. Lett. 99 (17) (2007) 177204.
[41] T. Espinosa-Ortega, I.A. Luk'yanchuk, Y.G. Rubo, Magnetic properties of graphene quantum dots, Phys. Rev. B 87 (20) (2013) 205434.
[42] M. Kabir, T. Saha-Dasgupta, Manipulation of edge magnetism in hexagonal graphene nanoflakes, Phys. Rev. B 90 (3) (2014) 035403.
[43] H.P. Heiskanen, M. Manninen, J. Akola, Electronic structure of triangular, hexagonal and round graphene flakes near the Fermi level, New J. Phys. 10 (10) (2008) 103015.
[44] Y. Sun, Y. Zheng, H. Pan, J. Chen, W. Zhang, L. Fu, et al., Magnetism of graphene quantum dots, npj Quantum Mater. 2 (1) (2017) 5.
[45] J. Jiang, W. Lu, J. Bernholc, Edge states and optical transition energies in carbon nanoribbons, Phys. Rev. Lett. 101 (24) (2008) 246803.
[46] B. Huang, F. Liu, J. Wu, B.-L. Gu, W. Duan, Suppression of spin polarization in graphene nanoribbons by edge defects and impurities, Phys. Rev. B 77 (15) (2008) 153411.
[47] J. Kunstmann, C. Özdoğan, A. Quandt, H. Fehske, Stability of edge states and edge magnetism in graphene nanoribbons, Phys. Rev. B 83 (4) (2011) 045414.
[48] Y. Niimi, T. Matsui, H. Kambara, K. Tagami, M. Tsukada, H. Fukuyama, Scanning tunneling microscopy and spectroscopy of the electronic local density of states of graphite surfaces near monoatomic step edges, Phys. Rev. B 73 (8) (2006) 085421.
[49] J. Fernández-Rossier, Prediction of hidden multiferroic order in graphene zigzag ribbons, Phys. Rev. B 77 (7) (2008) 075430.

[50] S.S. Rao, S.N. Jammalamadaka, A. Stesmans, V.V. Moshchalkov, J.V. Tol, D.V. Kosynkin, et al., Ferromagnetism in graphene nanoribbons: split versus oxidative unzipped ribbons, Nano Lett. 12 (3) (2012) 1210–1217.
[51] C. Tao, L. Jiao, O.V. Yazyev, Y.-C. Chen, J. Feng, X. Zhang, et al., Spatially resolving edge states of chiral graphene nanoribbons, Nat. Phys. 7 (2011) 616.
[52] Y. Li, Z. Zhou, P. Shen, Z. Chen, Spin gapless semiconductor – metal – half-metal properties in nitrogen-doped zigzag graphene nanoribbons, ACS Nano 3 (7) (2009) 1952–1958.
[53] C. Ma, X. Shao, D. Cao, Nitrogen-doped graphene nanosheets as anode materials for lithium ion batteries: a first-principles study, J. Mater. Chem. 22 (18) (2012) 8911–8915.
[54] Y. Ma, A.S. Foster, A.V. Krasheninnikov, R.M. Nieminen, Nitrogen in graphite and carbon nanotubes: magnetism and mobility, Phys. Rev. B 72 (20) (2005) 205416.
[55] P. Esquinazi, D. Spemann, R. Höhne, A. Setzer, K.H. Han, T. Butz, Induced magnetic ordering by proton irradiation in graphite, Phys. Rev. Lett. 91 (22) (2003) 227201.
[56] W. Min, H. Wei, B.C.-P. Mary, L. Chang Ming, Magnetism in oxidized graphenes with hydroxyl groups, Nanotechnology 22 (10) (2011) 105702.
[57] P.O. Lehtinen, A.S. Foster, A. Ayuela, A. Krasheninnikov, K. Nordlund, R.M. Nieminen, Magnetic properties and diffusion of adatoms on a graphene sheet, Phys. Rev. Lett. 91 (1) (2003) 017202.
[58] A.M. Black-Schaffer, Importance of electron-electron interactions in the RKKY coupling in graphene, Phys. Rev. B 82 (7) (2010) 073409.
[59] E.J.G. Santos, A. Ayuela, D. Sanchez-Portal, Universal magnetic properties of sp(3)-type defects in covalently functionalized graphene, New J. Phys. 14 (2012) 13.
[60] D.C. Elias, R.R. Nair, T.M.G. Mohiuddin, S.V. Morozov, P. Blake, M.P. Halsall, et al., Control of graphene's properties by reversible hydrogenation: evidence for graphane, Science 323 (5914) (2009) 610–613.
[61] X. Hong, K. Zou, B. Wang, S.H. Cheng, J. Zhu, Evidence for spin-flip scattering and local moments in dilute fluorinated graphene, Phys. Rev. Lett. 108 (22) (2012) 226602.
[62] D.W. Boukhvalov, M.I. Katsnelson, sp-Electron magnetic clusters with a large spin in graphene, ACS Nano 5 (4) (2011) 2440–2446.
[63] A. O'Hare, F.V. Kusmartsev, K.I. Kugel, A. Stable, "Flat" form of two-dimensional crystals: could graphene, silicene, germanene be minigap semiconductors? Nano Lett. 12 (2) (2012) 1045–1052.
[64] A. O'Hare, F.V. Kusmartsev, K.I. Kugel, Stable forms of two-dimensional crystals and graphene, Phys. B: Cond. Matter 407 (11) (2012) 1964–1968.
[65] T. Tang, F. Liu, Y. Liu, X. Li, Q. Xu, Q. Feng, et al., Identifying the magnetic properties of graphene oxide, Appl. Phys. Lett. 104 (12) (2014) 123104.
[66] J. Chen, W. Zhang, Y. Sun, Y. Zheng, N. Tang, Y. Du, Creation of localized spins in graphene by ring-opening of epoxy derived hydroxyl, Sci. Rep. 6 (2016) 26862.

CHAPTER 7

Experimental observation of low-dimensional magnetism in graphene nanostructures

Minghu Pan and Hui Yuan

School of Physics, Huazhong University of Science and Technology, Wuhan, P.R. China

Chapter Outline

7.1 Introduction

Ever since the first isolation of graphene [1], it has attracted enormous attention in science and technology. The simple honeycomb atomic structure and the unique electron structure with linear band dispersion at the Fermi level bring about many novel physical phenomena predicted and observed in this material [2–4]. While an ideal graphene is a purely sp^2 carbon material and not supposed to show any magnetism, many of the graphene nanostructures show some hints of magnetism. This reality makes graphene nanostructures very attractive for prospective applications in the field of spintronics. Unlike metallic spintronic materials, the negligible spin–orbit coupling in carbon-based graphene may bring about long spin coherence length [5]. All these exotic properties provide graphene highly prospective applications in spintronics devices. More interestingly and importantly, it is possible to tune the spin-transport properties through various applied stimuli. It was suggested that zigzag graphene nanoribbons (ZGNRs) can be triggered to have a half-metallic state by external electric fields [6]. It means that the spin transport could be controlled efficiently by an electric

Spintronic 2D Materials.
DOI: https://doi.org/10.1016/B978-0-08-102154-5.00006-0

field which is preferred when considering the compliance with current electronic technology.

In this chapter, we will introduce the mechanism of the magnetism in graphene nanostructures and review the experimental observation of the magnetism in these nanostructures and their edge states. First, we will go over a brief introduction of the theoretical model and its specific consequences for describing the electronic structure and magnetic properties of graphene-based materials. Here, the possible scenarios for the magnetism in graphene nanostructures are illustrated by a theoretical model based on the mean-field Hubbard Hamiltonian. Then, the magnetism in graphene nanostructures in three occasions: (1) finite graphene nanofragments, (2) one-dimensional (1D) graphene edges and graphene nanoribbons (GNRs), and (3) disorder in graphene nanostructures introduced by defects, is reviewed. In all cases, the theoretical analysis of the magnetism is provided before the introduction of the experimental observations. In the second case, the magnetism in GNRs and their edges will be covered in more detail for the reason that they received special attention in the scientific community for their better prospective applications in spintronics devices. Finally, the perspectives of the magnetism in graphene nanostructures are summarized in Section 7.6.

7.2 Electron–electron interaction in graphene

Graphene has a simple honeycomb atomic structure consists of pure carbon atoms which result in linear band dispersion at two specific points (K and K') in first Brillouin zone at the Fermi level (see Fig. 7.1) [2–4]. The orbitals of the four-valence electrons of each carbon atom in this honeycomb lattice undergo hybridization, forming three sp^2 bonds and one p_z orbital. The sp^2 orbitals form the σ band, which contains three localized electrons, thus does not contribute to conduction. The valence band, or π band, is generated from the bonding configuration among the p_z orbitals while the conduction band (π^*) is from the antibonding configuration. It is worth noting that the linear dispersion around K and K' points is protected by the lattice structure symmetry, and thus is robust against long-range hopping processes [8]. Another important fact is that graphene can also be considered as a two-dimensional (2D) example of the broad class of sp^2 carbon materials under a unifying concept includes polycyclic aromatic molecules, fullerenes, carbon nanotubes, and graphite.

In a conventional noncorrelative model, the electric system of graphene can be described by the nearest-neighbor tight-binding Hamiltonian as

$$\mathscr{H}_0 = -t \sum_{\langle i,j \rangle,\sigma} \left[c_{i\sigma}^{\dagger} c_{j\sigma} + \text{h.c.} \right] \tag{7.1}$$

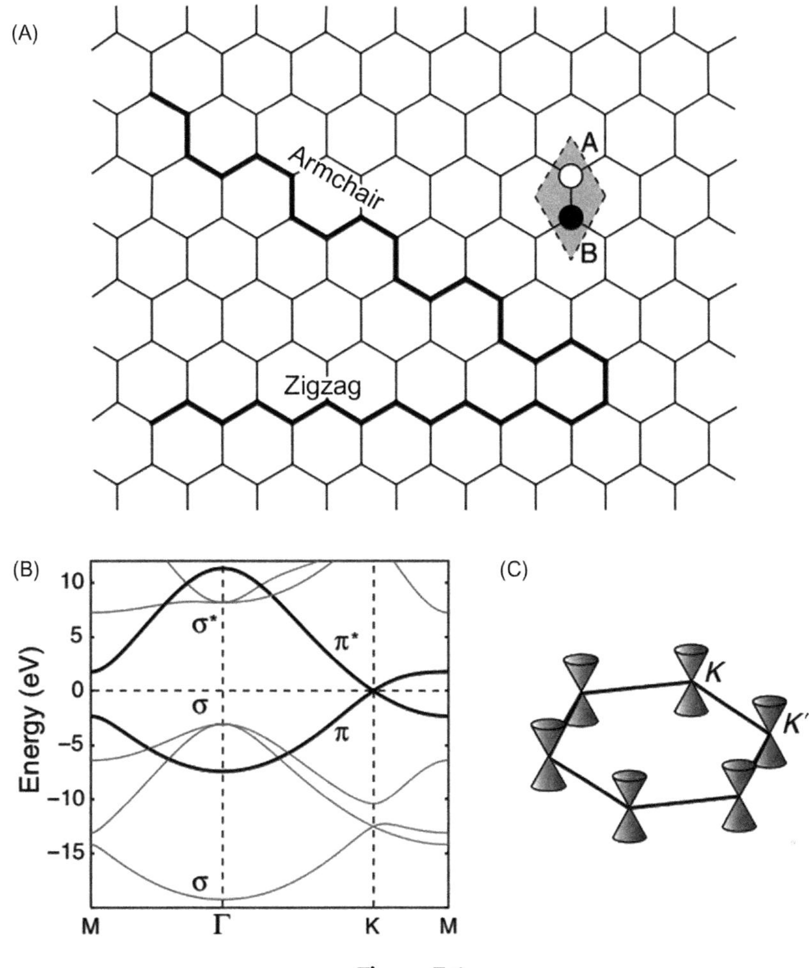

Figure 7.1

(A) Two-dimensional crystalline lattice of graphene. The shaded area denotes the unit cell of graphene containing two carbon atoms which belong to the two sublattices of graphene, A (*empty circle*) and B (*filled circle*). The two high-symmetry directions of graphene lattice, armchair and zigzag, are highlighted. (B) Band structure of graphene was obtained using first-principles calculations. The bands are labeled according to their symmetry. The π-symmetry bands responsible for the low-energy electronic properties of graphene are highlighted. The zero energy corresponds to the Fermi level. (C) The low-energy part of the band structure of graphene involves two inequivalent "Dirac cone" features in the corners (points *K* and *K*′) of the hexagonal Brillouin zone [7].

Here, the operators $c_{i\sigma}$ and $c_{i\sigma}^{\dagger}$ annihilate and create an electron with spin σ at site i, respectively. The notation $\langle \bullet, \bullet \rangle$ stands for the pairs of nearest-neighbor atoms; "h.c." is the Hermitian conjugate counterpart. The hopping integral t defines the energy scale of the Hamiltonian. The value of $t \approx 2.7$eV is well-established by previous density functional theory (DFT) calculations. In graphene system, each sp^2 carbon atom contributes one p_z orbital and one π electron. The system is thus half-filled. The spectrum of the eigenvalues of tight-binding Hamiltonian matrix exhibits electron-hole symmetry, that is, it is symmetric with respect to Fermi energy. In other words, if the Fermi energy E_F is set to 0, in a neutral graphene system for each eigenvalue <0 corresponding to an occupied (bonding) state, there is an unoccupied (antibonding) state with $\varepsilon^{\star} = -\varepsilon$. The states with $\varepsilon = 0$ are called zero-energy states (also referred to as nonbonding or midgap states).

The nearest-neighbor tight-binding model has been proved to be efficiently accurate to describe the electronic structure of graphene and its nanostructures. However, this model does not include any magnetism. So it could not explain the onset of magnetism in graphene nanostructures. To understand the magnetism in graphene and its nanostructures, electron–electron interactions have to be included. The Hubbard-model Hamiltonian could be used to investigate the magnetism in graphene and its nanostructures. The Hubbard-model Hamiltonian can be viewed as a correction of the standard nearest-neighbor tight-binding Hamiltonian. It can be written as

$$\mathcal{H} = \mathcal{H}_0 + \mathcal{H}' \tag{7.2}$$

in which the first term $\mathcal{H}_0$ is the nearest-neighbor tight-binding Hamiltonian that is given by Eq. (7.1). And the second term $\mathcal{H}'$ corresponds to the electron–electron interactions that are introduced through the repulsive onsite Coulomb interaction

$$\mathcal{H}' = U \sum_i n_{i\uparrow} n_{i\downarrow} \tag{7.3}$$

where $n_{i\sigma} = c_{i\sigma}^{\dagger} c_{i\sigma}$ is the spin-resolved electron density at site i; the parameter $U > 0$ defines the magnitude of the onsite Coulomb repulsion. Again, this model considers only the short-range Coulomb repulsion: only the repulsion of the two electrons that occupy the p_z orbital of the same atom is included. However, even this simplified term is not trivial from computations point of view. To overcome this difficulty, the mean-field approximation is applied.

$$\mathcal{H}'_{mf} = U \sum_i \left(n_{i\uparrow} \langle n_{i\downarrow} \rangle + \langle n_{i\uparrow} \rangle n_{i\downarrow} - \langle n_{i\uparrow} \rangle \langle n_{i\downarrow} \rangle \right) \tag{7.4}$$

Here, a spin-up electron at site i interacts with the average spin-down electron population $\langle n_{i\downarrow} \rangle$ at the same site and vice versa. This problem can then be solved self-consistently by iterative method: first assume some initial values of $\langle n_{i\sigma} \rangle$ which can be chosen randomly, following by the computation of the matrix element of the Hamiltonian matrix, its diagonalization and then calculate the updated spin densities. Repeat this process iteratively until all values of $\langle n_{i\sigma} \rangle$ are converged. The final solution provides the local spin density

$$M_i = \frac{\langle n_{i\uparrow} \rangle - \langle n_{i\downarrow} \rangle}{2} \quad (7.5)$$

at each atom i. And the total spin of the system is

$$S = \sum_i M_i \quad (7.6)$$

For a given graphene structure both local and total spins depend on the dimensionless parameter U/t. It should be noticed that in certain cases the broken-symmetry (antiferromagnetic) solutions can be obtained only if the initial guess of $\langle n_{i\sigma} \rangle$ values breaks the spin-spatial symmetry [9].

7.3 Magnetism in finite graphene fragment

Let us first consider the magnetism in finite graphene fragments. These graphene fragments are also referred to as nanoflakes, nanoislands, or nanodisks. They can be viewed as zero-dimensional variations of graphene system. For their finite size and number of total carbon atoms, some information about the solutions and thus the magnetic properties of these systems can be determined just by the analysis of the symmetry of these fragments.

Graphene could be viewed as a benzenoid system. The honeycomb lattice of graphene is a bipartite lattice, that is, it can be composited by two mutually interconnected sublattices A and B (see Fig. 7.1A). And each atom belonging to sublattice A is connected to the atoms in sublattice B only and vice versa. The spectrum of the tight-binding Hamiltonian of a honeycomb system can be analyzed using the benzenoid graph theory [10]. This theory is able to predict the number of zero-energy states of the nearest-neighbor tight-binding Hamiltonian in a "counting rule" fashion. The number of such states is equal to the graph's nullity

$$\eta = 2\alpha - N \quad (7.7)$$

where N is the total number of sites and α is the maximum possible number of nonadjacent sites, that is, the sites which are not the nearest neighbors to each other. The onset of magnetism in the system is determined by the competition of the exchange energy gain and the kinetic-energy penalty associated with the spin polarization of the system. The gain in exchange energy is due to the exchange splitting of the electronic states subjected to spin polarization [11].

$$\Delta s = \epsilon_\uparrow - \epsilon_\downarrow = \frac{U}{2} \sum_i n_i^2 \quad (7.8)$$

where $\sum_i n_i^2$ is the inverse participation ratio, a measure of the degree of localization of the corresponding electronic state. The kinetic-energy penalty is proportional to the energy of this state. Thus the zero-energy states undergo spin polarization at any $U > 0$ irrespective of their degree of localization. The spin polarization can be viewed as a mechanism to avoid the instability induced by the low-energy electrons in the system.

Now, let us consider the spin configuration in these states. The total spin of a bipartite system described by the Hubbard model can be determined by Lieb's theorem [12]. This theorem states that in the case of repulsive electron–electron interactions ($U > 0$), a bipartite system at half-filling as the ground state is characterized by the total spin

$$S = \frac{1}{2}|N_A - N_B| \tag{7.9}$$

where N_A and N_B are the numbers of sites in sublattices A and B, respectively. The ground state is unique and the theorem holds in all dimensions without the necessity of a periodic lattice structure. Importantly, the two counting rules are linked by the relation $\eta \geq |N_A - N_B|$. From this relation, it can be easily derived that no magnetism will be introduced to a graphene system if zero-energy states are avoided by the nearest-neighbor tight-binding model of a benzenoid system.

If we apply these counting rules to the finite graphene fragments, it is clear that the magnetism in these graphene fragments is determined by their shape and size. Let us first consider three simple examples of nanometer-sized graphene fragments as shown in Fig. 7.2 [7]. From the single-orbital physical models' point of view, only the connectivity of the π-electron conjugation network is important. In the simplest case, these π-electron networks shown can be realized in the corresponding all benzenoid polycyclic aromatic hydrocarbon molecules with the edges of the fragments passivated by hydrogen atoms.

The graphene fragment shown in Fig. 7.2A is equivalent to the coronene molecule. For this fragment, the number of sites belonging to the two sublattices is equal, $N_A = N_B = 12$. The number of nonadjacent sites is maximized when all atoms belonging to either of the two sublattices are selected, that is, $\alpha = 12$. Thus both the number of zero-energy states η and the total spin S are 0. As expected, the mean-field solution of Hubbard model for this fragment does not reveal any magnetism. Also, the absence of zero-energy states means a bandgap. Further, the tight-binding model predicts a wide bandgap of $1.08t \approx 3.0$ eV for this graphene molecule.

The second graphene fragment (triangulane) shown in Fig. 7.2B has a triangular shape. Unlike coronene, the two sublattices of this triangular fragment are no longer equivalent: $N_A = 12$ and $N_B = 10$. The unique choice maximizing the number of nonadjacent sites is

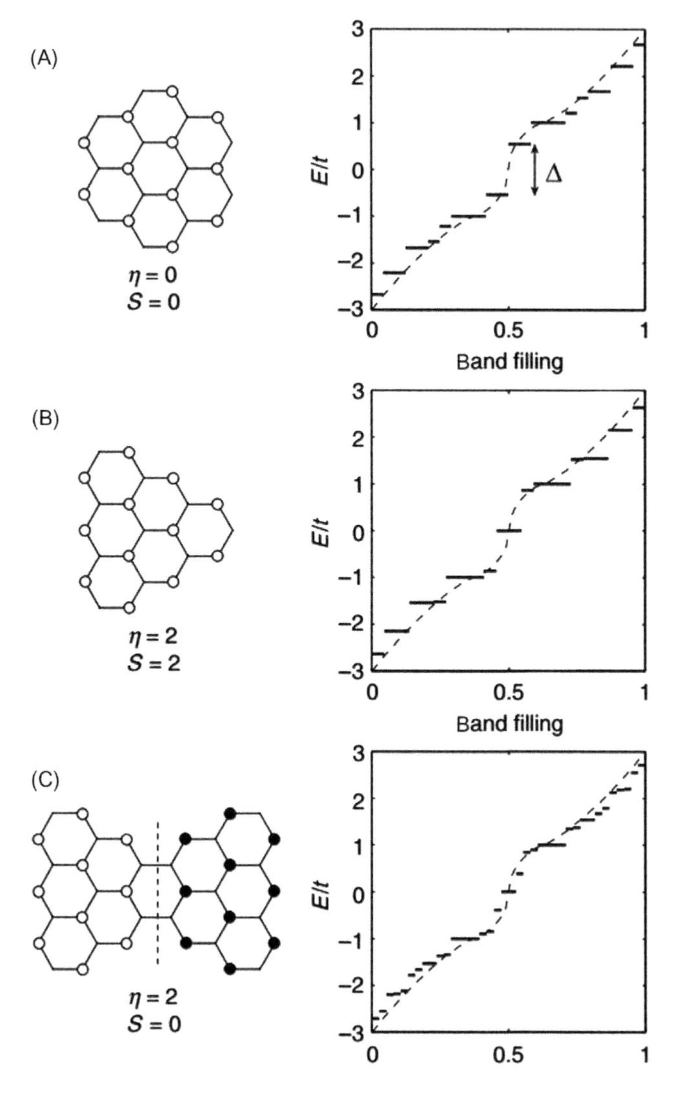

Figure 7.2
Atomic structures and tight-binding energy spectra of three graphene fragments: (A) coronene, (B) triangulane, and (C) a bowtie-shaped fragment ("Clar's goblet"). Nonadjacent sites are labeled by circles. Empty and filled circles correspond to sublattice A and sublattice B, respectively. Tight-binding energies are plotted as a function of band filling. Dashed line corresponds to the energy spectrum of ideal graphene.

achieved by selecting the atoms belonging to the dominant sublattice A, that is, $\alpha = N_A = 12$. Thus the benzenoid graph theory predicts the presence of two zero-energy states on sublattice A. Lieb's theorem predicts the $S = 1$ (spin-triplet) ground state or, equivalently, a magnetic moment of $2\mu_B$ per molecule. The two low-energy electrons

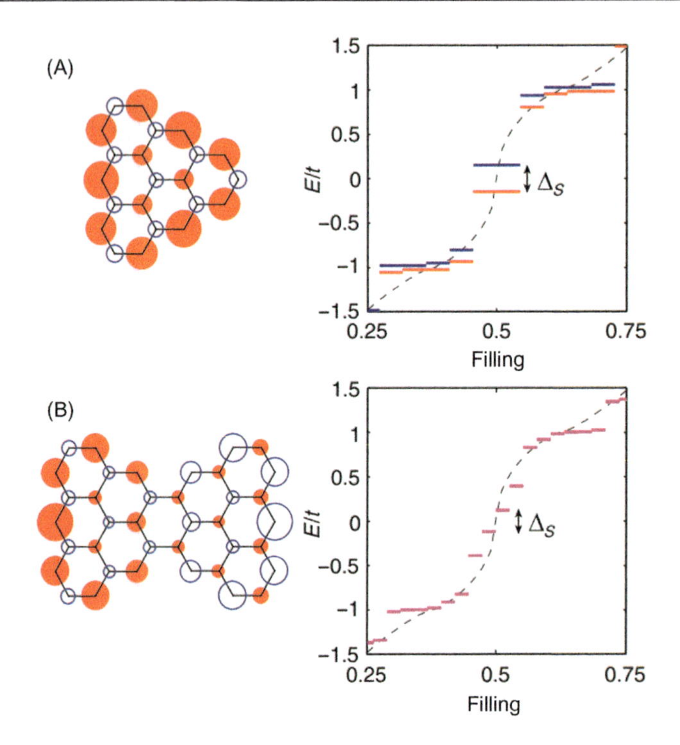

Figure 7.3

Local magnetic moments and spin-resolved energy levels obtained from the mean-field calculations of Hubbard model for (A) triangulane and (B) the bowtie-shaped graphene fragment ($U/t = 1.2$). Area of each circle is proportional to the magnitude of the local magnetic moment at each atom. Filled and empty circles correspond to spin-up and spin-down densities. Energy levels are plotted as a function of band filling. Dashed line corresponds to the energy spectrum of ideal graphene. Spin-up and spin-down channels are distinguished. In case of bowtie fragment, the energies in the two channels are identical [7].

populate a pair of zero-energy states according to Hund's rule, that is, their spins are oriented parallel to each other. The mean-field results of the Hubbard model for this system at half-filling are shown in Fig. 7.3A. One can see that spin polarization lifts the degeneracy of the zero-energy electronic states and opens an energy gap, $\Delta s = 0.30t \approx 0.8\text{eV}$. The system is stabilized by spin polarization. Most of the spin-up electron density localized on the atoms in sublattice A (see Fig. 7.3A) originates from the two electrons populating the nonbonding states. However, one can notice an appreciable amount of spin-down density on the atoms in sublattice B which is compensated by an equivalent contribution of the spin-up density in sublattice A. The occurrence of the induced magnetic moments is a manifestation of the spin-polarization effect which is related to the exchange interaction of the fully populated states with the two unpaired electrons.

The third bowtie-shaped graphene molecule shown in Fig. 7.2C composed of two triangulane fragments sharing one hexagon. For this system, Lieb's theorem predicts the spin-singlet ground state ($N_A = N_B = 19$). However, the maximized number of nonadjacent sites is $\alpha = 20$. As indicated in Fig. 7.2C, the set of sites with the maximized number of nonadjacent sites involves the atoms belonging to both sublattice A and sublattice B in the left and right parts of the structure, respectively. These atoms are marked differently in the figure. Hence, there are $\eta = 2 \times 20 - 38 = 2$ zero-energy states that are also confirmed by the tight-binding calculation. These two zero-energy states must undergo spin polarization. Thus they are spatially segregated in the two triangular parts of the molecule [13]. To satisfy the spin-singlet ground state, the two zero-energy states have to be populated by two electrons with oppositely oriented spins. In other words, the ground electronic configuration breaks spin-spatial symmetry and exhibits antiferromagnetic ordering. This result can be verified by the mean-field calculations of Hubbard model as shown in Fig. 7.3B.

The counting rules can be applied to larger graphene fragments as well. It is easy to derive that for triangular fragments with edges cut along the zigzag direction, the average magnetic moment per carbon atom decays with increasing the system size [14]. It has been shown theoretically that above some critical size, the system undergoes a transition into a broken-symmetry antiferromagnetic state [14]. The critical size itself depends strongly on the value of U/t.

Now, consider the realization of the magnetic graphene fragments in practice. It is not surprising that such magnetic systems are more reactive than nonmagnetic polyaromatic molecules. Thus triangulane itself has never been isolated. However, its chemical derivatives (see Fig. 7.4) have been successfully synthesized [15,16].

To verify the spin-triplet ground state of these chemical compounds, electron spin resonance (ESR) measurements were performed. ESR or electron paramagnetic

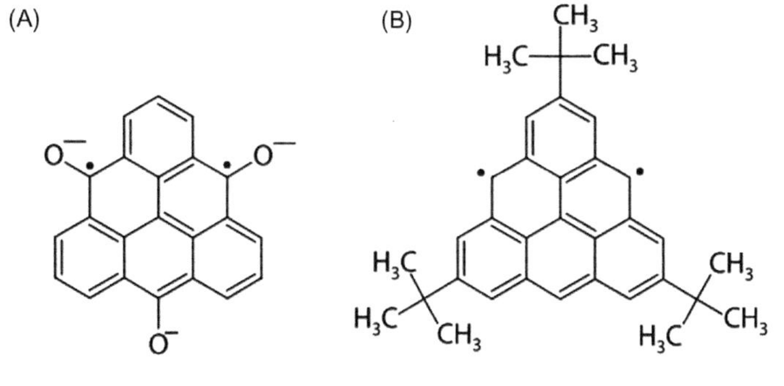

Figure 7.4
Chemical derivatives of triangulane that have been successfully synthesized [15,16].

resonance (EPR) spectroscopy is a method to study the spin states of unpaired electrons in materials. In the ESR measurement, an external magnetic field with strength B_0 is applied. Due to the Zeeman effect, the electrons gain specific energy depending on their magnetic moment

$$E = m_s g_e \mu_B B_0 \tag{7.10}$$

where m_s is the electron magnetic moment, g_e is the electrons g-factor, $g_e = 2.0023$ for the free electron [17], and μ_B is the Bohr magneton. For free electrons, the spin quantum number is $S = 1/2$, and the magnetic components $m_s = \pm(1/2)$. (see Fig. 7.5). Therefore the energy separation between the two spin states is $\Delta E = g_e \mu_B B_0$. An unpaired electron can move between the two energy levels by either absorbing or emitting a photon of energy $h\nu$, such that the resonance condition, $h\nu = \Delta E$, is obeyed. Here, h is Plank's constant, and ν is the photon frequency. This leads to the fundamental equation of ESR spectroscopy:

$$h\nu = g_e \mu_B B_0 \tag{7.11}$$

In the ESR measurement, a laser is used as an incident light source to activate the electrons in the material. When the resonance condition is fulfilled, the unpaired electrons can move between the two spin states. Since typically more electrons are in the lower state due to the Maxwell–Boltzmann distribution, there is a net absorption of energy. Usually, the absorption of the light is monitored to form the spectrum with the fixed laser photon frequency and the switching of the magnitude of the applied magnetic field [18].

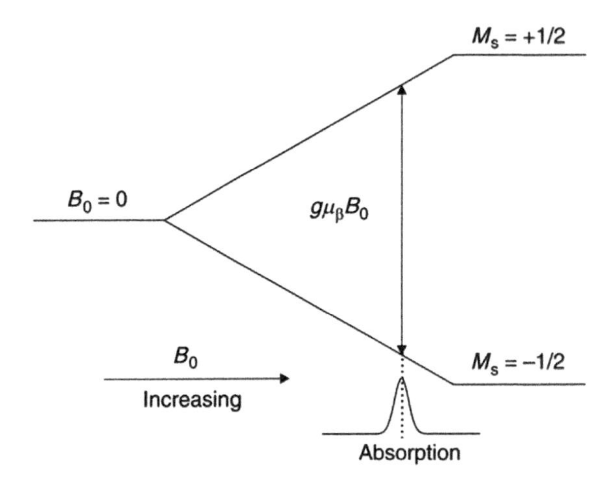

Figure 7.5
The magnetic-field-dependent splitting of the energy levels for a single unpaired electron [18].

When ESR measurement is applied to materials with multiple unpaired electrons, the interaction between electrons should be considered. For the molecules shown in Fig. 7.4, the ground state of the electron is a spin triplet state. The Hamiltonian could be expressed as:

$$\mathcal{H} = g\mu_B B_0 \cdot S + D\left(S_z^2 - \frac{1}{3}S^2\right) + E\left(S_x^2 - S_y^2\right) \tag{7.12}$$

Here, D and E are energy parameters along the primary axes [19]. If the primary axis of the molecules considered here is parallel to the z-axis of the system, then $E = 0$ since the molecules have uniaxial structures. The energy splitting and corresponding absorption spectrum are illustrated in Fig. 7.6.

Fig. 7.7 illustrates the first derivative of the microwave absorption in the ESR measurement of the molecule is shown in Fig. 7.4B [16]. The ESR measurement was performed at 123K with the microwave frequency of 9.6190 GHz. The ESR spectrum of the sample shows a superposition of spectra from spin-doublet states and a typical fine-structure spectrum of a triplet state with an axial symmetry with random orientations of the molecules [19]. The spin Hamiltonian parameters were determined by the spectral simulation to be $S = 1$, $g = 2.003$ (g-factor), $|D|/hc = 0.0073\ \text{cm}^{-1}$, and $|E|/hc \sim 0\ \text{cm}^{-1}$, which are attributed to triangulene, a genuine hydrocarbon structure

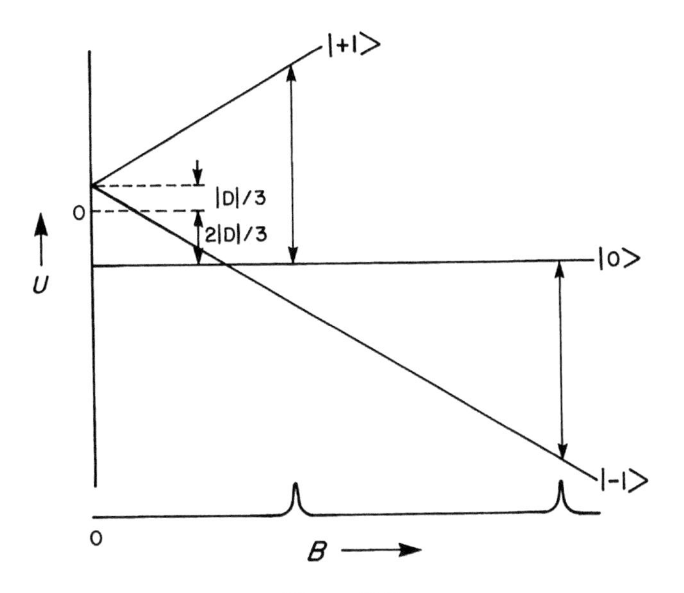

Figure 7.6
The state energies as a function of applied magnetic field B for a spin-triplet system of spin $S = 1$ and $B_0 \| z$, shown for $D > 0$ and $E = 0$. The two primary transitions with $\Delta M_s = \pm 1$ are indicated for a constant frequency spectrum [19].

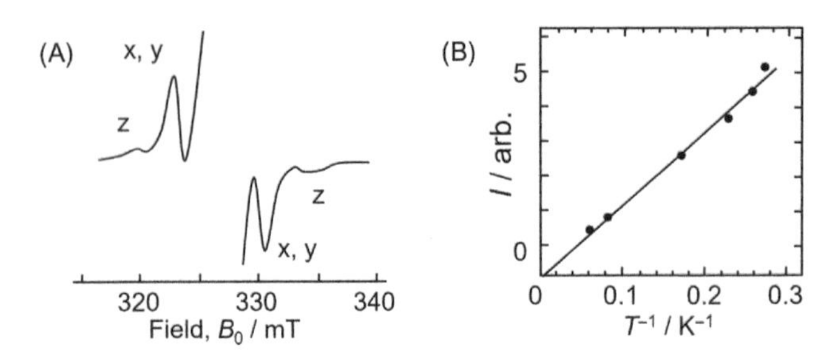

Figure 7.7
(A) Observed triplet-state ESR spectrum of the SED is 9.6190 GHz. (B) Plot of the molecule shown in Fig. 7.4B with monoradical species in a frozen toulene matrix at 123K; *x*, *y*, and *z* denote the canonical absorption peaks. The microwave frequency used is 9.6190 GHz. (B) A plot of the temperature dependence of the triplet signal intensity *I* versus $1/T$ from 3.7K to 16K [16].

with a threefold rotation axis. The linear dependence of the triplet signal intensity I on $1/T$ (Fig. 7.7B) shows that the observed triplet is the electronic ground state with other states located far above. ESR measurement could also provide the information on the spin distribution in this molecule [15,16]. Since D value is primary from the electron–electron dipole interaction, the mean interspin distance r_0 could be derived from the expression: [16]

$$D = \frac{3\mu_0}{16\pi}\frac{(g\mu_B)^2}{r_0^3} \tag{7.13}$$

This estimation gave $r_0 = 5.64\mathring{A}$ for the molecule is shown in Fig. 7.4B. It is consistent with the size of the molecular plane and indicated the spin density distribution in the triangulene.

Despite the confirmation of the triplet ground state in triangulene derivatives, the direct magnetism measurement is still lacking. Besides, the successful synthesis of larger magnetic triangular molecules has not been achieved so far. In principle, the finite graphene fragments can be considered as indirect proof of edge magnetism in graphene systems. Moreover, these graphene fragments provide a prospective to predefine magnetic interactions in designing graphene nanostructures so that spintronics devices such as spintronic logic gates could be realized [13].

7.4 Edges in graphene nanoribbons

For GNRs with infinite lengths, counting rules are no longer applicable. GNRs can be easily modeled as 1D period strips of graphene. Like carbon nanotubes, the band structure of GNRs is highly dependent on their orientations and widths. There are two

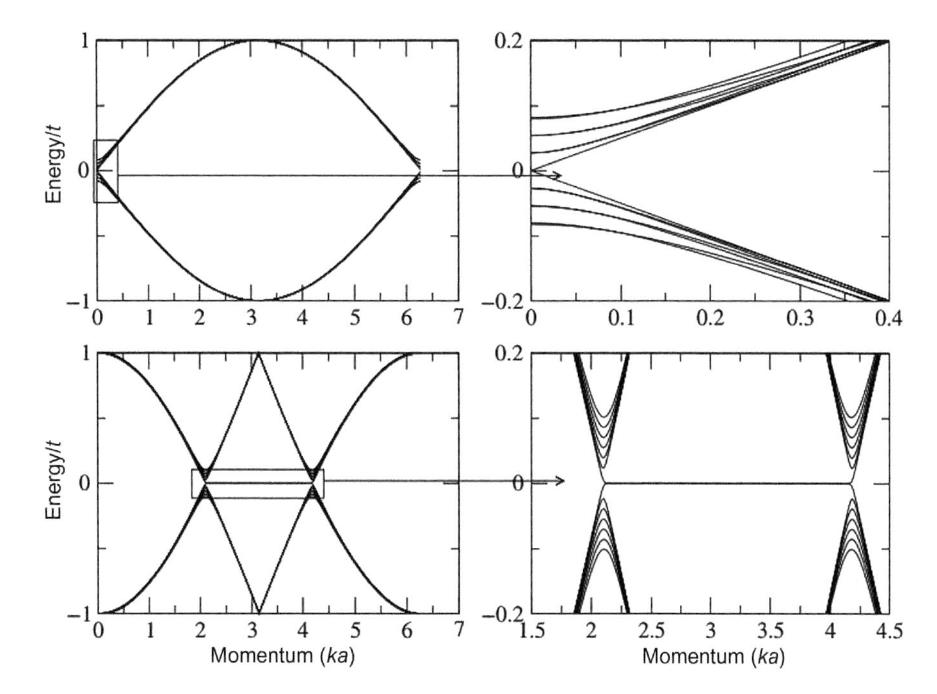

Figure 7.8
Electronic dispersion for graphene nanoribbons. Left: energy spectrum, as calculated from the tight-binding equations, for a nanoribbon with armchair (top) and zigzag (bottom) edges. The width of the nanoribbon is $N = 200$ unit cells. Only 14 eigenstates are depicted. Right: zoom of the low-energy states shown on the right [4].

high-symmetry crystallographic directions in graphene, armchair, and zigzag, as shown in Fig. 7.1A. The direction (also referred as chirality) of GNR is defined by the direction of their cutting edge. For instance, GRNs with cutting edge along zigzag direction are called zigzag nanoribbons. Fig. 7.8 shows the energy spectrum of an armchair and zigzag nanoribbons with a width of $N = 200$ unit cells.

In general, armchair nanoribbons are either semimetallic or semiconducting as predicted by nearest-neighbor tight-binding model [20–23], and the two situations alternate as the nanoribbon's width increases. The bandgap of semiconducting GNRs decreases with increasing width. In the case of metallic nanoribbons, two bands cross the Fermi level at the Γ point. No magnetic ordering is predicted in this case.

On the other hand, the zigzag ribbons present a band of zero-energy modes that is absent in the armchair case. This band at zero energy is corresponding to the surface states near the edges of the GNRs. These states decay quickly in the bulk. The high density of zero-energy states suggests a possibility of magnetic ordering. As discussed in Section 7.2, the onset of magnetism could be viewed as a mechanism to lift the degeneracy of the zero-energy states.

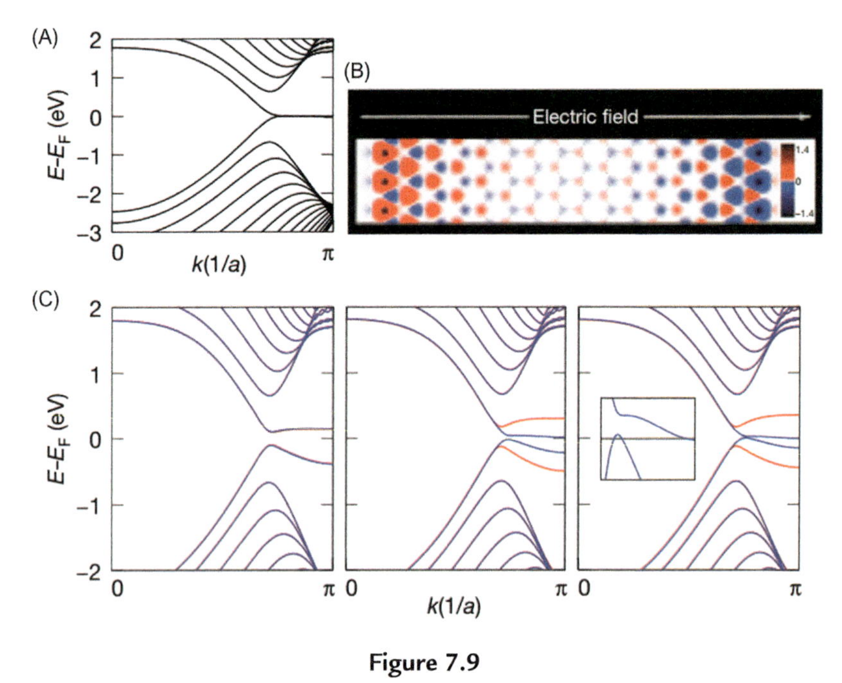

Figure 7.9

Electronic structures of graphene nanoribbons. In all figures, the Fermi energy (E_F) is set to 0. (A) The spin-unpolarized band structure of (A) 16-ZGNR. (B) The spatial distribution of the charge difference between α-spin and β-spin ($\rho_\alpha(r) - \rho_\beta(r)$) for the ground state when there is no external field. The graph is the electron density integrated in the *z* direction, and the scale bar is in units of $10^{-2}|e|\text{Å}^{-1}$. (C) From left to right, the spin-resolved band structures of a 16-ZGNR with $E_{ext} = 0.0$, 0.05, and 0.1 VÅ^{-1}, respectively. The red and blue lines denote bands of α-spin and β-spin states, respectively. Inset, the band structure with $E_{ext} = 0.1\ \text{VÅ}^{-1}$ in the range $|E - E_F| < 50$ meV and $0.7\pi \leq ka \leq \pi$ (the horizontal line is E_F) [6]. *ZGNR*, zigzag graphene nanoribbons.

Fig. 7.9 illustrates the band structure of a zigzag GNR with a width of $N = 16$ (16-ZGNR) with/without the consideration of spin polarization [6]. Interestingly, when the spin polarization is considered, localized magnetic moments display ferromagnetic ordering along the zigzag edge while the magnetic moments at the opposite edge is antiparallel, as shown in Fig. 7.9B [6]. Thus the net magnetic moment of a zigzag nanoribbon keeps zero in agreement with Lieb's theorem ($N_A = N_B$). The band structure with the consideration of spin polarization is shown in the left plot of Fig. 7.9C. The spin polarization opens a bandgap across the whole flat-band segment, turning the system into a semiconductor although it almost shows no effects on the electronic states at higher energies. Importantly, the band structures for the two spin channels are equivalent, but spin-spatial symmetry is broken (see Fig. 7.9B).

A very important and pioneering idea about the spin polarization in GNRs is that the spin polarization and the corresponding band structure could be tuned by an external electric field so that half-metallicity can be induced in ZGNRs [6]. The half-metallicity refers to the coexistence of a metallic state for electrons with one-spin orientation and an insulating state with the opposite. Fig. 7.9C shows the spin-resolved energy spectrum of electrons under different external electric field. The electric field was applied to the ZGNR in the way indicated in Fig. 7.9B. It is clear that as the magnitude of the external electric field increases, the bandgap increases for α-spin while decreases for β-spin. When the electric field is strong enough, the bandgap for β-spin is closed which means the nanoribbons are metallic for β-spin. However, the bandgap for α-spin is still open. This means that under this certain electric field bias, this ZGNR is turned to half-metal.

The half-metallicity of the ZGNRs originates from the fact that the applied electric fields induce energy-level shifts of opposite signs for the spatially separated spin-ordered edge states. Such separate and opposite energy shifts are made possible by the localization of the edge states around E_F. As illustrated in Fig. 7.10A, electrons with α-spin occupied the valence band on the left-hand edge, while electrons with β-spin occupied the valence band on the right. When E_{ext} is applied, the electrostatic potential is raised on the right side and lowered on the left side as E_{ext} (>0) increases. Correspondingly, the energies for localized edge states on the right side are shifted upwards and those on the left side downwards. As a result, the conduction band for the electrons with β-spin on the left edge approaches to the valence band for the electrons with β-spin on the right edge. Finally, when E_{ext} is sufficiently large, the bandgap closes for the electrons with β-spin. For electrons with α-spin, the opposite procedure takes place, the bandgap for the electrons with α-spin increases with the increase of E_{ext}. Eventually, states of only one-spin orientation at E_F are left (Fig. 7.10B). The density-of-states (DOS) evolution during this process is illustrated in Fig. 7.8C. The occupied β-spin states in the middle of a 16-ZGNR are the tails of the localized β-spin states on the right side and the unoccupied β-spin states are from the left side, so that occupied and unoccupied β-spin states in the middle of the ZGNR move oppositely to close the gap.

It is worth mentioning that no direct proof of antiparallel edge magnetism in pristine GNRs without doping the edges has been reported so far. However, the presence of localized low-energy states at edges of GNRs has been verified using scanning tunneling microscopy (STM) [24,25].

Because of its high spatial resolution and sensitivity to local electron states, STM is a powerful tool to investigate the edge states of GNRs. Fig. 7.11 demonstrates the edge states in an (8,1) chirality graphene nanoribbon located on a gold substrate [24]. Fig. 7.11A shows the atomically resolved topography of an (8,1) GNR with a measured width of 19.5 ± 0.4 nm. The GNR shown in Fig. 7.11A has an (8,1) chirality (equivalent to $\theta = 5.8°$), and the resulting structural model for this GNR is shown in Fig. 7.11B. It is interesting to see that the edge of the nanoribbon is mostly zigzag. The local electronic structures of GNR edges

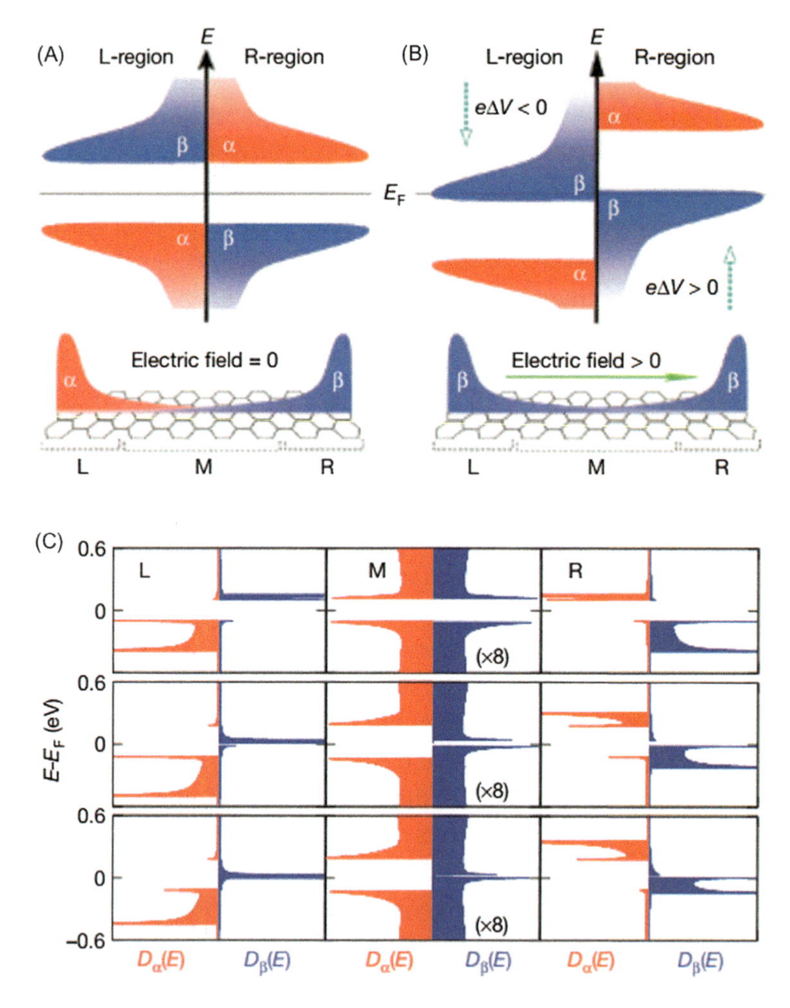

Figure 7.10

Origin of half-metallicity. (A) Schematic density-of-states diagram of the electronic states of a ZGNR in the absence of an applied electric field. Top: the occupied and unoccupied localized edge states on the left side (L-region as defined at the bottom) are the α-spin and β-spin states, respectively, and vice versa on the right side (R-region) with the same energy gap for both sides. Bottom: schematic diagram of the spatial spin distribution of the highest occupied valence band states without an external electric field. (B) Top: with a transverse electric field, the electrostatic potential on the left edge is lowered ($e\Delta V < 0$), whereas the one on the right edge is raised ($e\Delta V > 0$). Correspondingly, the energies of the localized states at the left edge are decreased and those of the localized states at the right edge are increased. Bottom: the resulting states at E_F are only β-spin. (C) From left to right, the local DOS of a-spins and β-spin (ordinate) of a 16-ZGNR as a function of energy (abscissa) for atoms in the L, M, and R regions shown in A, respectively. From top to bottom, $E_{ext} = 0.0$, 0.05, and 0.1 $V\mathring{A}^{-1}$, respectively. The local DOS in the middle panels is enlarged eightfold for clarity. For $E_{ext} = 0.1\ V\mathring{A}^{-1}$, the van Hove singularities of the β-spin in the M and R regions are above the E_F by 5 meV, and all states at E_F are of β-spin. *DOS*, density of states; *ZGNR*, zigzag graphene nanoribbons.

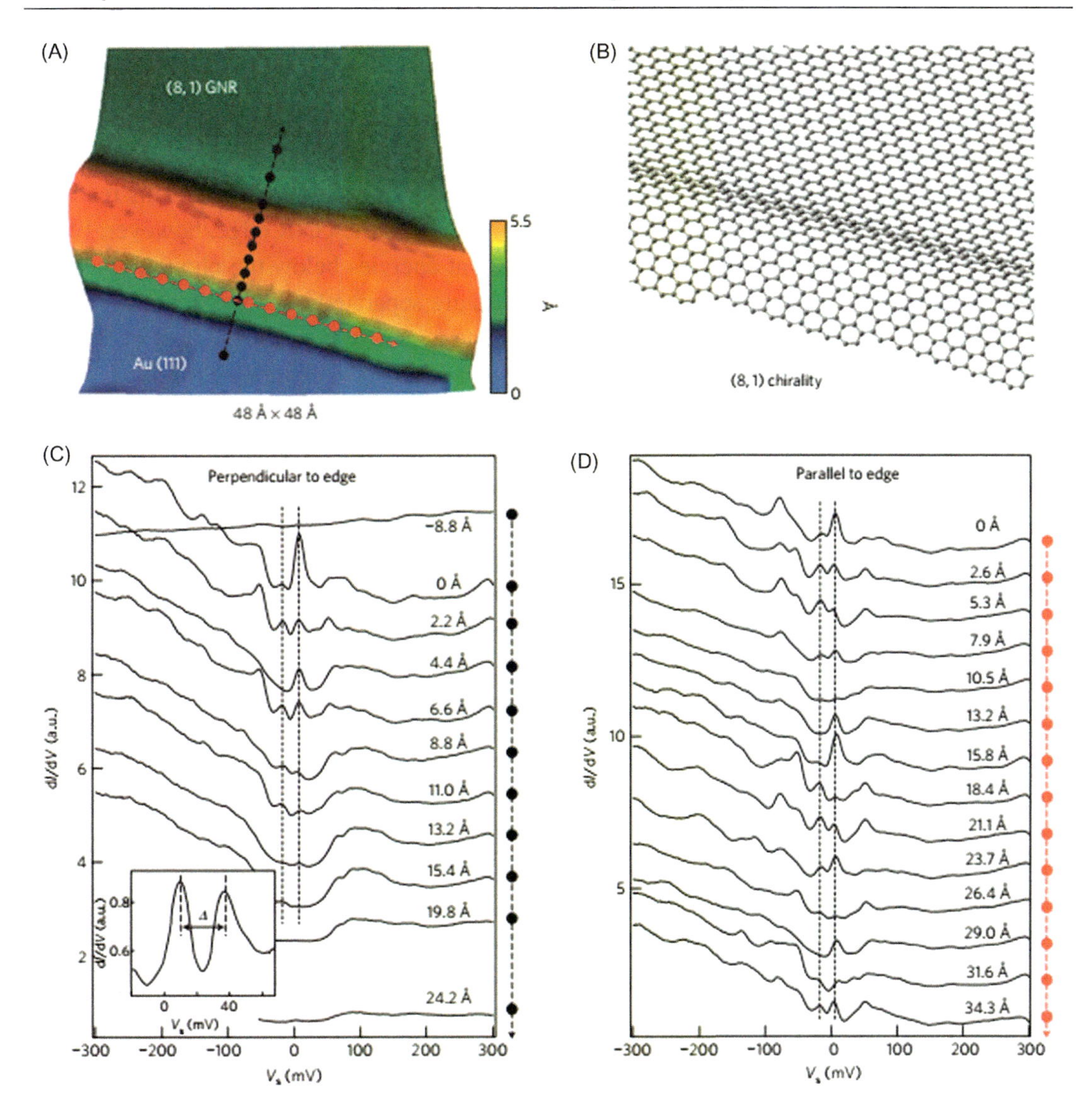

Figure 7.11

Edge states of GNRs. (A) Atomically resolved topography of the terminal edge of an (8,1) GNR with measured width of 19.5 ± 0.4 nm. (B) Structural model of the (8,1) GNR edge shown in (A). (C) dI/dV spectra of the GNR edge shown in (A) measured at different points (*black dots*, as shown) along a line perpendicular to the GNR edge at $T = 7$K. Inset shows a higher resolution dI/dV spectrum for the edge of a (5,2) GNR with width of 15.6 ± 0.1 nm. The dashed lines are guides to the eye. (D) dI/dV spectra measured at different points (*red dots*, as shown) along a line parallel to the GNR edge shown in (A) at $T = 7$K [24]. *GNR*, graphene nanoribbon.

were investigated using scanning tunneling spectrum (STS) technology, in which dI/dV measurement reflects the energy-resolve local density of states (LDOS). Fig. 7.11C and D shows dI/dV spectra obtained at different positions (as marked) near the edge of the (8,1) GNR as shown in Fig. 7.11A. The dI/dV spectra measured within 24 Å of the GNR edge typically show a broad gap-like feature having an energy width of $\sim$130 meV. This feature disappears further into the interior of the GNR. Very close to the GNR edge, however, two peaks rise up within the elastic tunneling region (i.e., at energies below the phonon-assisted inelastic onset) and straddle zero bias. The two peaks are separated in energy by a splitting of $\Delta = 23.8 \pm 3.2$ meV. Similar energy-split edge-state peaks have been observed in other the clean chiral GNRs. For example, the inset in Fig. 7.11C shows a higher resolution spectrum exhibiting energy-split edge-state peaks for a (5,2) GNR having a width of 15.6 nm and an energy splitting of $\Delta = 27.6 \pm 1.0$ meV. The two edge-state peaks are often asymmetric in intensity (depending on a specific location in the GNR edge region), and their midpoint is often slightly offset from $Vs = 0$ (within a range of ± 20 meV). As seen in the spectra of Fig. 7.11C, the amplitude of the peaks grows as the test point moves closer to the terminal edge of the GNR, before falling abruptly to 0 when the carbon/gold terminus is crossed. The edge-state spectra also vary as one moves parallel to the GNR edge, as shown in Fig. 7.11D. The edge-state peak amplitude oscillates along the direction parallel to the GNR edge with an approximate 20 Å period, consistent to the 21 Å periodicity of an (8,1) edge.

The spectroscopic features observed correspond closely to the spatial- and energy-dependence predicted for 1D spin-polarized edge states coupled across the width of a chiral GNR. It was found that the experimental spectroscopic edge-state data for the (8,1) GNR is in agreement with Hubbard-model Hamiltonian calculations for $U = 0.5t$. The two small peaks in the center of STS spectra are corresponding to the low-energy states at the nanoribbon edges. These states are generated by the splitting of the zero-energy states. As illustrated in Fig. 7.9A and C, without consideration of spin, the electrons ate GNR edges exhibit a flat band at zero-energy which is corresponding to a van Hove singularity in the DOS at zero energy. The introduction of spin induces an energy split of the zero-energy states and creates a small bandgap, as a result, the van Hove singularity splits to two small peaks around zero energy. The theoretical bandgap of 29 meV is very close to the experimentally observed value of 23.8 ± 3.2 meV. The observed energy-split spectroscopic peaks thus provide evidence for the formation of spin-polarized edge states in pristine GNRs. We are further able to compare the spatial dependence of the calculated edge states with the experimentally measured STS results.

An important factor that affects the magnetic edge states in GNRs is the edge reconstruction. Fig. 7.12 compares the edge states of a pristine (3,1) edge and 5-7 reconstructed edge in a (3,1) chirality GNR. Fig. 7.12A shows the topography of a (3,1) chirality GNR. The width of this GNR is 6.4 nm. Fig. 7.12B presents a line profile of the

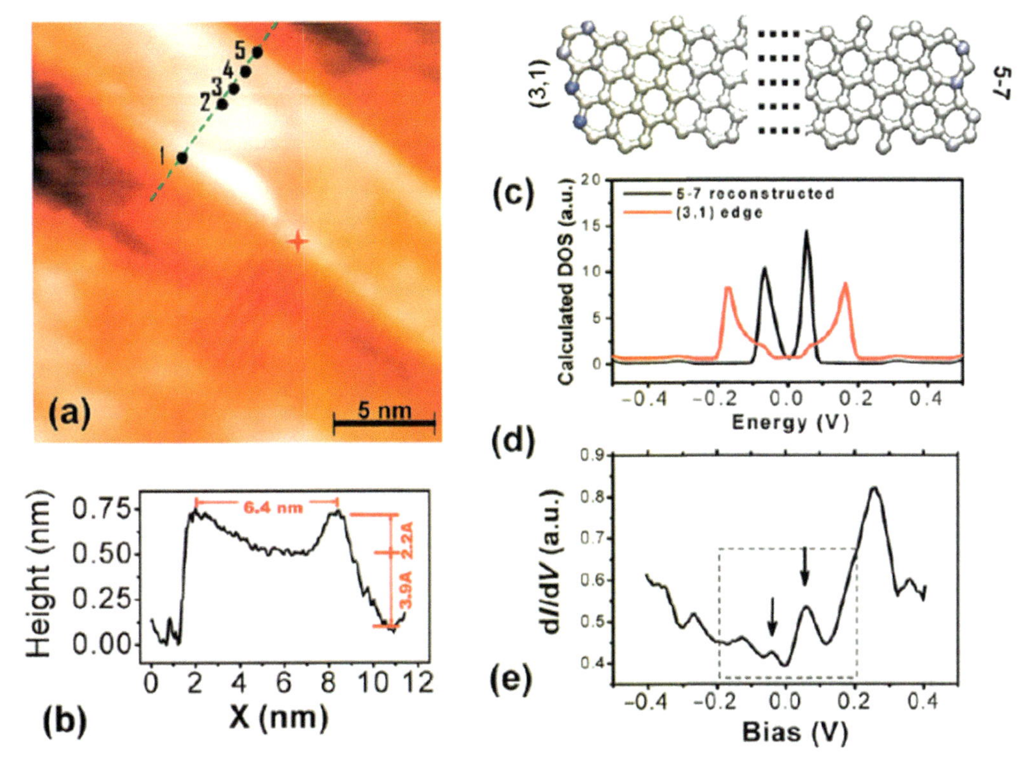

Figure 7.12
Tunneling spectra measured at a GNR edge. (A) A GNR with a width of 6.4 nm and a height of 6 Å. (B) Line profile of height measurement across the ribbon showing a "bright edge," which is about 2.4 Å higher than the plain terrace of the ribbon. (C) Ball – stick model of proposed GNR structure, where the left edge is a pristine (3,1) edge. The right edge of the ribbon is a reconstructed structure composed of 5 – 7 member rings. (D) The calculated electronic structure for the (3,1) (*red*) and for the 5-7 reconstructed edge of a GNR (*black*) obtained by using a model with an identical width as the GNR in panel A. (E) d*I*/d*V* spectrum obtained at the edge of the GNR in panel A. *GNR*, graphene nanoribbon.

height across the GNR. For the STM tests the local tunneling current, both the possible curve at the edge and higher LDOS could contribute to the seemly higher height at the edges. On investigating the edge states of this GNR, the possible STS spectra were calculated. Fig. 7.12C illustrates the ball-stick model of a GNR with a pristine (3,1) edge on one side and a 5-7 reconstructed on the other. The calculated LDOS at both edges were compared in Fig. 7.12D. The energy splitting in the lowest energy states in the pristine (3,1) edge is much larger than the splitting in the reconstructed edge. This means that the edge reconstruction has a strong impact on the energy splitting. Fig. 7.12E shows the d*I*/d*V* spectrum collected at the edge of the GNR indicating by the red star at the bottom edge. The measured split is 92 meV which is close to the theoretical value of the 5-2 reconstructed edge.

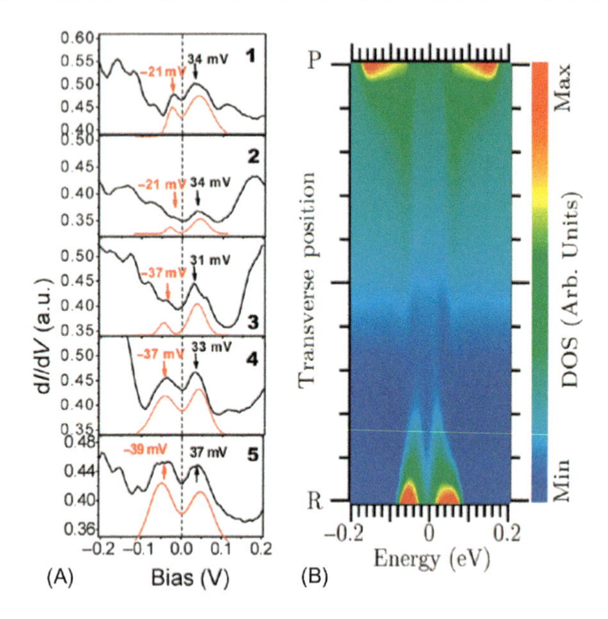

Figure 7.13

Two spin-split peaks around the Fermi energy taken perpendicularly to the axis of a graphene ribbon. (A) Spatial dependence of the tunneling spectra across the ribbon width, *dI*/*dV* spectra obtained at five different positions (as marked in Fig. 7.12A) in the low-energy regime. The tunneling current 50 pA, sample bias 300 mV. The red curves are Gaussian fittings to the peaks. Although all spectra show two peaks around the Fermi energy, the energy separation between the peaks and the intensity of the peaks varies according to the positions on the ribbon (1, 2, 3, 4, and 5). (B) Color mapping for the calculated LDOS obtained by moving perpendicularly away from the pristine (3,1) edge (top) to the 5-7 reconstructed edge (bottom). *LDOS*, local density of states.

The spacious distribution of the energy split is included in Fig. 7.13. Fig. 7.13A shows the STS spectra at five different points along the vertical direction across the GNR indicated with black points in Fig. 7.12A. On one edge of the ribbon, the two peaks are separated by a splitting of 55 meV (positions 1and 2). These two peaks are asymmetric both in intensity and in energy positions. The middle point has a 6-meV offset from zero bias. The energy positions of the two peaks remain the same (positions 1 and 2 in Fig. 7.12A) and just the intensity decreases as we move toward the middle of the ribbon. When moving to the other side to the opposite terminal edge from the middle, the amplitude of the two peaks again grows higher. The peaks become more symmetric and the energy separation jumps from 68 meV in the middle to 76 meV at the edge. Fig. 7.13B shows a series of calculated LDOS are shown as a color mapping as a function of transverse position with a 2.5-Å interval along the perpendicular direction away from the pristine (3,1) edge (P) to (5,7) reconstructed edge (R). If we compare Fig. 7.13A and B, it is possible that the GNR investigated has a (3,1) pristine edge at the top and a 5-7 reconstructed edge at the bottom. This configuration of both edges was also confirmed by an atomic resolved STM topography measurement and theoretical simulation [25].

Another factor should be considered for the magnetism of graphene nanostructures is temperature. Magnetic order in low-dimensional systems is particularly sensible to thermal fluctuations. In particular, the Mermin–Wagner theorem excludes long-range order in 1D magnetic systems (such as the magnetic graphene edges) at any finite temperatures. According to DFT calculations [9], above the crossover temperature $T_x \approx 10\mathrm{K}$, weak magnetic anisotropy does not play any role and the spin correlation length $\xi \propto T^{-1}$. However, below T_x the spin correlation length grows exponentially with decreasing temperature. At $T = 300\mathrm{K}$ the spin correlation length $\xi \approx 1$ nm. From a practical point of view, this means that the dimensions of spintronic devices based on the magnetic zigzag edges of graphene and operating at normal temperature conditions are limited to several nanometers. At present, such dimensions are rather challenging for device fabrication.

7.5 Magnetism induced by defects in graphene nanostructures

Magnetism in graphene nanostructures can also be induced by defects [11,26–29]. Concerning the onset of magnetism in a practical graphene material, the types and distribution of defects must also be considered. The defects in graphene could be introduced by radiation damage. High-energy particles (e.g., proton) in the radiation may induce several different defects. Fig. 7.14 shows two typical types of defects: hydrogen chemisorption and carbon vacancy. Fig. 7.14A shows a defect where proton (indicated by a *red triangle*) chemisorbed by a carbon atom in graphene lattice. This chemisorption results

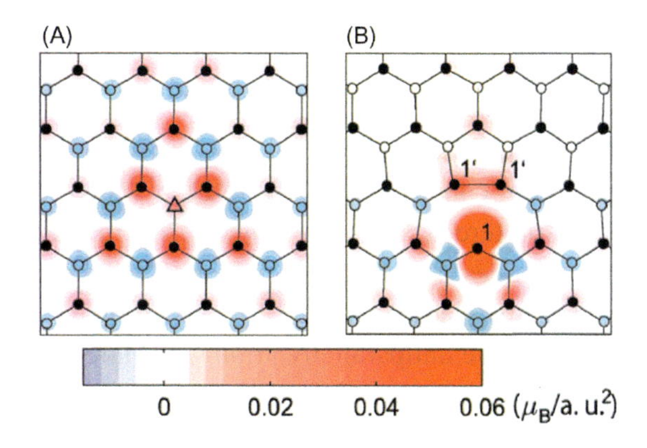

Figure 7.14
Spin-density projection (in $\mu_B/\mathrm{a.u}^2$) on the graphene plane around (A) the hydrogen chemisorption defect (△) and (B) the vacancy defect in the A-sublattice. Carbon atoms corresponding to the sublattice A (○) and to the sublattice B (●) are distinguished [27].

in the carbon rehybridization into the sp^3-state so that this carbon atom no longer contributes its p_z orbital to the π-electron system. Fig. 7.14B shows another type of defect where a carbon atom is knocked out by the radiated particle, so that a carbon vacancy is left on the graphene lattice. In this case, one p_z-orbital is eliminated together with the knocked-out carbon atom. Thus in the point of view that one p_z-orbital is removed from π-system of graphene, hydrogen chemisorption and vacancy of carbon atoms are equivalent. As a result, these two types of defects are also referred to as p_z-vacancies.

Now, let us consider a p_z-vacancy is created on a graphene nanostructure originally with total $2N$ ($N = N_A = N_B$) carbon atoms. Suppose this defect occupies A-cite and turns the sp^2 bond of this carbon atom to an sp^3 bond, then the total atom at A-cite becomes to $N_A = N - 1$. According to the counting rule and Lieb's theorem, this defect induces $\eta = 2N_B - (2N - 1) = 1$ zero-energy state and a spin of $S = (1/2)|N_A - N_B| = 1/2$ to this graphene nanostructure. This means that the introduction of this defect turns the graphene nanostructure to ferromagnetic. This result has been confirmed by both the first-principle calculations [27–29] and mean-field analysis of Hubbard model [11,26].

It is worth noticing that these defects induced zero-energy states are quasilocalized sates [30,31]. In the hydrogen chemisorption case, the spin-resolved DOS shows a sharp peak near the Fermi level. The peak is fully split by exchange and the system is characterized by a magnetic moment of 1 μ_B at any defect concentration. The distribution of spin density around the defective site clearly shows a $\sqrt{3} \times \sqrt{3}R30$ degrees superstructure (Fig. 7.14A) [27]. For carbon vacancies, the situation is more complicated. Due to the σ-symmetry dangling-bond in this defect, a local nonbonding state also presents together with the quasilocal state. This dangling-bond state contributes 1 μ_B to the total magnetic moment of the defect. On the other hand, the quasilocalized is suppressed due to the self-doping effect related to the structural reconstruction of the vacancy. The overall magnetic moment per vacancy defect varies from 1.12 μ_B to 1.53 μ_B for defect concentrations ranging from 20% to 0.5% [27]. The spin density distribution around a vacancy is shown in Fig. 7.14B.

In realistic graphene materials, the defects are randomly distributed. Defects could occupy both sublattices at arbitrary local concentrations. From Lieb's theorem, the total magnetic induced in a graphene nanostructure is $M = |N_A - N_B| = \left|N_A^d - N_B^d\right|$, where N_A^d and N_B^d are the number of defects in sublattices A and B, respectively. This means that the electron spins of the defects in the same sublattice are parallel to each other while the spins populated at the other sublattice are antiparallelly arranged. Moreover, the quasilocalized states induced by the defects at complementary sublattices interact with each other [11,26]. When these states are close enough, their interaction is so strong that the degeneracy of the zero-energy states are lifted leading to weakly bonding and antibonding states. In this case, the defect-induced magnetic moments are quenched [32,33]. Besides, the types of defects can be more complex. For instance, interstitial defects could be created when the knocked-off carbon atoms are attracted in the interface between the neighboring graphene layers; single-atom vacancies may aggregate producing extended defects; different functionalities other than hydrogen could also be absorbed by the carbon atoms; etc. All these factors affect the magnetism in graphene nanostructures, making an analysis of the magnetism of graphene nanostructure very complicated. It is predicted that the magnetism induced by hydrogen chemisorption in graphene nanostructures could only be realized with a very low concentration of hydrogen chemisorption due to the interaction between defects [33].

The realization of magnetism in graphene nanostructures without magnetic element doping has been achieved in GNRs from chemical synthesis [34]. These chemical approaches are benefit for mass productivity of GNRs. Depending on their synthesis method, GNRs can sometimes be referred to as chemically converted GNRs (CCGNRs). Technically speaking, GNRs are generally more pristine than CCGNRs from the point of view of a carbon π-electron system. GNRs could be typically synthesized by potassium splitting of multiwalled carbon nanotubes (MWCNTs) while CCGNRs are made from a reduction of oxidative unzipped MWCHTs, for example, by hydrazine. These two chemical approaches make the characteristics of these two types of GNRs extremely different. GNRs are far more conductive, almost never monolayered, and hydrogen atoms likely edge the ribbons [35]. On the other hand, CCGNRs are heavily oxidized both on the planes and at the edges due to the strong oxide conditions. They are poorly conductive, but very well exfoliated [36–38]. Fig. 7.15 compares the magnetization of GRNs and CCGNRs [34]. Clearly, s-shape in the magnetization curve in both GNRs and CCGNRs indicates the ferromagnetism (FM) state at low temperatures. GNRs even show FM orientation at room temperature (RT) (see Fig. 7.15A). At low temperatures, the magnetization of GNRs and CCGNRs can be fitted with a Langevin function $L(x)$ which can be expressed as:

$$M = M_0\left(coth\left(\frac{\mu\mu_0 H}{k_B T}\right) - \left(\frac{\mu\mu_0 H}{k_B T}\right)^{-1}\right) \tag{7.14}$$

Here, μ is the magnetic moment of the material, μ_0 is the permittivity of vacuum, k_B is the Boltzmann constant, and M_0 is the saturation magnetization. Langevin function typically describes the disordered magnetic behavior such as superparamagnetism. The fitting of magnetization of GNRs at 5K gives the saturation magnetization $M_0 = 0.25\text{emu}/\text{g}$, and the magnetic moment of $\mu = 34.27(5)\mu_B$, suggesting clustering of the $S = 1/2$ spins. Also, at low temperatures, the hysteresis in the magnetization of GNRs can be clearly identified (see top left inset of Fig. 7.15A). A coercivity (H_C) of ~ 0.002 T is achieved in GNRs at 5K. All these features clearly confirm the ferromagnetic state existence in GNRs up to RT.

In the measurement of CCGNRs, however, the ferromagnetic state only exists in low temperatures. When the temperature is higher than 10K, only a linear magnetization

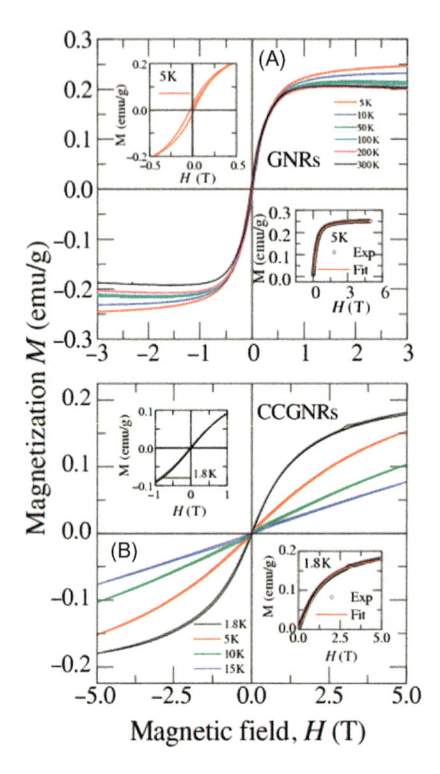

Figure 7.15

(A) Isothermal magnetization ($M-H$) observed on GNRs collected for $\mu_0 H$ varying from -5 to $+5$ T at various temperatures 5K, 10K, 50K, 100K, 200K, and 300K. The upper left inset in this figure shows an enlarged version of the $M-H$ curve measured at 5K. The lower right inset of this figure represents the Langevin fit (*solid curve*) to the experimental data observed at 5K using Eq. 7.14. (B) Isothermal magnetization ($M-H$) observed on CCGNRs collected for H varying from -5 to $+5$ T at various temperatures 1.8K, 5K, 10K, and 15K. The top left inset of this figure shows the enlarged version of 1.8K $M-H$ curve. The lower right inset of this figure represents the Langevin fit (*solid curve*) to the 1.8-K experimental data using Eq. 7.14. In the GNRs, a clear ferromagnetic character can be seen even at room temperature (300K). However, in the CCGNEs case, a clear $M-H$ open loop is only visible at 1.8K [34].

indicating paramagnetic state can be identified, This means the Curie temperature for CCGNRs is very low (<10K). At lower temperatures, the magnetization shows an s-shape superparamagnetic feature (left top inset of Fig. 7.15B). Fitting magnetization measured at 1.8K with Langevin function Eq. (7.14) gives a magnetic moment of $\mu = 7.04(6)\mu_B$, suggesting clustering of $S = 1/2$ spins like in GNRs case. Compared to GNRs, the magnetism of CCGNRs fades a lot: both saturate magnetization, and Curie temperature drops dramatically in CCGNRs. This elimination of magnetism in CCGNRs may be explained as the majority of the edges being reconstructed, passivated or closed upon the oxidative unzipping process [39].

It is worth pointing out that the ferromagnetic states and antiferromagnetic states exist simultaneously in GNRs. In by field-cooled magnetization measurement of GNRs, a negative shift of the M loops with increasing cooling field could be observed. This phenomenon, often referred as "negative exchange bias," has been commonly observed in systems exhibiting coexisting magnetic phases such as FM and antiferromagnetism [34]. Another problem for the magnetism of these GNRs and CCGNRs is that the origin of the magnetism is still an open question. Theoretical calculation suggests that the clustering of spins observed in the magnetization was possibly generated from the carbon adatom agglomeration other than randomly distributed hydrogen chemisorption [33,40].

Apparently, a big problem for the application of defect-induced magnetism in graphene nanostructures is the fine control of the defect distribution. Some theoretic calculations suggest that the chemisorbed hydrogen atoms tend to occupy locations that can quench the magnetic states on graphene [32,33]. And the edge reconstruction also prefers to the nonmagnetic structures [39]. Moreover, high defect density and thermal treatment of graphene may also destroy the fantastic physical properties of graphene, reducing the carrier mobility and spin coherent length dramatically [41]. To induce magnetism with fine control to graphene nanostructures without ruining its electric properties is still a big challenge.

7.6 Summary and outlook

This chapter illustrates the theoretical scenarios in magnetism of graphene nanostructures and attempts to observe the magnetism experimentally. According to the mean-field Hubbard-model Hamiltonian, the magnetism in graphene nanostructures can be understood thoroughly as a result of the electron–electron interaction. The spin and electron states of the graphene nanostructures can be tested with ESR and STM, which provide indirect evidence of the magnetic configurations in these graphene nanostructures. Also, the direct observation of the magnetization of nanostructures was achieved in defect-induced magnetism in GNRs. Although a lot of prototype applications have been proposed [13,42–44], a number of challenges have to be coped with. The most important problem is, despite the fierce computational and theoretical studies of the physics of magnetic graphene edges, little direct experimental evidence has been reported so far. One big problem that lies in the field is the manufactory of the graphene nanostructures with fine control of the edge configuration with atomic precision. On consideration of the temperature, Curie temperature has to be elevated to enhance spin correlation length. New concepts on spintronic devices using graphene nanostructures remain to be further developed for any practical spintronics applications

References

[1] K.S. Novoselov, A.K. Geim, S.V. Morozov, D. Jiang, Y. Zhang, S.V. Dubonos, et al., Science 306 (5696) (2004) 666.
[2] A.K. Geim, K.S. Novoselov, Nat. Mater. 6 (2007) 183.
[3] M.I. Katsnelson, Mater. Today 10 (1) (2007) 20.
[4] A.H. Castro Neto, F. Guinea, N.M.R. Peres, K.S. Novoselov, A.K. Geim, Rev. Mod. Phys. 81 (1) (2009) 109.
[5] F. Zhai, L. Yang, Appl. Phys. Lett. 98 (6) (2011) 062101.
[6] Y.-W. Son, M.L. Cohen, S.G. Louie, Nature 444 (2006) 347.
[7] O.V. Yazyev, Rep. Prog. Phys. 73 (5) (2010) 056501.
[8] J.C. Slonczewski, P.R. Weiss, Phys. Rev. 109 (2) (1958) 272.
[9] O.V. Yazyev, M.I. Katsnelson, Phys. Rev. Lett. 100 (4) (2008) 047209.
[10] S. Fajtlowicz, P. John, H. Sachs, On Maximum Matchings and Eigenvalues of Benzenoid Graphs, 2005.
[11] J.J. Palacios, J. Fernández-Rossier, L. Brey, Phys. Rev. B 77 (19) (2008) 195428.
[12] E.H. Lieb, Phys. Rev. Lett. 62 (10) (1989) 1201.
[13] W.L. Wang, O.V. Yazyev, S. Meng, E. Kaxiras, Phys. Rev. Lett. 102 (15) (2009) 157201.
[14] J. Fernández-Rossier, J.J. Palacios, Phys. Rev. Lett. 99 (17) (2007) 177204.
[15] G. Allinson, R.J. Bushby, J.-L. Paillaud, M. Thornton-Pett, J. Chem. Soc. [Perkin 1] (4) (1995) 385.
[16] J. Inoue, K. Fukui, T. Kubo, S. Nakazawa, K. Sato, D. Shiomi, et al., J. Am. Chem. Soc. 123 (50) (2001) 12702.
[17] B. Odom, D. Hanneke, B. D'Urso, G. Gabrielse, Phys. Rev. Lett. 97 (3) (2006) 030801.
[18] G.R. Eaton, S.S. Eaton, D.P. Barr, R.T. Weber, in: G.R. Eaton, S.S. Eaton, D.P. Barr, et al. (Eds.), Quantitative EPR: A Practitioners Guide, Springer, Vienna, Vienna, 2010, p. 1.
[19] J.A. Weil, J.R. Bolton, Electron Paramagnetic Resonance, John Wiley & Sons, 2006.
[20] K. Nakada, M. Fujita, G. Dresselhaus, M.S. Dresselhaus, Phys. Rev. B 54 (24) (1996) 17954.
[21] L. Brey, H.A. Fertig, Phys. Rev. B. 73 (23) (2006) 235411.
[22] M. Ezawa, Phys. Rev. B 73 (4) (2006) 045432.
[23] N.M.R. Peres, A.H. Castro Neto, F. Guinea, Phys. Rev B 73 (19) (2006) 195411.
[24] C. Tao, L. Jiao, O.V. Yazyev, Y.C. Chen, J. Feng, X. Zhang, et al., Nat. Phys. 7 (2011) 616.
[25] M. Pan, E.C. Girão, X. Jia, S. Bhaviripudi, Q. Li, J. Kong, et al., Nano. Lett. 12 (4) (2012) 1928.
[26] H. Kumazaki, D.S. Hirashima, J. Phys. Soc. Japan 76 (6) (2007) 064713.
[27] O.V. Yazyev, L. Helm, Phys. Rev. B 75 (12) (2007) 125408.
[28] P.O. Lehtinen, A.S. Foster, Y. Ma, A.V. Krasheninnikov, R.M. Nieminen, Phys. Rev. Lett. 93 (18) (2004) 187202.
[29] E.J. Duplock, M. Scheffler, P.J.D. Lindan, Phys. Rev. Lett. 92 (22) (2004) 225502.
[30] V.M. Pereira, F. Guinea, J.M.B. Lopes dos Santos, N.M.R. Peres, A. H. Castro Neto, Phys. Rev. Lett. 96 (3) (2006) 036801.
[31] W.-M. Huang, J.-M. Tang, H.-H. Lin, Phys. Rev. B 80 (12) (2009) 121404.
[32] O.V. Yazyev, Phys. Rev. Lett. 101 (3) (2008) 037203.
[33] D.W. Boukhvalov, M.I. Katsnelson, A.I. Lichtenstein, Phys. Rev. B 77 (3) (2008) 035427.
[34] S.S. Rao, S. Narayana Jammalamadaka, A. Stesmans, V.V. Moshchalkov, J. van Tol, D.V. Kosynkin, et al., Nano. Lett. 12 (3) (2012) 1210.
[35] D.V. Kosynkin, W. Lu, A. Sinitskii, G. Pera, Z. Sun, J.M. Tour, ACS Nano. 5 (2) (2011) 968.
[36] D.V. Kosynkin, A.L. Higginbotham, A. Sinitskii, J.R. Lomeda, A. Dimiev, B.K. Price, et al., Nature 458 (2009) 872.
[37] T. Shimizu, J. Haruyama, D.C. Marcano, D.V. Kosinkin, J.M. Tour, K. Hirose, et al., Nat. Nanotechnol. 6 (2010) 45.
[38] A. Sinitskii, A.A. Fursina, D.V. Kosynkin, A.L. Higginbotham, D. Natelson, J.M. Tour, Appl. Phys. Lett. 95 (25) (2009) 253108.
[39] J. Kunstmann, C. Özdoğan, A. Quandt, H. Fehske, Phys. Rev. B 83 (4) (2011) 045414.

[40] C. Gerber Iann, V. Krasheninnikov Arkady, S. Foster Adam, M.N. Risto, New J. Phys. 12 (11) (2010) 113021.
[41] X. Liang, B.A. Sperling, I. Calizo, G. Cheng, C.A. Hacker, Q. Zhang, et al., ACS Nano. 5 (11) (2011) 9144.
[42] F. Zou, L. Zhu, K. Yao, Sci. Rep. 5 (2015) 15966.
[43] M. Wimmer, İ. Adagideli, S. Berber, D. Tománek, K. Richter, Phys. Rev. Lett. 100 (17) (2008) 177207.
[44] W.Y. Kim, K.S. Kim, Nat. Nanotechnol. 3 (2008) 408.

CHAPTER 8

Magnetic topological insulators: growth, structure, and properties

Liang He[1], Yafei Zhao[1], Wenqing Liu[1,2] and Yongbing Xu[1,3]

[1]York-Nanjing Joint Center (YNJC) for Spintronics and Nanoengineering, School of Electronics Science and Engineering, Nanjing University, Nanjing, P.R. China, [2]Department of Electronic Engineering, Royal Holloway University of London, Egham, United Kingdom, [3]Department of Electronic Engineering, The University of York, York, United Kingdom

Chapter Outline

Spintronic 2D Materials.
DOI: https://doi.org/10.1016/B978-0-08-102154-5.00010-2

8.1 Introduction

The recently discovered topological insulator (TI), which a novel state of quantum matter with enormous potential for electronic and spintronic applications, has promoted the development of condensed matter physics. TI is a class of quantum material featuring with an energy gap in its bulk band structure and unique Dirac-like metallic states on the surface. The gapless surface states of the TI are protected by time-reversal symmetry (TRS), and the spin of the Dirac-like surface states is tightly locked to the momentum. Thus such nontrival surface states can be utilized to perform dissipationless spin transport.

However, it is equally important to break the TRS of TIs to realize novel physical phenomena, such as quantum Hall effect [1], quantum spin Hall effect (SHE) [2], quantum anomalous Hall effect (QAHE) [3,4], magnetoresistance switch effect [5], hedgehog-like spin textures [6], the giant magneto-optical Kerr effect [7], and so on. Intuitively, doping TIs with transition metal (TM) elements is one simple route to break TRS [8]. Another way is through the exchange interaction with an adjacent ferro- or ferri-magnetic material, namely, magnetic proximity effect. Also, due to the strong spin–orbital coupling (SOC), TIs possess giant SHE, which means the spin current per unit charge current density in TIs is much greater than that in any other heavy metals, even for nonideal TI films in which the surface states coexist with bulk conduction [9,10]. This enables the large spin-transfer torque (STT) to the adjacent ferromagnetic materials and opens a pathway toward electrically manipulating magnetism at room temperatures (RTs). In this chapter, we will discuss the growth, the structure and various properties of the TIs with the emphases on the magnetism and various spin-related effects.

8.2 Crystal structure and thin film grown by molecular-beam epitaxy

8.2.1 Crystal structures

Three-dimensional (3D) TIs are V–VI compound semiconductors, such as Bi_2Se_3, Bi_2Te_3, and Sb_2Te_3, and their derivatives. All of them have a tetradymite-type crystal lattice, as shown in Fig. 8.1 in the case of Bi_2Se_3 [11]. It has a trigonal axis (threefold rotational symmetry), defined as the z-axis, a binary axis (twofold rotational symmetry); defined as the x-axis, a bisectrix axis (in the reflection plane); and defined as the y-axis. Each Bi_2Se_3 layer consists of two Bi layers and three Se layers in the sequence of Se–Bi–Se–Bi–Se along the z-direction, normally called a quintuple layer (QL). Each TI's unit cell has the space group $D_{3d}^5(R\overline{3}m)$ and contains three such QLs, and the neighboring QLs are held

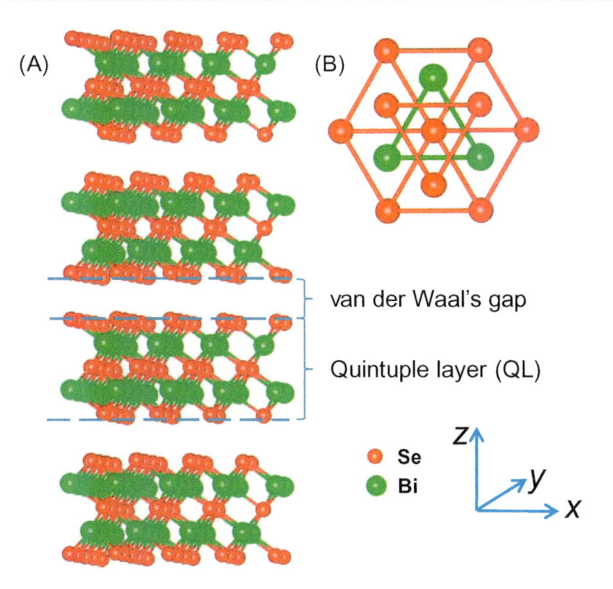

Figure 8.1
(A) 3D view of Bi_2Se_3 crystal structure. Five atomic layers of Se–Bi–Se–Bi–Se form a QL, and there are van der Waals gaps among QLs. (B) Top view along the *z*-direction of the lattice. Each atomic layer sits on the others' center of triangle lattices, with the stacking order of A–B–C–A–B–C, etc. [11]. *QL*, quintuple layer.

together by a weaker van der Waals (VDW) force. The VDW gap between QLs gives the structure a distinctly anisotropic character similar to other 2D materials, such as TM dichalcogenide materials, III–VI layered semiconductors, hexagonal boron nitride (h-BN), black phosphorus, graphene, and so on [12–15].

8.2.2 Thin film grown by molecular-beam epitaxy

Experimental realization of TI materials is mainly by wet chemical synthesis [16,17], chemical vapor deposition (CVD) [18–20], Bridgman method [21–23], and molecular-beam epitaxy (MBE) [24–28]. Among them, MBE technology has tremendous advantages, including convenient physical deposition without conceiving the complexity of chemistry in CVD, ability to precisely control growth thickness down to atomic layer and doping concentration, growth of heterogeneous and superlattices structures for electronic applications, and wafer-scale growth method. In general, MBE deposition is realized by depositing individual elements onto the substrates, where they form the desired compound structure. The quality of the deposited films is affected by the substrate temperature, the flux ratio of the components, and more importantly, the substrates. This chapter will further discuss the effect of different substrates on the growth quality. Other growth factors reported in previous review especially on the growth of TI [11].

8.2.2.1 Film grown on conventional semiconductors

The growth mechanism of TIs is different from that of the conventional MBE growth for covalent or ionic bonded structures due to the existence of VDW gaps among the QLs. Therefore there is a special name called VDW epitaxy for the growth of these kinds of layered materials.

For VDW epitaxy, it releases constraint in the lattice matching required for mostly epitaxial growth of common semiconductors and their heterostructures. Thus a series of substrates with large lattice mismatch can be used for the growth of TIs, and epitaxial films have been obtained from various reports [29]. Such as, large atomically flat terraces have been observed in Bi_2Se_3 grown on Si (111) and Bi_2Te_3 grown on GaAs (111)B, as shown in Fig. 8.2. Meanwhile the lattice constants extracted by reflection high-energy electron diffraction (RHEED) demonstrate that the films are fully relaxed after finishing the first QL and the lattice constant reaches the bulk value.

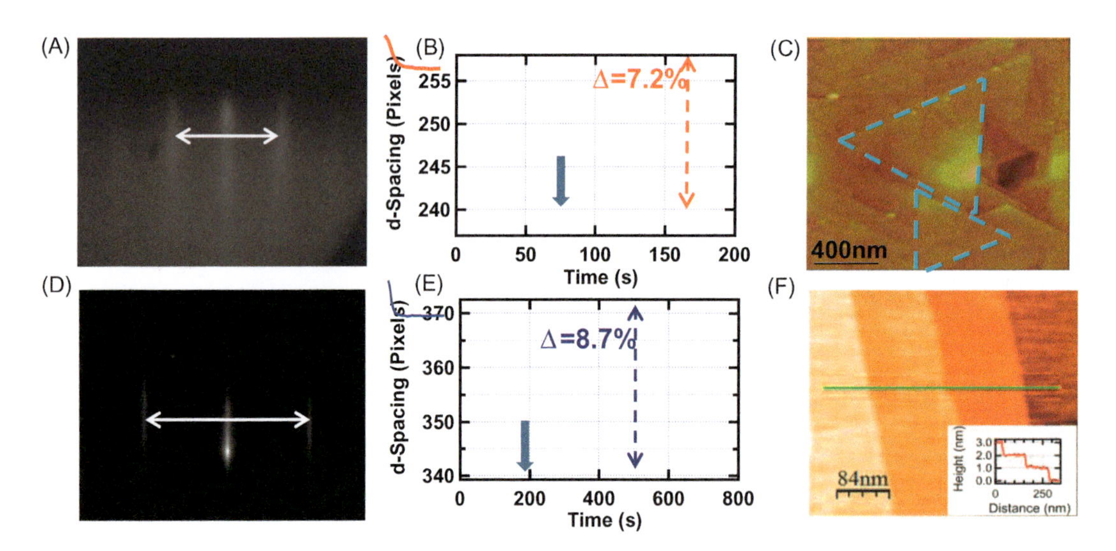

Figure 8.2

d-Spacing evolutions. (A) and (D) typical RHEED images of an as-grown Bi_2Se_3film on Si (111) and a Bi_2Te_3 film on GaAs (111)B, respectively. The double arrows between the two first-order stripes represent the *d*-spacing, which is inversely proportional to the lattice constant. (B and E) The *d*-spacing evolution as a function of growth time. The solid blue arrows indicate the growth time of the first QL. The *d*-spacing reaches a constant value after it, suggesting that the lattice is almost fully relaxed to that of the TI films, (C and F) atomic force microscope (AFM) image of an as-grown Bi_2Se_3 film on Si (111) and a Bi_2Te_3 film on GaAs (111)B, respectively [11]. *QL*, quintuple layer; *RHEED*, reflection high-energy electron diffraction; *TI*, topological insulator.

However, it should be pointed out that although the VDW epitaxy growth makes the substrates' lattice constants are not so important, the substrates still somehow affects the structure and the overall quality of the film. Kou et al. reported an improved surface morphology and higher mobility using lattice-matched substrate of CdS [30]. Moreover, the presence of dangling bonds on a clean surface of those conventional semiconductors could cause the growth of an unwanted interfacial layer at the interface, which lowers the quality of TI films [31,32]. Intuitively, the passivation of the dangling bond would mean a better way. Experimentally, people have demonstrated high-quality TI films grown on passivated Si and GaAs substrates by H and Se atoms [33–35]. However, it would be much better to grow TIs directly onto 2D materials, whose surfaces are fully saturated, to avoid the complex of passivation.

8.2.2.2 Films grown on two-dimensional materials

2D materials, such as graphene, MoS_2, h-BN, and mica are suitable for VDW epitaxy due to the fact that there are no dangling bonds associated with the surface [36–38]. In addition, lattice mismatch and crystalline defect of the substrates would not be problematic due to the lack of covalent bonding across the VDW gaps of the two materials. Such as, an excellent crystallinity Bi_2Se_3 film (Fig. 8.3) has been epitaxially grown on a single layer of

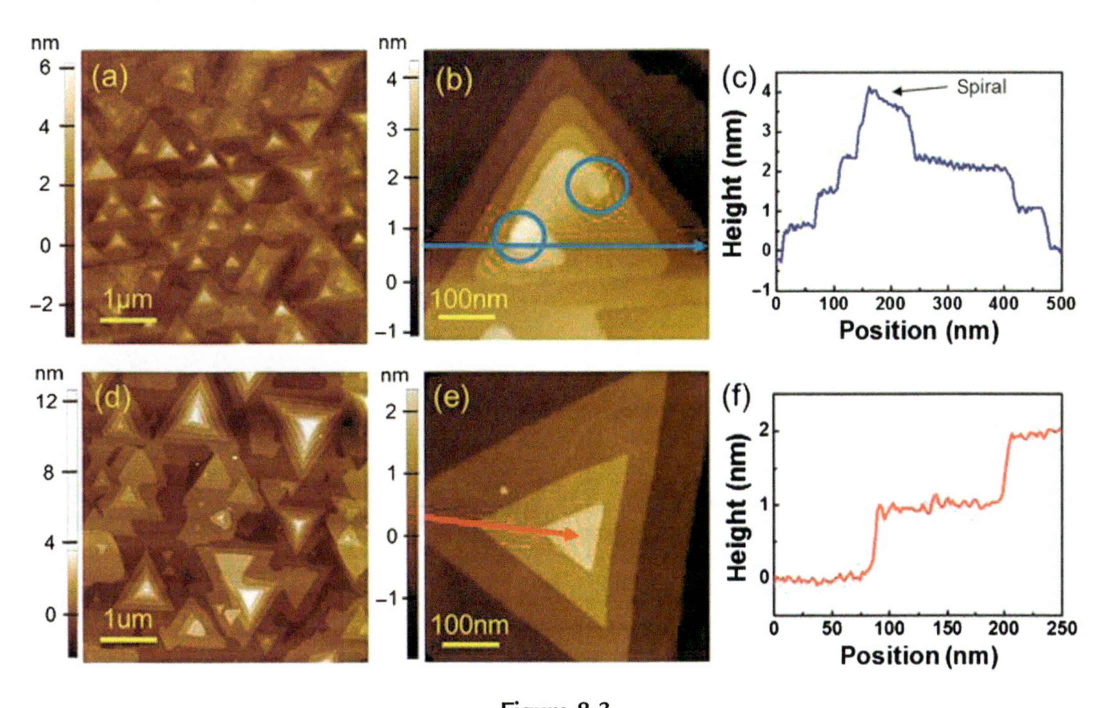

Figure 8.3
Surface morphologies of 30 QL Bi_2Se_3 films (A)–(C) on the Al_2O_3 (0001) substrate and (D)–(F) on MoS_2/Al_2O_3 (0001) [36]. *QL*, quintuple layer.

MoS_2 via VDW epitaxy in a planar geometry [36], with the increased size of the triangular-shaped Bi_2Se_3 domains and 2–3 times enhancement in mobility, leading to the observation of Shubnikov–de Haas oscillations in transport measurements.

Similarly, based on VDW epitaxial growth, ultrahigh quality of Bi_2Se_3/In_2Se_3 superlattices can be directly integrated onto mica (without surface dangling bonds) [37]. Such as, Fig. 8.4A shows the RHEED patterns of the mica, Bi_2Se_3, and In_2Se_3. It is found that the sharp streaks in insets (ii) and (iii) reveal the smooth surface morphology of the deposited Bi_2Se_3 and In_2Se_3, and RHEED oscillations have also been observed in the growth of Bi_2Se_3. Meanwhile, the evolution of the reciprocal lattice parameters indicates that the lattice of Bi_2Se_3 is fully relaxed after the first QL. And the superlattices have demonstrated good uniformity and no strain, as shown in Fig. 8.4B.

Moreover, large-area high-quality Bi_2Se_3 thin films with excellent surface coverage and crystallinity have been grown on h-BN, by two-step MBE method [39]. Namely, at the first step, 1–2 QLs of TIs were deposited at low temperature, and then annealed under Se flux for 1 h; at the second step, growth was then obtained at higher temperatures. The film quality was verified by AFM and transmission electron microscope (TEM). Giant terrace of Bi_2Se_3 was achieved with only one monolayer step. And the atomically sharp interface between Bi_2Se_3 and BN was observed.

In summary, high-quality TIs can be grown epitaxially on most common semiconductor substrates and/or 2D materials by VDW epitaxy on large scale. This lays the foundation for future TI-based device applications.

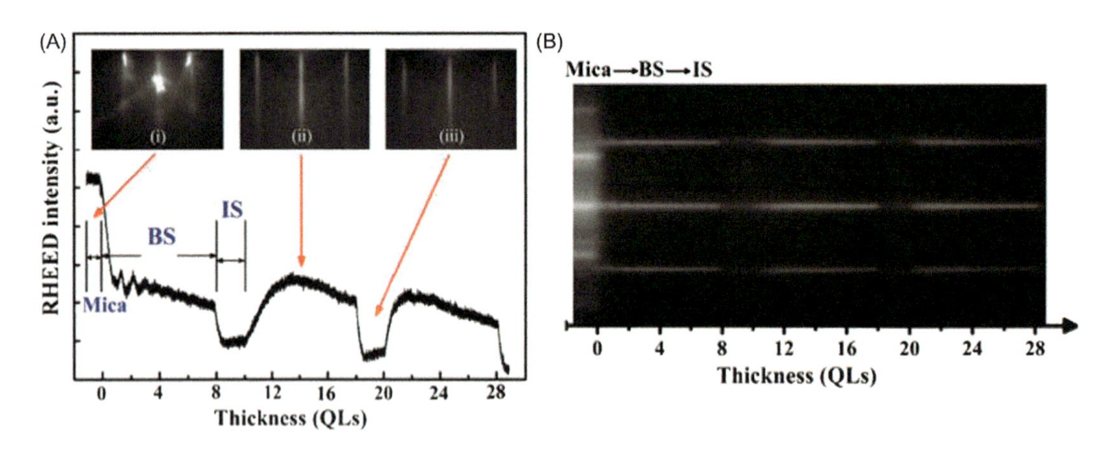

Figure 8.4
(A) Variation of the RHEED specular-beam intensity. The insets (i), (ii), and (iii) show typical RHEED patterns of the mica, Bi_2Se_3, and In_2Se_3 surfaces, respectively. (B) Intensity linescan across the diffraction streaks [37]. *RHEED*, reflection high-energy electron diffraction.

8.3 Magnetic topological insulator

8.3.1 Transition metal–doped topological insulator

Doping TIs with TM elements is intuitively one simple route to break TRS or to introduce ferromagnetic order, such as V-, Cr-, Mn-, and Fe-doped TIs [40–46]. Experimental realization of TM-doped TI materials is mainly by MBE and Bridgman method. MBE technique was usually used to grow high-quality single-crystalline Cr–Bi2Se3, Cr–Bi2Te3, and Cr– $(Bi_xSb_{1-x})_2Te_3$ by codeposition method [47]. And, Bridgman method (the required elements are mixed together according to a certain stoichiometric ratio, then mixed powders were ground and placed in quartz ampoule) was widely used to grow bulk TM-doped TIs, such as Fe- and Mn-doped Bi_2Se_3 bulk samples [48].

8.3.2 Magnetic properties

Using MBE, Kou et al. studied the magnetic properties in epitaxial $Bi_{2-x}Cr_xSe_3$ $(0 \leq x \leq 0.2)$ films [41]. It is found that with Cr doping, the samples turn into ferromagnetic. Fig. 8.5A exhibits the hysteresis loops of a 50 QL $Bi_{1.9}Cr_{0.1}Se_3$ thin film at various temperatures. Clear hysteresis loops at various temperatures can be traced up to

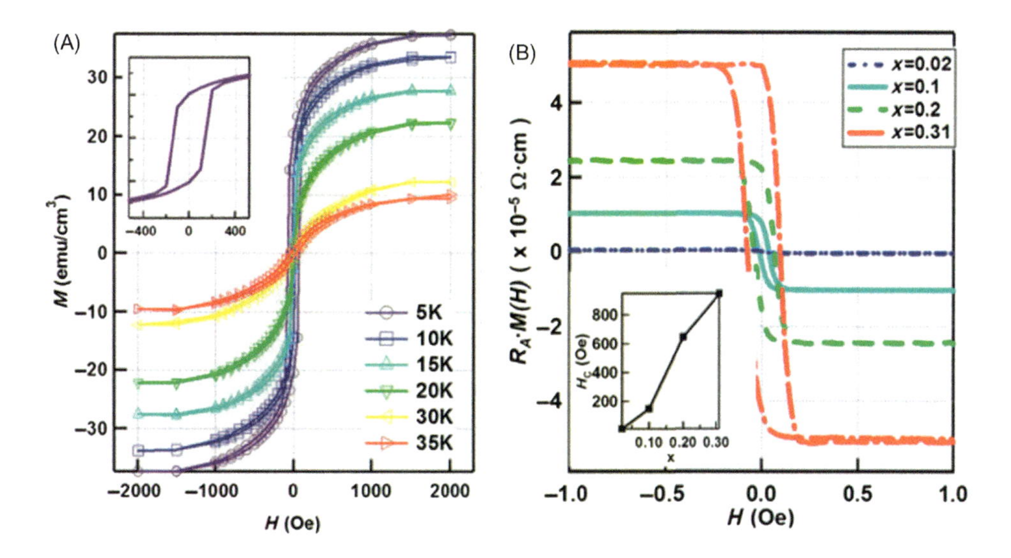

Figure 8.5

(A) $M-H$ loops of the 50 QL $Bi_{2-x}Cr_xSe_3$ thin film at 5K, 10K, 15K, 20K, 30K, and 35K. Inset: zoom-in hysteresis loop at 5K. B field-dependent AHE (R_A, M) with Cr doping concentration ($x = 0.02$, 0.1, 0.2, and 0.3). Linear ordinary Hall component of R_o, H has been subtracted from the Hall resistance. Inset: the evolution of coercivity field H_C under different Cr compositions at 1.9K [41]. *AHE*, anomalous Hall effect; *QL*, quintuple layer.

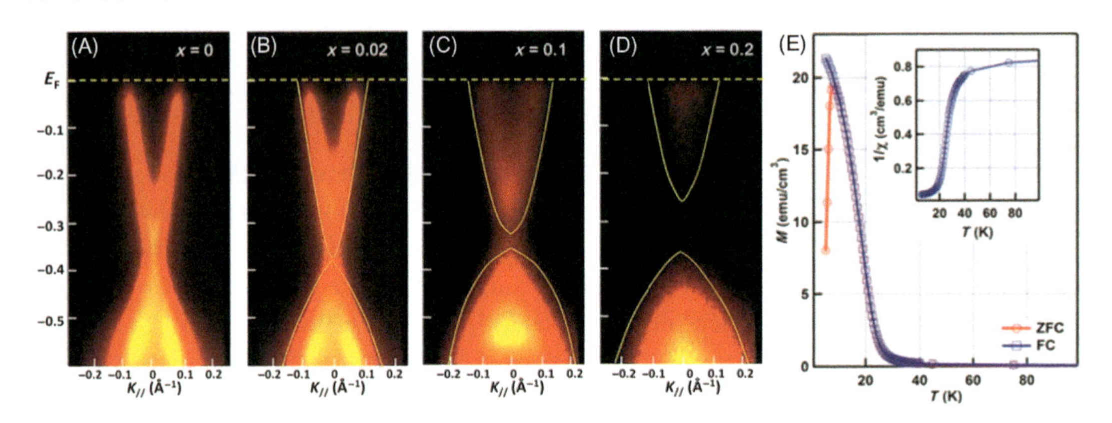

Figure 8.6
ARPES intensity maps of uncapped 50 QL $Bi_{2-x}Cr_xSe_3$ thin films with (A) $x = 0$, (B) $x = 0.02$, (C) $x = 0.1$, and (D) $x = 0.2$ on Si (111). The surface will also open a larger bandgap with higher Cr concentration as illustrated by the solid yellow lines. All data are taken using 52 eV photons under the temperature of 10K.(E) Magnetization versus temperature for 50 QL $Bi_{1.9}Cr_{0.1}Se_3$ thin film sample under zero-field cooling (*red circles*) and field cooling (*blue squares*) conditions. Inset: this relation does not follow the linear Curie–Weiss law above T_c [41]. *ARPES*, angle-resolved photoelectron spectroscopy; *QL*, quintuple layer.

35K, with a coercivity field of 150 Oe at 5K. This suggests a Curie temperature (CT) of 35K. Moreover, the Hall resistance R_{xy} is nonlinear when the temperature is below the T_C, due to the anomalous Hall effect (AHE). After subtracting the linear ordinary Hall component, the field-dependent AHE demonstrates again clearly hysteresis loops for various Cr doping concentrations, as shown in Fig. 8.6B. Most importantly, from this we can see that the magnetic direction of those Cr-doped TIs is out of plane, parallel to the *z*-axis. And this is crucial, since only out-of-plane magnetization can break the time-reserve symmetry (TRS). As a result of the out-of-plane magnetic moment, the surface states deviate from the original Dirac fermion character and a gap is open. This gap opening monotonically increases with Cr concentration up to ~100 meV at 10K, as shown in Fig. 8.6.

Similarly, Liu et al. also observed that the bandgap opening systematically increasing as a function of Cr doping concentration, see Fig. 8.7 [49].

Moreover, Liu et al. provided direct evidence for the detailed magnetic states of the $Bi_{1.94}Cr_{0.06}Se_3$ epitaxial thin films by X-ray magnetic circularly dichroism (XMCD), as shown in Fig. 8.8 [50]. They found that the reduced magnetic moment with the increasing of the Cr doping concentration is due to the formation of the antiferromagnetic coupling of the Cr in different sites.

In addition, Zhang et al. also observed ferromagnetism in $Co_{0.08}Bi_{1.92}Se_3$ and Co cluster. They found that the Ruderman–Kittel–Kasuya–Yosida interaction and the spin structure of Co impurity determine the ferromagnetism of bulk $Co_{0.08}Bi_{1.92}Se_3$ [51] (Fig. 8.9).

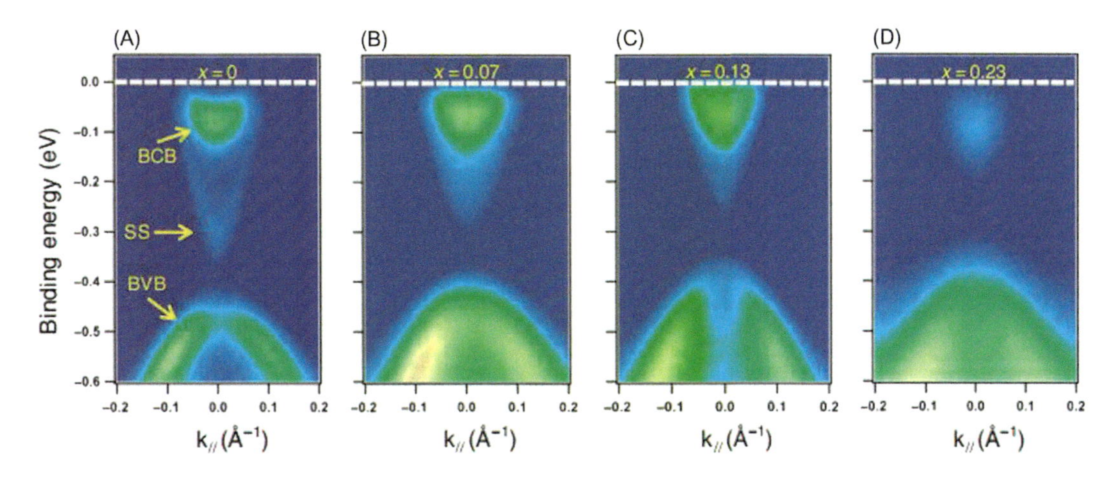

Figure 8.7

(A) ARPES spectrum of Bi_2Se_3 shows well-resolved Dirac-like SS. The SS, BCB, and BVB are marked by arrows. The ~100 meV gap near the Dirac point is induced by the hybridization between the two surfaces. In the $x = 0.07$ (B) and $x = 0.13$ films (C), the surface states become weaker and the gap becomes larger. (D) In the ARPES spectrum of the strongly magnetic $x = 0.23$ film, the surface states totally disappear [49]. *ARPES*, angle-resolved photoelectron spectroscopy; *BCB*, bulk conduction band; *BVB*, bulk valence band; *SS*, surface states.

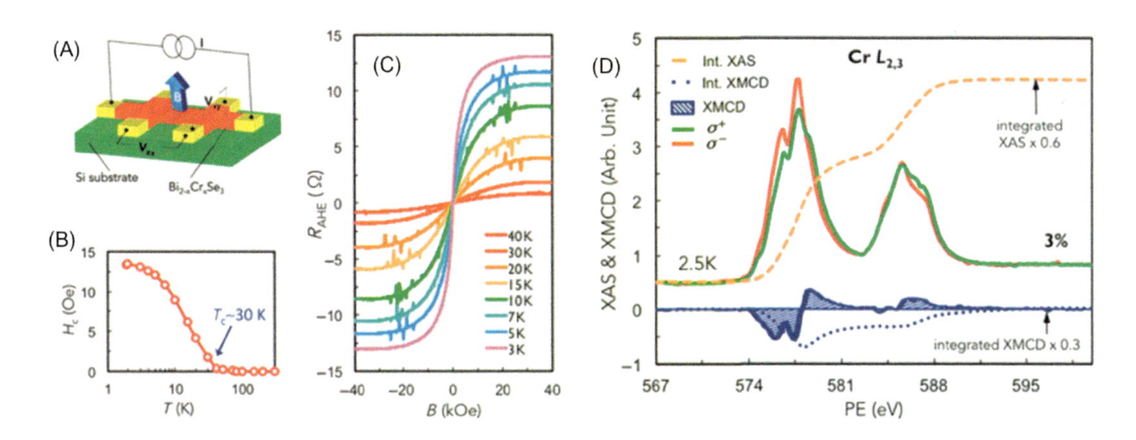

Figure 8.8

Electrical magneto-transport measurements. (A) Schematic diagram of the experimental setup of the AHE measurement. (B) H_c versus temperature of the $Bi_{1.94}Cr_{0.06}Se_3$/Si (111) thin film from 3K to 300K. (C) R_{AHE} versus magnetic field B of the thin film from 3K to 40K after background removal. (B) Typical pair of XAS and XMCD spectra of the $Bi_{1.94}Cr_{0.06}Se_3$/Si (111) thin film. Data are offset and scaled for clarity [50]. *AHE*, anomalous Hall effect; *XAS*, X-ray absorption spectroscopy; *XMCD*, X-ray magnetic circularly dichroism.

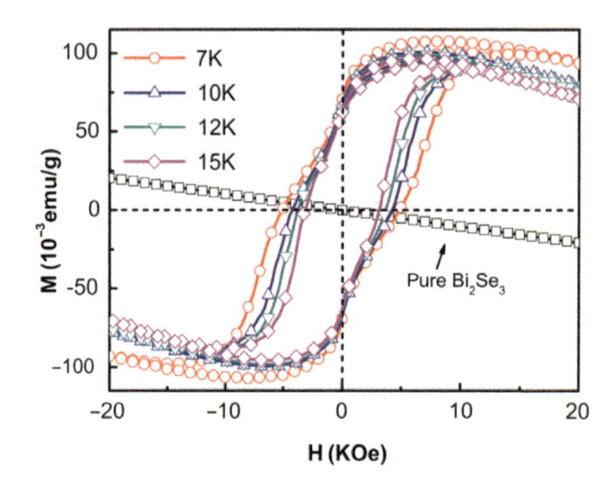

Figure 8.9
The M–H curves of $Co_{0.08}Bi_{1.92}Se_3$ along the in-plane direction at different temperatures. The M–H curves of Bi_2Se_3 at 7K [51].

8.3.3 Novel phenomena based on magnetic topological insulators

Magnetic TIs have many novel physical phenomena, and we focus on QAHE. QAHE is a quantized version of AHE. Since QAHE presents the realization of the quantum Hall effect (QHE) in zero magnetic fields, therefore the dissipationless quantum transport is expected. Chang et al. reported that for very thin TM-doped TI systems, it will induce the QAHE if the magnetic exchange field is perpendicular to the plane and overcomes the semiconductor gap. The QAHE was first experimentally observed in a 5 QL $Cr_{0.15}(Bi_{0.1}Sb_{0.9})_{1.85}Te_3$ film grown on $SrTiO_3(111)$ substrates [52]. The realization of QAHE may promote the development of low-power electronic devices.

Soon after that, Kou et al. also successfully observed QAHE in MBE-grown $(Cr_{0.12}Bi_{0.26}Sb_{0.62})_2Te_3$ sample with a thickness of 10 QL [53]. It is found that when the sample temperature falls below 85 mK, the sample reaches the QAHE regime, as shown in Fig. 8.10A and B. Moreover, it is observed that the Hall resistance R_{xy} reaches the quantized value of h/e^2 (25.8K) at $B = 0$ T, while the longitudinal resistance R_{xx} is nearly vanished when the film is magnetized. Compared to the 2D hybridized thin film, an additional weak field-dependent longitudinal resistance is observed in the 10 QL film.

Meanwhile, Checkelsky et al. using solid-dielectric and ionic-liquid gating, also successfully observed QAHE in Mn-doped $Bi_2Te_{3-y}Se_y$ sample [54].

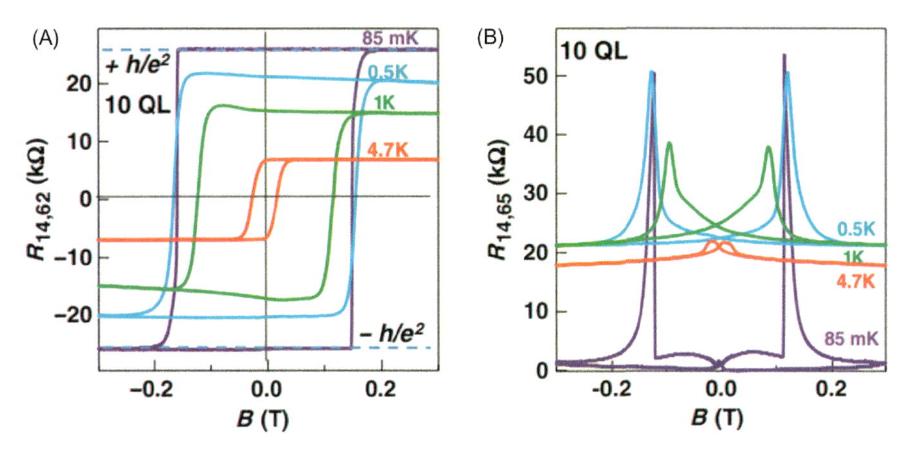

Figure 8.10

The QAHE in the $(Cr_{0.12}Bi_{0.26}Sb_{0.62})_2Te_3$ thin films with a thickness of 10 QL. (A) Hysteresis $R_{xy} - B$ curves of the 10 QL film at different temperatures. For $T < 85$ mK, R_{xy} attains the quantized value of h/e^2. (B) Butterfly-shaped $R_{xy} - B$ curves of the 10 QL film. In the QAHE regime, R_{xx} nearly vanishes at low fields [53]. *QAHE*, quantum anomalous Hall effect; *QL* quintuple layer.

8.4 Magnetic proximity effect in topological insulator–based heterojunctions

Ferromagnetism has been achieved in TM-doped TIs at low temperatures, and the highest T_C so far is only 35K. Obviously, this is far too low for RT applications of magnetic TIs. Pioneering theoretical work suggests that suitable ferromagnetic insulators (FMI) have the potential to achieve a strong and uniform exchange coupling in contact with TIs, which is called the magnetic proximity effect. This route can subsequently be divided into two ways in terms of ferromagnetic metal (FM) and FMIs, as illustrated in Fig. 8.11.

8.4.1 Topological insulators/ferromagnetic metal

Inducing magnetism into TI through the proximity effect was first tried on TIs/FM heterojunctions. Most FMs are selected from the TM groups with high T_C above RT. Recent work has reported about tuning the band structures of Bi_2Se_3 by depositing Fe over-layers observed by angle-resolved photoelectron spectroscopy (ARPES). The topological spin structure of the Bi_2Se_3 surface can be systematically modified depending on the thickness of the deposited Fe layers [55], due to their large Coulomb charge and significant magnetic moment. When the Fe layer is thick enough, odd multiples of Dirac fermions of

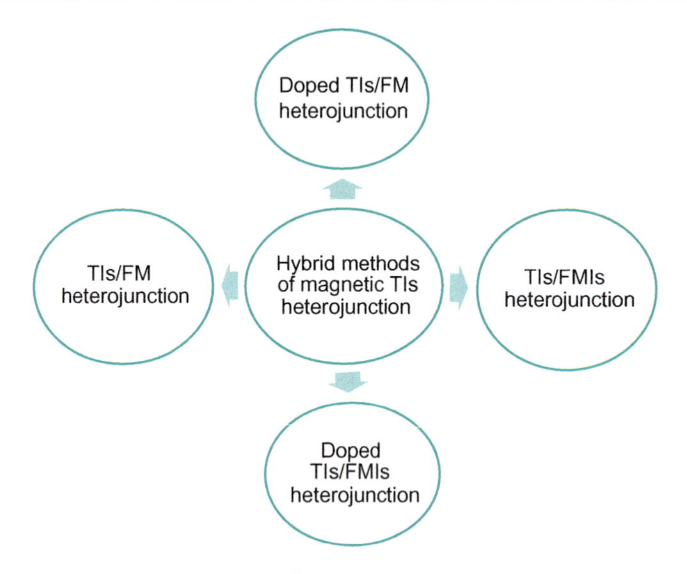

Figure 8.11
Classification of the magnetic proximity effect in TIs-based heterojunction: TIs/FM; TIs/FMIs; magnetically doped TIs/FM; and magnetically doped TIs/FMIs. *FM*, ferromagnetic metal; *TI*, topological insulator; *FMI*, ferromagnetic insulator.

Bi_2Se_3 are observed and TRS is broken in the presence of band hybridizations, as shown in Fig. 8.12. Furthermore, the electron mobility and quantum behaviors of topological surfaces have also been altered. More importantly, this TI/FM heterostructure has set the stage for creating exotic particles, such as axions or imaging monopoles on the surface.

Another similar work is about a detailed thickness-dependent study of Co films grown on Bi_2Se_3 substrate using both magneto-optic Kerr effect (MOKE) and element-specific XMCD measurements [56]. It has been found that the ultra-thin Co layer on Bi_2Se_3 is not ferromagnetically ordered when the Co layer thickness is below 1.2 nm. However, by inserting a spacer layer of 3 nm Ag between Co and Bi_2Se_3, the Co film becomes completely ferromagnetically ordered, as shown in Fig. 8.13. This suggests that the loss of the ferromagnetic order of the Co film at the Co/Bi_2Se_3 interface is due to the band intermixing between Co and Bi_2Se_3. With the added 3-nm Ag, the electron wave functions do not overlap, thus no band intermixing exists and the Co layer resumes ferromagnetic. Basically, this work suggests that the TI and FM layers do interact with each other, and the interaction range is about 1.2 nm. The short-range nature of magnetic proximity coupling with ferromagnetic layer allows the TIs' surface states to experience the ferromagnetic interactions, where the symmetry breaking happens right at the interface, rather than affecting the bulk states or introducing defects [57,58]. This also implies that the proximity effect only affects the magnetism of the surface states, which we will discuss further more later.

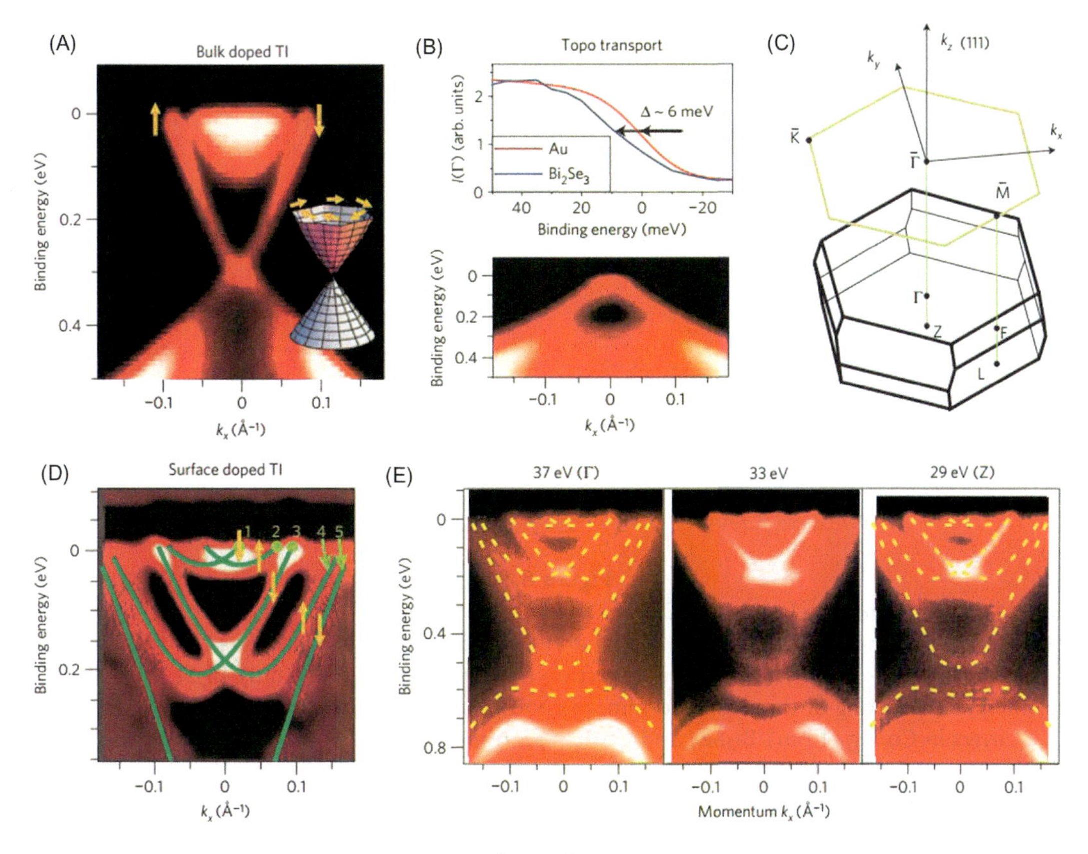

Figure 8.12

Iron deposition strongly modifies the topological surface. (A) Uniformly electron-doped Bi_2Se_3 has a single surface-state Dirac cone. (B) When the surface chemical potential of as-grown Bi_2Se_3 is lowered to the Dirac point by NO_2 deposition, observing a slight gap in the leading edge of ARPES intensity. (C) The hexagonal surface Brillouin zone of Bi_2Se_3 is drawn above a diagram of the 3D bulk Brillouin zone. (D) A second-derivative image of new surface states in Bi_2Se_3 (sample #1) after surface iron deposition is labeled with numerically predicted spin texture. (E) Low-energy features from *d* have no *z*-axis momentum dispersion, as obtained from the data taken with varying incident photon energy (37–29 eV), confirming the 2D character of the state [55]. *ARPES*, angle-resolved photoelectron spectroscopy.

8.4.2 Topological insulators/ferromagnetic insulators

Although TIs/FM heterostructure can introduce ferromagnetism into TIs surface, the conducting nature of FM layer shadows the conduction of the exotic surface states of TIs. Thus if we want to study the novel transport properties of the magnetic surface states, TIs/FMI heterostructure is needed.

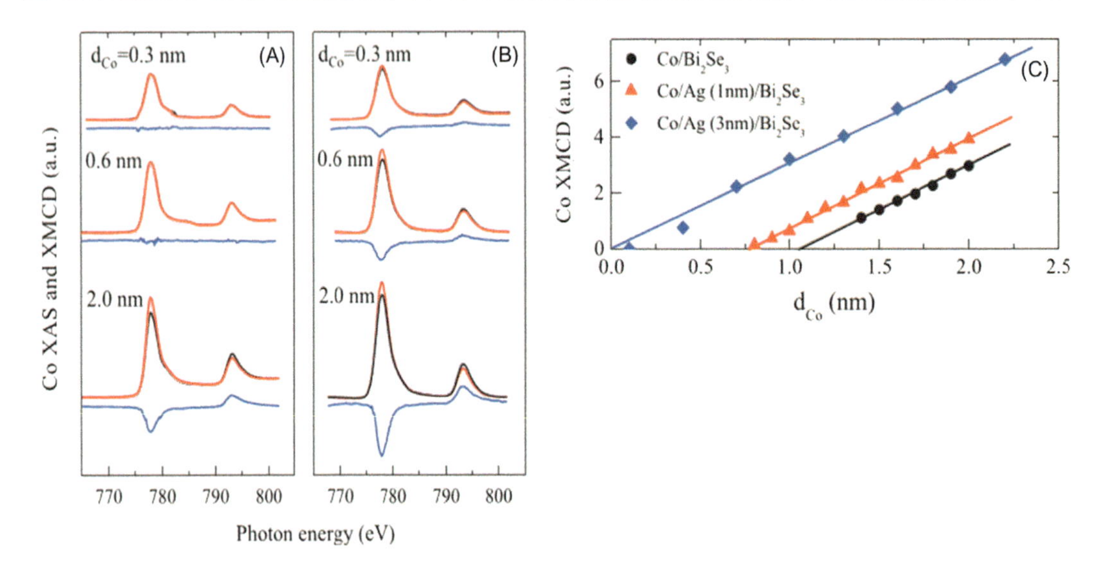

Figure 8.13

Co XAS (*black and red lines*) and XMCD (*blue lines*) of (A) Co/Bi_2Se_3 and (B) Co/Ag (3 nm)/Bi_2Se_3. Red and black lines are XAS at $B = +5000$ and $B = -5000$ Oe, respectively. (C) XMCD signal versus Co thickness. The zero Co intersection thickness for Co/Ag (3 nm)/Bi_2Se_3 shows a ferromagnetic ordering of the entire Co film, in contrast to the 0.8-nm and 1.2-nm magnetic dead layers in Co/Ag (1 nm)/Bi_2Se_3 and Co/Bi_2Se_3, respectively [56]. *XAS*, X-ray absorption spectroscopy; *XMCD*, X-ray magnetic circularly dichroism.

Experimentally Lang et al. demonstrated the magnetic properties of Bi_2Se_3 surface states in the proximity of high T_C (~550K) FMIs, YIG ($Y_3Fe_5O_{12}$) [59]. Proximity-induced butterfly- and square-shaped magnetoresistance loops are observed by magneto-transport measurements with out-of-plane and in-plane fields, respectively. And these can be correlated with the magnetization of the YIG substrate, as shown in Fig. 8.14. More importantly, a magnetic signal from the Bi_2Se_3 up to 130K is clearly observed by MOKE measurements. This suggested a much higher T_C in magnetic TIs.

Another work has reported the heterostructure of Bi_2Se_3 and EuS (T_C~16.6K), a ferromagnetic phase in Bi_2Se_3 has emerged, evidenced by magnetic and magneto-transport studies. And also, there is no observable defect in Bi_2Se_3 layers [60]. Further, an interface electron state at the junction of Bi_2Se_3/YIG was investigated by ARPES and XMCD [61]. The surface state of the Bi_2Se_3 film was directly observed and localized 3d spin states of the Fe^{3+} in the YIG film were confirmed. The proximity effect is likely described in terms of the exchange interaction between the localized Fe 3d electrons in the YIG film and delocalized electrons of the surface and bulk states in the Bi_2Se_3 film.

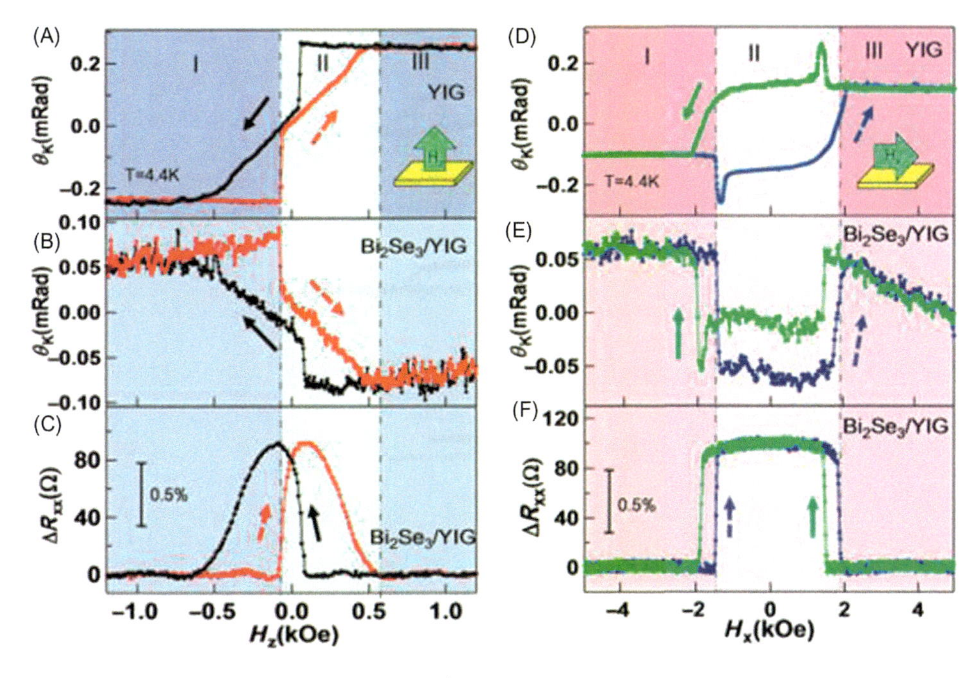

Figure 8.14

The comparison between MOKE and magneto-transport of Bi_2Se_3/YIG. The illustrated by a sharp increase in the polar MOKE of YIG (A) and correspondingly the sharp drop in the Bi_2Se_3/YIG (B), (C) WAL background–subtracted ΔMR shows a shape increase of resistance at ~90 Oe. In-plane magnetization of YIG (D) and Bi_2Se_3/YIG (E) measured by a longitudinal-mode MOKE setup. The peaks in the hysteresis loop may come from the domain nucleation. (F) Parabolic background–subtracted MR with in-plane field applied, clearly displaying two resistance states [59]. *MOKE*, magneto-optic Kerr effect; *YFO*, $Y_3Fe_5O_{12}$.

8.4.3 Doped topological insulators/ferromagnetic metal

Besides native TIs, doped TIs can also be used to look for proximity effect. Doping TM impurities into TIs can introduce a perpendicular ferromagnetic ordering, open a gap at the Dirac point, and produce massive Dirac fermions at the surface [62–64]. On the other hand, compared to the doping method, proximity effect has a number of advantages, including spatially uniform magnetization, better controllability of surface state, free from dopant-induced scattering, and preserving Tis' intrinsic crystalline structure, etc. [65]. It is easy to think of combining these two methods together to get stronger ferromagnetism in Tis. Here we have to note that the magnetic doping is to induce magnetism to the bulk and surface states together, while the proximity effect only works on the surface states. Therefore proximity effect can be used to enhance the magnetism of the surface states instead of bulk states.

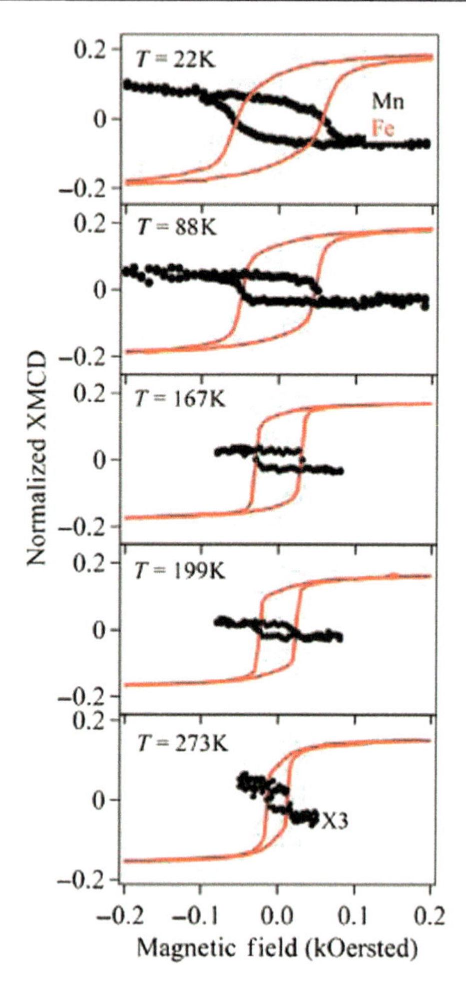

Figure 8.15

Mn and Fe XMCD hysteresis loops versus temperature. Element selective XMCD hysteresis loops for Fe (*black*) and Mn (*red*). Hysteresis curves are measured with the photon energy fixed at the maximum of the XMCD magnetic signal (L_3) and sweeping the magnetic field. Normalization is obtained through on-edge and off-edge measurements. The orientation of the Mn loops is opposite to that of the Fe loops, dictated by the antiparallel alignment of the magnetizations. The experimental geometry probes only the in-plane magnetization of both Fe and Mn. Both the magnetic signal (height of the loops) and the coercive field (width of the loops) increase when the temperature is decreased, suggesting a robust magnetic coupling. The Mn XMCD signal at RT, although clearly visible, is close to 0, while Fe magnetic signal, as expected for a 1-nm ferromagnetic film, is still large [66]. *XMCD*, X-ray magnetic circularly dichroism.

Experimentally, enhanced ferromagnetic ordering at ambient temperatures has been demonstrated in $Bi_{2-x}Mn_xTe_3$ through deposited Fe over-layer [66]. The magnetism of the doped Mn atoms has been investigated by XMCD from 22K to RT, as shown in Fig. 8.15.

And the results clearly demonstrate that the magnetic ordering of the surface states can be maintained up to RT, which is crucial for device applications. Also, the magnetic TI and the Fe over-layer are antiparallel. More importantly, recent findings indicate that the surface of a TI maintains its topological character in the presence of external magnetic perturbations (Fe over-layers), because that the magnetic direction of Fe is in-plane. Thus this result opens a new path to interface-controlled FM in TIs-based spintronic devices by changing the structure and the local environment of the TIs surfaces through the usage of deposited FM over-layers.

8.4.4 Doped topological insulators/ferromagnetic insulators

Finally, the proximity effect of doped TI/FMI has also been studied. For example, clearly enhancement of the T_C of Cr-doped Bi_2Se_3 has been observed in $Bi_{2-x}Cr_xSe_3$/YIG [67], since the magnetization direction of the YIG top layer is out of plane. Thus the FMI provides the TIs with a source of exchange interaction yet without removing the nontrivial surface state. By performing the elemental-specific XMCD, unequivocally observed an enhanced T_C of 50K in this magnetically doped TI/FMI heterostructure, as shown in Fig. 8.16. Moreover, this system also showed a larger (6.6 nm at 30K) but faster decreasing (by 80% from 30K to 50K) penetration depth compared to that of the dilute magnetic semiconductor (DMS), which could indicate a novel mechanism for the interaction between FMIs and the nontrivial TIs surface.

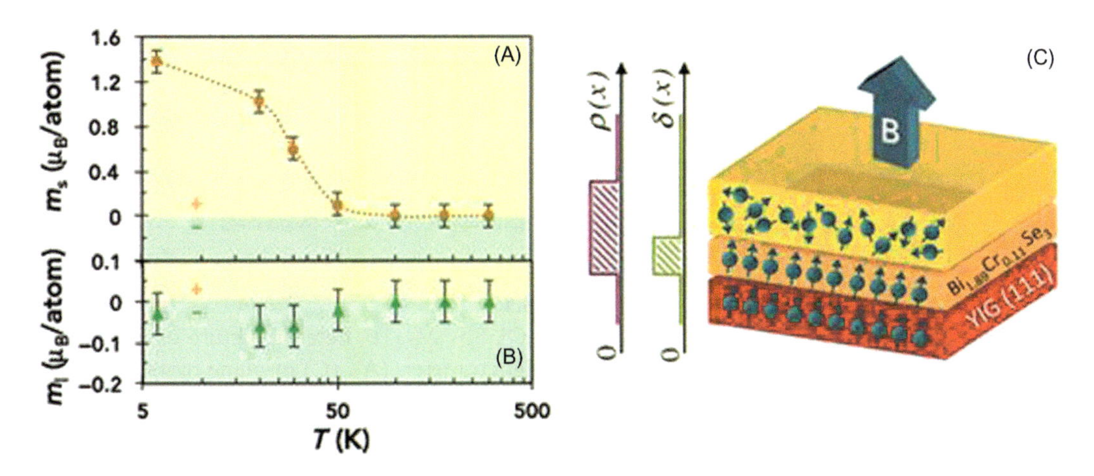

Figure 8.16

Enhanced magnetic ordering of $Bi_{1.89}Cr_{0.11}Se_3$ via the proximity effect. (A), (B) The m_s and m_l of Cr at 6K − 300K derived from the sum rules. The dashed line is a guide to the eye. (C) A schematic diagram of the model used to estimate the proximity length, showing the Cr distribution $\rho(x)$ and the ferromagnetically ordered Cr distribution $\delta(x)$ at given temperature [67].

In addition, $Sb_{2-x}V_xTe_3$/EuS heterostructure has demonstrated an enhancement of magnetism in Sb_2Te_3 which includes both proximity exchange coupling and magnetic doping, as shown in Fig. 8.17 [68]. But here, since the magnetism direction of EuS is

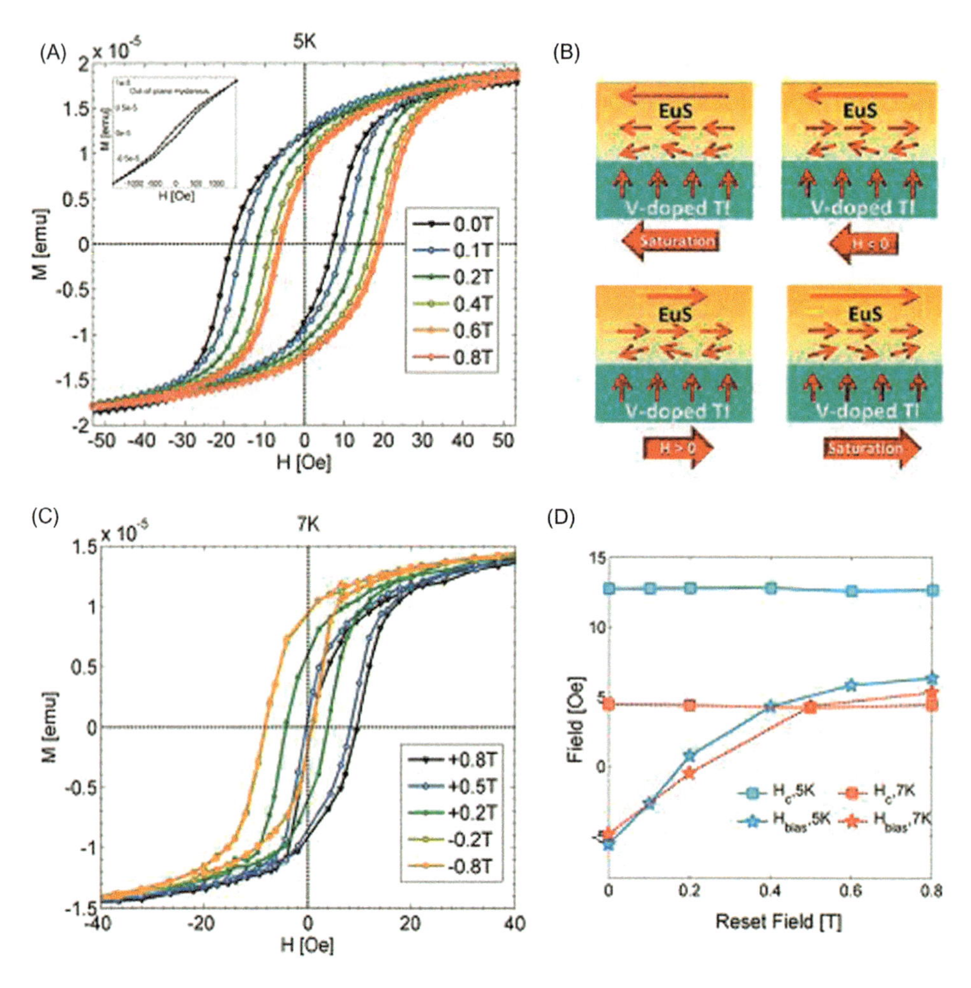

Figure 8.17

Magnetic measurements of a 2-nm EuS/10 QL $Sb_{1.9}V_{0.1}Te_3$ heterostructure using a superconducting quantum interference devices magnetometer. (A), (C) In-plane hysteresis at 5K (A) and 7K (C), respectively, showing a negative EB following a set field of −1 T which can be switched to positive bias by applying a positive resetting field. Inset of (A) is the out-of-plane magnetic hysteresis of the same sample, showing a finite remnant moment. (B) Schematic interfacial magnetic structure, where the interfacial EuS moments (*horizontal arrows*, the *long arrow* on the top of EuS means a saturated magnetization) are pinned by the exchange-coupled moments in the presence of a V-doped TI (*vertical arrows*). (D) EB and coercive field as a function of the in-plane resetting field at 5K and 7K, respectively [68]. *TI*, topological insulator; *QL*, quintuple layer.

in-plane, perpendicular to that of the V-doped Sb_2Te_3, the enhancement of T_C is moderate. The observed resetting field of a few Gausses below the T_C of EuS suggests that there is interplay between EuS and V-doped Sb_2Te_3. And the interplay between the proximity effect and doping in a hybrid heterostructure provides insights into the engineering of magnetic ordering.

8.5 Spin-transfer torque/spin–orbital torque in topological insulators

Similar to electric currents being carried by moving charges (electrons or holes), the spin current occurs due to moving spins. The spin current is a small quantity of angular momentum of the carriers, and this angular momentum can be transferred to the magnetic moment and alter it, if it is large enough. This is called STT [69–73]. The origin of the STT is the absorption of the itinerant flow of angular momentum components normal to the magnetization direction, as shown in Fig. 8.18 [73]. STT has been found to be both present and important in all known magnetic materials, including TM ferromagnets, magnetic semiconductors, and oxide ferromagnets.

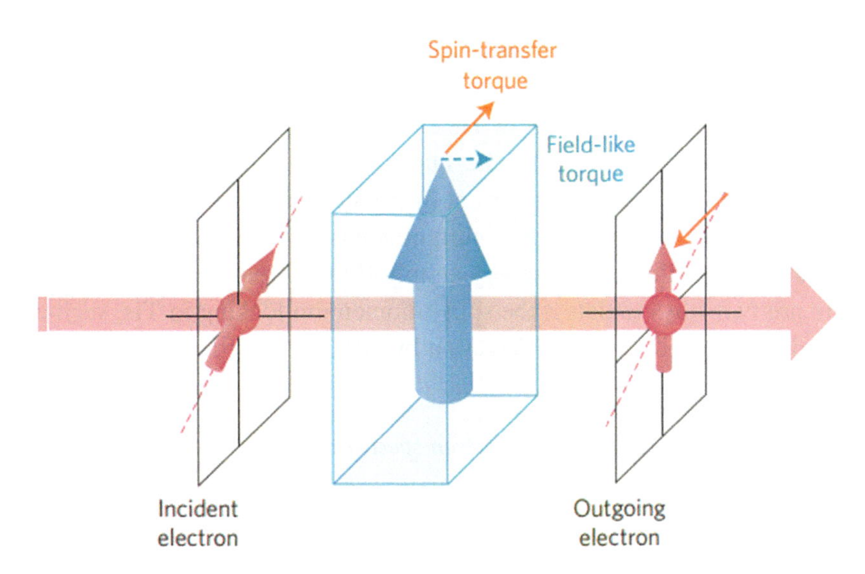

Figure 8.18
Illustration of current-induced torques. A spin-polarized current enters a ferromagnet. The interaction between the spin-polarized current and the magnetization causes a change in the spin direction of the outgoing electron compared with the incident electron. The difference in spin polarization causes torques on the ferromagnet. The torque in the plane of the incident and outgoing electron spin directions is called the STT and the torque perpendicular to that plane is called the field-like torque. The bold vertical arrow is the magnetization of the ferromagnetic layer [73]. *STT*, spin-transfer torque.

SOC is a relativistic interaction of a particle's spin with its motion inside a potential. A key example of this phenomenon is the SOC leading to shifts in an electron's atomic energy levels, due to electromagnetic interaction between the electron's magnetic dipole, its orbital motion, and the electrostatic field of the positively charged nucleus. The SOC is the one which causes SHE. Similar to Hall effect, SHE is a transport phenomenon consisting of spin accumulation on the lateral surfaces of a sample carrying electric current, without external magnetic field. The SOC acts as an internal effective magnetic field, which scatters electrons/holes with different spins into opposite directions.

STT-based magnetization witching has promoted the commercially viable magnetic random access memory (STT-MRAM) [74]. But the high current density required by STT effect has conflicted with energy-efficient applications. And it has been discovered that SOC in heavy-metal/ferromagnet heterostructures (HMFHs) can produce strong driven torques on an adjacent magnetic layer [75–79] via the SHE in the heavy metal or the Rashba–Edelstein effect in the ferromagnet [80–83]. In searching for materials to provide even more efficient spin–orbit torques [84–86], 3D TIs is considered as the prospect material for the STT and SOT due to its massless Dirac electrons have spins locked with their momenta [45,87–89].

8.5.1 Spin–momentum locking in topological insulators

Strong spin–orbit interaction and TRS in TIs enable the spin–momentum locking for the helical surface states. More importantly, the spin–momentum locking leads to a spin current induced simply by an electron current in the surface states. The surface states conduction is spin-polarized once electric current is passed through, and this spin polarization can be accordingly reversed by simply flipping the electric current direction. To date, there are two ways to detect the spin–momentum locking in TIs, such as photoelectron detection and electrical detection by ARPES and transport measurements, respectively.

8.5.1.1 Spin-resolved angle-resolved photoelectron spectroscopy

Spin-resolved ARPES was one of the most promising methods to study the spin-polarized surface states. It detects the energy and momentum of the emitted photoelectrons by exciting with ultraviolet photons from synchrotron light source or simply He lamps. Using the conservation of energy and momentum, the band structure of the materials can be directly mapped out. More importantly, it can also analyze the spin information of the photoelectrons, if a spin detector is installed [90–92].

Experimentally, Hsieh et al. measured the spin imaging and momentum resolved spectroscopy of $Bi_{2-x}Ca_xSe_3$. Here the doping with Ca atoms is to tune the Fermi level to below the conduction band minimum, since most Bi_2Se_3 are heavily electron doped due to a large quantity of Se vacancies. They have revealed a spin–momentum locked Dirac cone carrying a nontrivial Berry's phase that is nearly 100% spin-polarized. This exhibits a tunable topological fermion density in the vicinity of the Γ point and can be driven to the long-sought topological spin transport regime, as shown in Fig. 8.19 [93]. The observed topological nodal state is shown to be protected even up to 300K. And, this demonstration of RT topological order and nontrivial spin texture in Bi_2Se_3.

In addition, Hasan et al. measured the topological quantum numbers and invariants of a series of BiTI $(S_{1-x}Se_x)_2$ samples by ARPES. In this system, as more S replaces Se, the SOC decreases. Eventually, when it is weak enough, the samples change from the topological nontrivial state into the trivial state [94]. They have demonstrated the spin–momentum locking in the surface states, and the observed spin-polarized

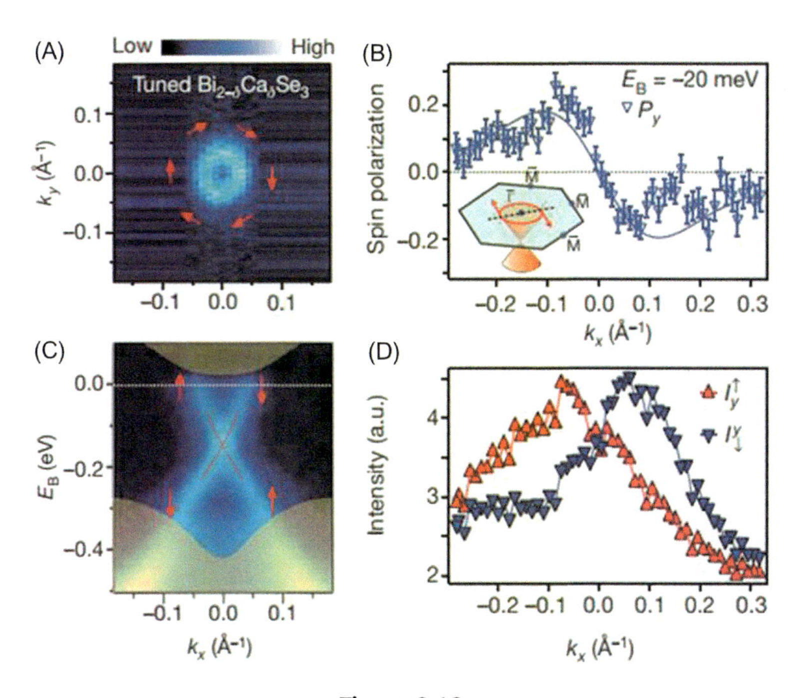

Figure 8.19

Detection of spin – momentum locking of spin-helical Dirac electrons in Bi_2Se_3 using spin-resolved ARPES. (A) ARPES intensity map at Fermi energy of the (111) surface of tuned stoichiometric $Bi_{2-x}Ca_xSe_3$. Red arrows denote the direction of spin projection around the Fermi surface. (C) ARPES dispersion of tuned $Bi_{2-x}Ca_xSe_3$ along the k_x cut. The dotted red lines are guides to the eye. The shaded regions in care our projections of the bulk bands of pure Bi_2Se_3 onto the (111) surface. (B) Measured *y*-component of spin-polarization along the Γ–M direction at $E_B = -20$ meV, which only cuts through the surface states. Inset, schematic of the cut direction. (D) Spin-resolved spectra obtained from the *y*-component spin polarization data [93]. *ARPES*, angle-resolved photoelectron spectroscopy.

edge/surface channel is crucial for future applications in spintronic and computing technologies possibly at RT.

8.5.1.2 Electrical detection

Another way to detect the spin–momentum locking is to detect the spin current on the surface. Using a Fe/Al_2O_3 tunneling contact, Jonker et al. have successfully detected a potential change on the contact as the magnetization direction changes by external magnetic field, as shown in Fig. 8.20 [95]. This voltage is proportional to the projection of the spin polarization onto the contact magnetization, which is determined by the direction and magnitude of the charge current. And its sign is that expected from spin–momentum locking rather than Rashba effects. More importantly, this voltage change scales inversely with Bi_2Se_3 film thickness, suggesting that it is associated with the surface conduction. Similar data are obtained from two different ferromagnetic contacts, demonstrating that these behaviors are independent of the details of the ferromagnetic contacts. These results

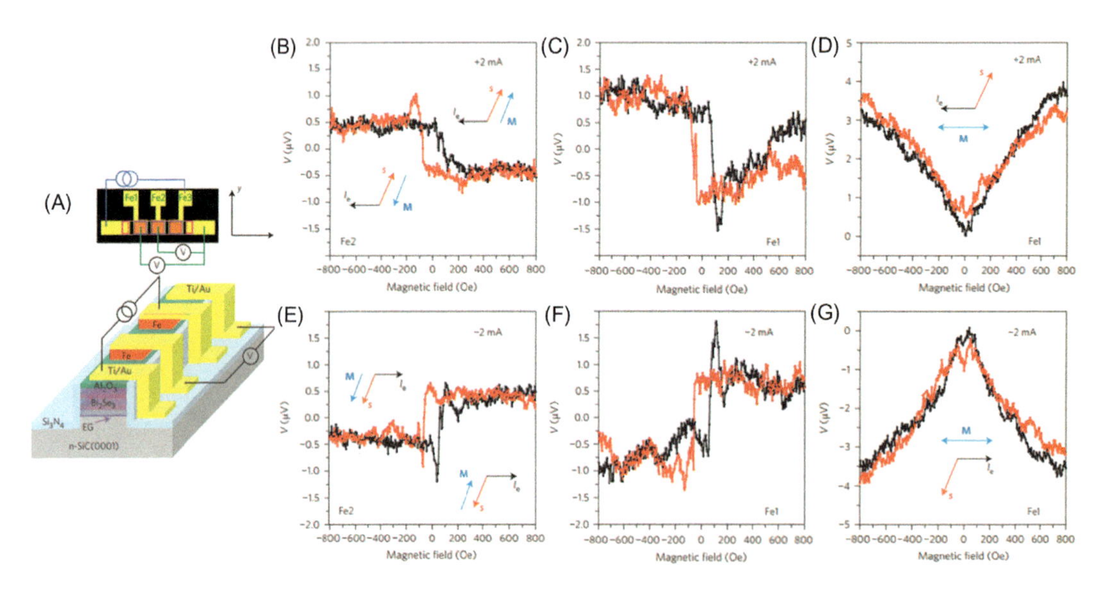

Figure 8.20

TI spin polarization detected as a voltage with Fe/Al_2O_3 contacts. The contact layout of (A) is used and the Bi_2Se_3 film is 45 nm thick. (A) and (E) Field dependence of the voltage measured using contact Fe2 as the detector with the magnetization collinear with the induced TI spin for bias currents of +2 mA and −2 mA, respectively. (C) and (F) Similar data for contact Fe1 as the detector. (D) and (G) Field dependence of the voltage measured using contact Fe1 as the detector with the magnetization orthogonal to the induced TIs spin for bias currents of +2 mA and −2 mA, respectively. Black traces correspond to positive magnetic field sweeps (from negative to positive values) and red traces correspond to negative field sweeps. The relative orientations of I_e, induced TIs spin polarizations and detector magnetization M is shown in the insets. All data were obtained at $T = 8$K [95]. *TI*, topological insulator.

demonstrate direct electrical access to the TIs' surface-state spin system and enable utilization of its remarkable properties for future applications.

Moreover, Tang et al. using a Co/Al_2O_3 ferromagnetic tunneling contact have successfully demonstrated the electrical detection of spin-polarized surface states conduction in the $(Bi_{0.53}Sb_{0.47})_2Te_3$ film [96] (Fig. 8.21). By changing the directions of both the magnetic field and the electric current, reversible voltage (resistance) hysteresis was observed up to 10K, in which the high-resistance state (HRS) and low-resistance state (LRS) were obtained from the relative orientation between the Co magnetization and the spin polarization of the topological surface states. And, these electrical transport results show direct evidence of

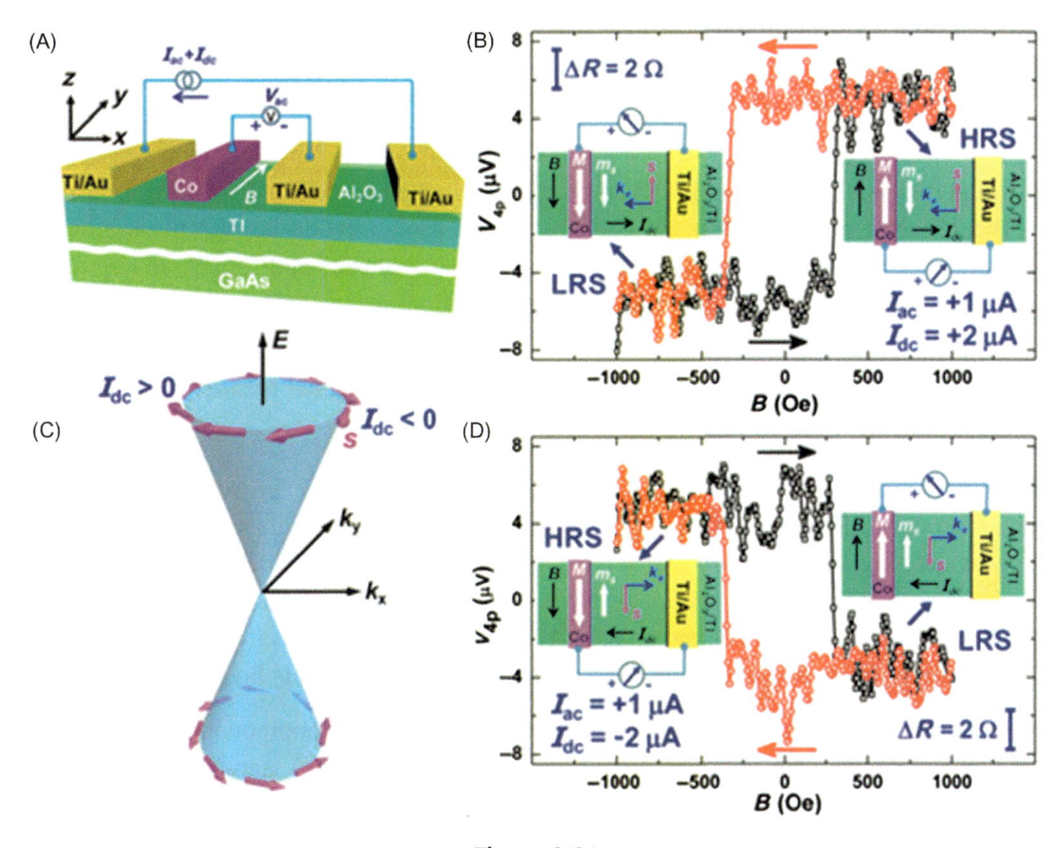

Figure 8.21

Electrical detection of the spin-polarized surface states conduction in $(Bi_{0.53}Sb_{0.47})_2Te_3$. (A) Schematic illustration of the device structure with one ferromagnetic tunneling Co/Al_2O_3 contact for spin detection. (C) Schematic illustration of the helical spin texture of the surface states in TI: clockwise spin texture (left-handed chirality) above the Dirac point while counterclockwise spin texture (right-handed chirality) below the Dirac point. (B) and (D) The measured voltage (resistance) at $T=1.9$K as the in-plane magnetic field is swept back and forth 1.9K as the in-plane magnetic field is swept back and forth [96]. *TI*, topological insulator.

spin–momentum locking feature of the topological surface states enabled by the strong spin–orbit interaction and the TRS in TIs. However, the short mean-free path and phase coherence length in TIs and the terrace-like TIs surface morphology limit the spin detection efficiency (1.02%) for the surface states conduction in $(Bi_{0.53}Sb_{0.47})_2Te_3$. This value can be further enhanced by tuning the Fermi level and increasing the surface states conduction ratio.

In addition, quaternary TIs with insulating bulk are highly desirable to achieve better spin-valve effect when integrated with ferromagnets [97]. They can not only overcome the metallic bulk conduction caused by native imperfections in Bi_2Se_3, but also maintaining the high surface mobility to ensure the surface-dominated conduction. Thus spin-valve transistors are fabricated based on $BiSbTeSe_2$ with enhanced surface mobility. And, the output of spin-valve transistors exhibits a dominant step-like behavior when sweeping the magnetic field to change the magnetization orientation of the $Ni_{21}Fe_{79}$ electrode.

Most importantly, the HRS and LRS can be switched when reversing the direction of the d.c. The TI-based spin-valve transistors enable the current direction-dependent switching of HRS and LRS, allowing for the applicability in magnetic sensors and spin-logic circuits, and showing the potential use of TIs as innovative current-driven spin generators.

8.5.2 Theoretically predicted spin-transfer torque/spin–orbital torque in topological insulators

The surface spin current due to SOC will exert torques on adjacent ferromagnetic layer, which is so-called STT or SOT effect. Theoretically, it has been predicted that the STT acting on the magnetization of a free ferromagnetic (F) layer can be found within N/TIs/F vertical heterostructures. This originates from strong SOC on the surface of a 3DTI and from charge current becoming spin polarized in the direction of transport as it flows perpendicularly from the normal metal (N) across the bulk of the TIs layer [98] (Fig. 8.22).

Moreover, the STT vector has both in-plane and perpendicular components that are comparable in magnitude to conventional torque in F/I/F (where I stands for insulator) magnetic tunnel junctions (MTJs), while not requiring additional spin polarizing F layer with fixed magnetization. Meanwhile, such heterostructures could exploit strong interfacial SOC without requiring a perfectly insulating bulk whose unintentional doping in the present experiments obscures the topological properties anticipated for lateral transport along the TIs surface, which makes it advantageous for spintronics applications [99–101].

Meanwhile, using the nonequilibrium Green's function (NEGF) method, the current-induced STT in graphene-based graphene/TIs/F heterostructures have also been investigated [102].

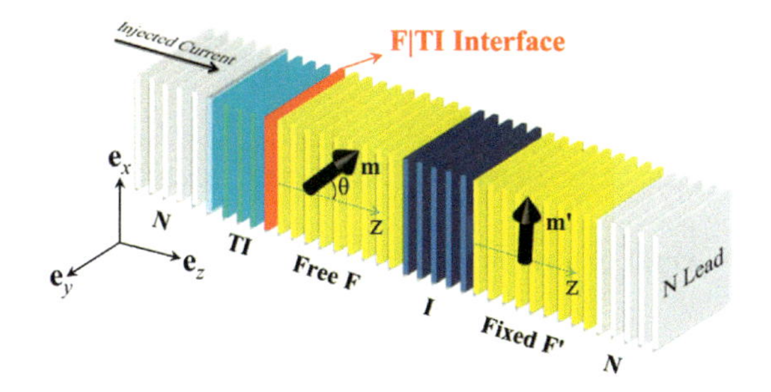

Figure 8.22

The proposed device for the detection of unconventional STT driven by perpendicular charge transport through TI slab consists of the free and reference (with fixed magnetization) ferromagnetic layers forming a conventional MTJ, where the free F layer is additionally capped by a thin layer of the TIs material. Both the free and reference ferromagnetic (F) layers are initially magnetized along the *x*-axis, while the unpolarized injected charge current owing perpendicularly through the TI layer will become spin-polarized along the *z*-axis. Such spin-polarized current, together with additional torque driven by the surface SOC, will switch the magnetization of the F layer which is then detected as the change in the tunneling magnetoresistance of F/I/F′MTJ [98]. *MTJ*, magnetic tunnel junction; *SOC*, spin–orbital coupling; *STT*, spin-transfer torque; *TI*, topological insulator.

It is found that the charge current becomes spin polarized, when it flows from the left leads (N) across the quantum spin Hall insulator layer (TI), and induces STT on the magnetization of the Ferromagnetic region, as shown in Fig. 8.23. The STT per unit of the bias voltage can reach $0.8e/4\pi$ which is comparable in magnitude to the conventional F/N/F junctions. The features of the STT can be explained through the analysis of the band structure of the right ferromagnetic graphene electrode. It is found that the STT obtained in this nanostructure is immune to the changes in the geometry of the quantum spin Hall insulator. That means it stays constant while introduces an asymmetric square notch or random nanopores in the quantum spin Hall insulator. This offers an efficient method to change the magnetization direction of magnetic materials based on TIs.

8.5.3 Spin–orbital torque in topological insulators

Soon, after the experimental observation of SOT on the interface of heavy metal/Py heterostructure, magnetism procession due to the SOT of TIs has also been found in TI/Py heterostructure. Experimentally it has been demonstrated that the charge current flowing in-plane in a thin Bi_2Se_3 film at RT can exert a very strong torque on the adjacent ferromagnetic permalloy ($Ni_{81}Fe_{19}$) thin film (Fig. 8.24A), with a direction consistent with that expected from the topological surface state [103].

Moreover, spin torque ferromagnetic resonance technique has been used to evaluate the strength of current-induced torque, as shown in Fig. 8.24B and D. The effective spin current

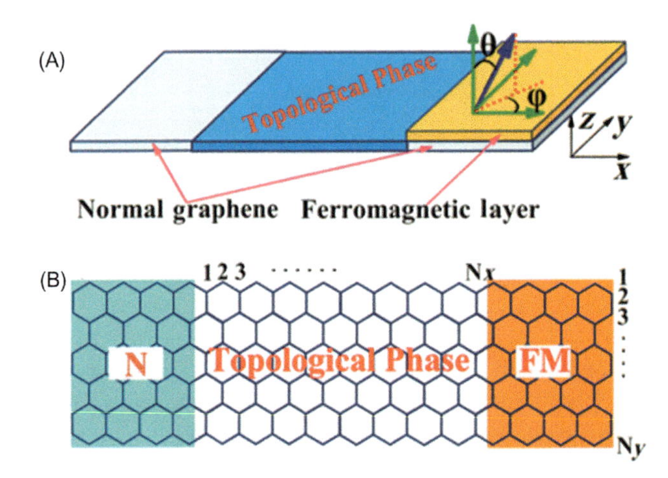

Figure 8.23
(A) Schematic illustration of the N/TI/F heterostructures: the graphene-based TIs is connected to a normal graphene electrode and a ferromagnetic electrode. The magnetization direction (denoted by *blue arrow*) of the ferromagnetic electrode is assumed to be tunable. (B) Schematic view of the graphene nanoribbon device with graphene-based TIs connected with two semiinfinite leads. Te size of the scattering region is determined by the ribbon length N_x and ribbon width N_y [102]. *TI*, topological insulator.

conductivities ($\sigma_{s,\parallel}$) and spin torque ratio ($\theta_{s,\parallel}$) of TIs are listed in Table 8.1. It is found that the strength of the torque per unit charge current density in Bi_2Se_3 is greater than for any source of STT measured so far, even for nonideal TIs films in which the surface states coexist with bulk conduction. The spin Hall angle is larger than 1. This suggests that the SOT exerted by TI is enormous larger than that from normal heavy metals, whose spin Hall angle is normally smaller than 1. The reason is that TIs possess a spin–momentum locked surface state, in which an electron can transfer its spin–momentum multiple times to the adjacent ferromagnetic layers.

This implies that TIs could enable very efficient electrical manipulation of magnetic materials at RT, for memory and logic applications.

Indeed, Fan et al. have experimentally demonstrated the magnetization of a ferromagnetic layer can be switched by TIs with an electric current passing through, due to the giant spin–orbital torque, in a $(Bi_{0.5}Sb_{0.5})_2Te_3/(Cr_{0.08}Bi_{0.54}Sb_{0.38})_2Te_3$ heterostructure [106]. The effective magnetization switching induced by an in-plane current was tested by AHE resistance, as shown in Fig. 8.25. By increasing the current in TI layer or switching the external magnetic field, the AHE resistance can change sign. This is a piece of solid evidence for the current-induced magnetization switch. Most importantly, it finds that the effective field to current ratio, as well as the spin Hall angle tangent, is nearly two orders of magnitude larger than those reported in HMFHs so far. This can also be seen as the critical current density

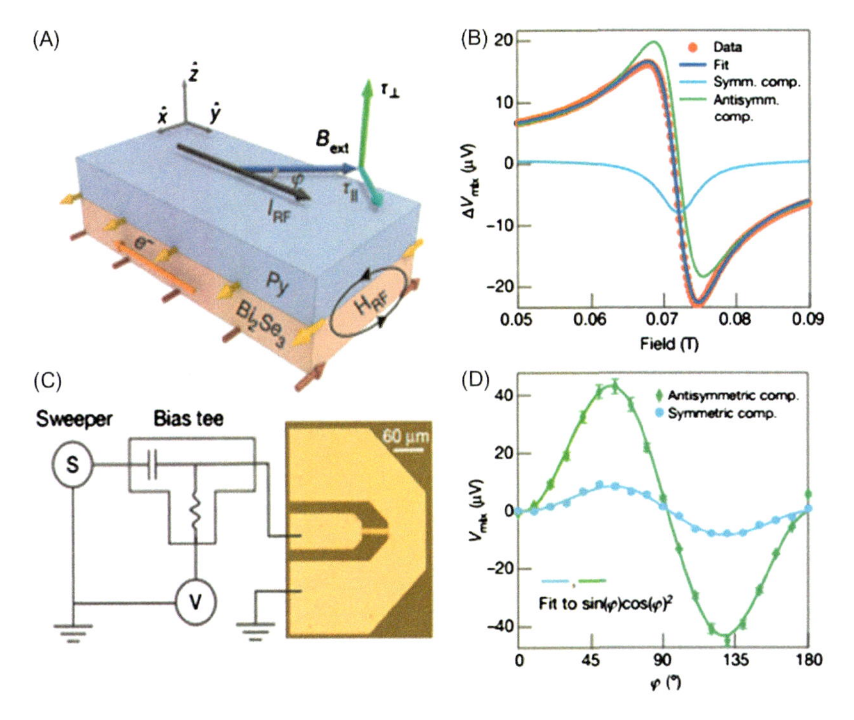

Figure 8.24

The sample geometry and ST-FMR measurement used in the measurement. (A) Schematic diagram of the layer structure and coordinate system. The yellow and red arrows denote spin–momentum directions. (B) Measured ST-FMR resonance at room temperature with microwave frequency 8 GHz for an 8-nm Bi_2Se_3/16-nm permalloy sample. (C) Depiction of the circuit used for the ST-FMR measurement and the sample contact geometry. (D) Measured dependence on the magnetic field angle φ for the symmetric and antisymmetric resonance components for a different sample (8 nm Bi_2Se_3/16 nm permalloy device) [103]. *ST-FMR*, spin torque ferromagnetic resonance.

Table 8.1: Comparison of RT effective spin current conductivities $\sigma_{s,\parallel}$ and spin torque ratio $\theta_{s,\parallel}$ for Bi_2Se_3 with other materials. $\theta_{s,\parallel}$ is dimensionless and the units for $\sigma_{s,\parallel}$ are $10^5\hbar/2e\Omega^{-1}m^{-1}$

Parameter	Bi_2Se_3 [103]	Pt [67]	β-Ta [99]	Cu(Bi) [104]	β-W [105]
$\theta_{s,\parallel}$	2.0–3.5	0.08	0.15	0.24	0.3
$\sigma_{s,\parallel}$	1.1–2.0	3.4	0.8	—	1.8

required for switching is below 8.9×10^4 A cm^{-2} at 1.9K, while the current density in normal STT-MRAM is in the order of 1×10^7 A cm^{-2}. Thus TIs are good SOT source materials.

This giant SOT, together with the current-induced switching behavior, suggests that TIs could potentially be the materials/structures to generate STTs with efficiency beyond

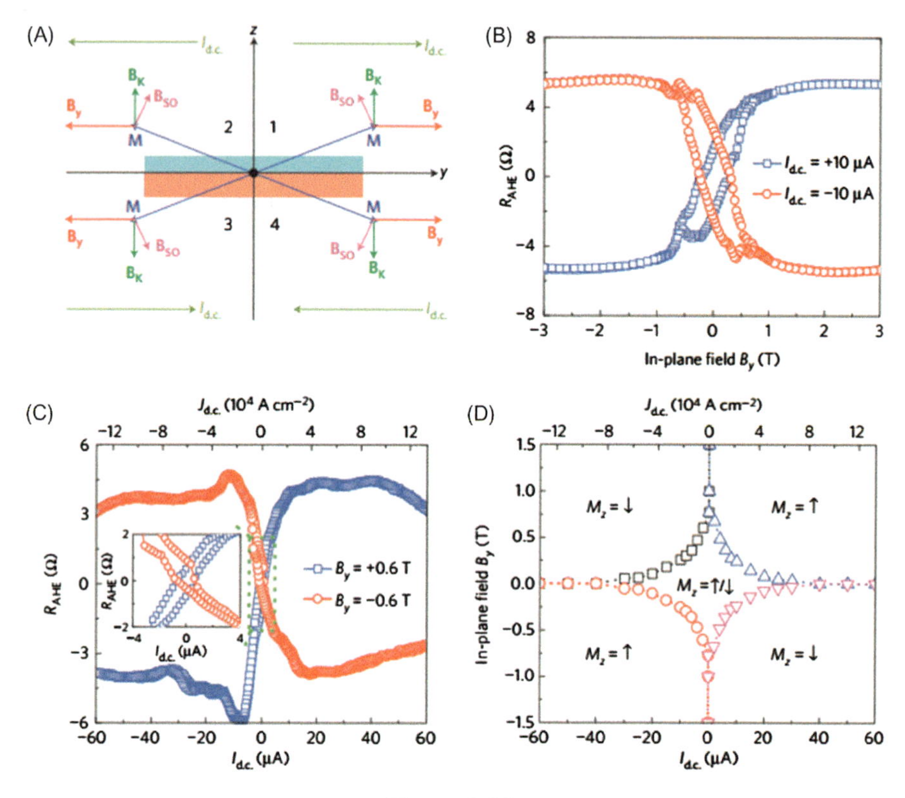

Figure 8.25

Magnetization switching due to the SOT induced by an in-plane DC. (A) Schematic of the four stable magnetization states (panels 1–4) when passing a large d.c., $I_{d.c.}$, and applying an in-plane external magnetic field, B_y, in the $\pm y$-directions. The effective spin–orbit field B_{SO} induced by the d.c. and the anisotropy field B_K are both considered. (B) The AHE resistance R_{AHE} as a function of the in-plane external magnetic field when passing a constant d.c. with $I_{d.c.} = +10\ \mu A$ and $I_{DC} = -10\ \mu A$ along the Hall bar, respectively, at 1.9K. (C) Current-induced magnetization switching in the Hall bar device at 1.9K in the presence of a constant in-plane external magnetic field with $B_y = +0.6$ T and $B_y = -0.6$ T, respectively. Inset: Expanded scale to show the hysteresis windows. (D), Phase diagram of the magnetization state in the presence of an in-plane external magnetic field B_y and a $I_{d.c.}$. The dashed lines and symbols (obtained from experiments) represent switching boundaries between the different states [106]. *AHE*, anomalous Hall effect; *SOT*, spin–orbital torque.

today's HMFHs. The use of the giant SOT revealed in TI/Cr-doped TIs bilayer for RT applications will probably require a search for high-T_C magnetic TIs or other alternative RT insulating (or high resistivity) magnetic materials that can couple efficiently with TIs. Therefore this finding in the TI/Cr-doped TIs heterostructure may spur further work on heterostructures that integrate TIs and magnetic materials, and may potentially lead to the innovation of novel SOT devices.

8.5.4 Magnetic random access memory based on topological insulators

Further, the transport properties of TI-based MRAM devices have been simulated by the quantum transport model [107]. The model captures the effects of spin–momentum locking, Klein tunneling, and coupled spin dynamics. In the model, the designed STT device consists of a thin-layer TI coupled to the top ferromagnetic layer, as shown in Fig. 8.26.

The channel length is $L = 20$ nm. The electrostatic potential of the TI thin film is controlled by a back gate through a high-k insulator such as $SrTiO_3$ (STO) [108]. In recent experiments, thin films of TIs have been grown on STO whose dielectric constant is over 100. The top surface is in close coupling with a 1-nm ferromagnetic thin film. As shown in Fig. 8.26B, the coupling of electrons in TI surface channel to top magnetic layer is a spin flipping process described by a self-energy term modeled with the self-consistent Born approximation (SCBA) [109,110] and the strength of exchange interaction is obtained from Ref [111]. Spin transport in the TI surface state is modeled by the NEGF formalism, which is coupled to the Landau–Lifshitz–Gilbert equation that describes magnetic dynamics of the soft ferromagnetic layer, as shown in Fig. 8.26C.

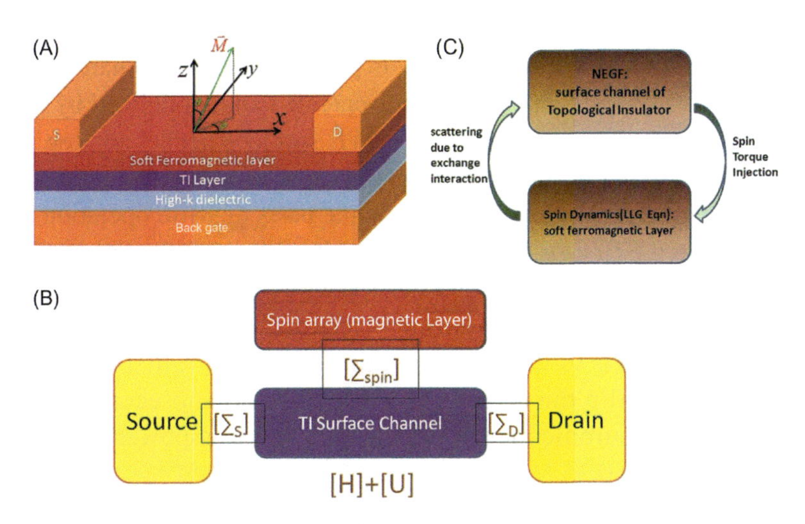

Figure 8.26

(A) Schematic structure of a TI-based spintronic device. Nonmagnetic metal is used as the source/drain contact. The magnetization of the ferromagnetic layer is described by angle θ and φ in spherical coordinates. The easy axis is chosen as $+y$-direction. (B) Quantum transport model of the device: The channel is described by the Hamiltonian H and electrostatic potential U, the interaction of the channel with the source, drain, and top magnetic layer is represented by corresponding self-energy terms. (C) Simulation method. The LLG equation and NEGF are used to describe the spin dynamics of the magnetic layer and electron transport in the TI layer. They are coupled through exchange interaction of spin [107]. *LLG*, Landau–Lifshitz–Gilbert; *NEGF*, nonequilibrium Green's function; *TI*, topological insulator.

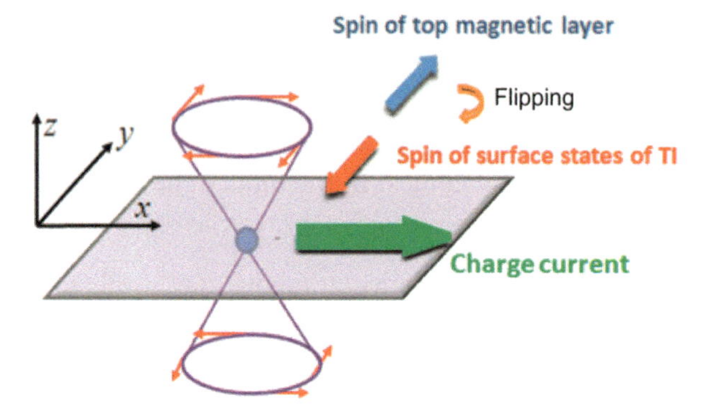

Figure 8.27
Switching mechanism of the TIs-based STT device. The surface state of TI is helical Dirac fermions; its spin polarization is perpendicular to the direction of the charge current, due to the nature of spin–momentum locking in TIs. The STT is injected into the top magnetic layer due to exchange interaction, resulting in the flipping of magnetization [107]. *STT*, spin-transfer torque; *TI*, topological insulator.

The switching mechanism of the TI-based STT device is schematically shown in Fig. 8.27. Because the surface state of TIs is Dirac fermions, its spin polarization is perpendicular to the direction of momentum k. The spin density is proportional to the charge current, and the spin polarization of TIs' surface channel can be generated and modulated by a charge current. Due to the exchange interaction between TIs surface channel and top magnetic layer, an STT will be exerted into the top magnetic layer, results in the possibility of flipping its magnetization. This mechanism forms the basis of write operation in the design of the proposed TI-based STT-MRAM. And the results show that the current at the surface of TIs is intrinsically spin polarized. Through spin scattering processes, the magnetization of the magnetic layer can be read by measuring the charge current in the TIs. The spin torque injection from the TIs to the magnetic thin film is controlled by the charge current and is efficient to flip the magnetization in several nanoseconds. The simulation results of the TI-based spintronic memory device suggest low-power operation with a power supply voltage of $\sim$0.35 V.

On the other hand, Liu et al. proposed another memory device called SOT-MRAM. The MTJ cell is the core part of a bit-cell in SOT-MRAM as in STT-MRAM [112]. However, to eliminate the shortcomings of STT-MRAM, its cell has an additional terminal to separate the (unidirectional) read and the (bidirectional) write path, which are perpendicular to each other [65]. The terminals comprise a bit line, a write line, a source line, and a word line, as shown in Fig. 8.28B. The word line is used to access the required cell during the read operation, while in the write operation, the current flows between the source line and the write line. The direction of the current affects the magnetization of the free layer and hence, the value stored in the bit-cell. If the current flows from the source line to the write line,

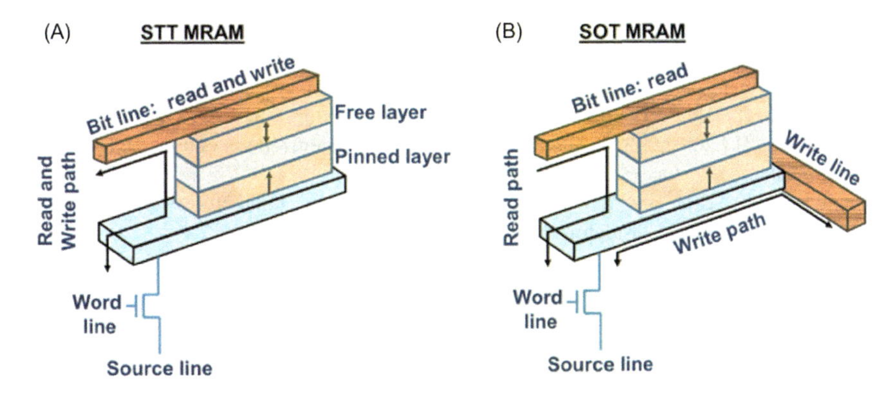

Figure 8.28

Illustration of the path of write and read currents in (A) STT-MRAM and (B) SOT-MRAM. SOT-MRAM uses separate paths for reading and writing information, as shown by the arrows [73]. *MRAM*, magnetic random access memory; *SOT*, spin–orbital torque; *STT*, spin-transfer torque.

the MTJ resistance will be low. To achieve a high MTJ resistance, the current needs to flow from write line to source line (high potential for the write line) [113].

This new designed SOT-MRAM could potentially have low-power consumption and faster switching. The use of three-terminal MTJ-based on spin–orbit torque is also beneficial in isolating the read and the write line, which could dramatically increase the lifetime of the insulator layer in the MTJ tri-layers, since no large writing current will flow through it. Because of all the advantages of SOT-MRAM, it can potentially be a prominent memory technique in the near future.

8.6 Summary and outlook

The study of 3D TIs of V–VI compound chalcogenide semiconductors is one of the most exciting new fields to explore new physics and applications. In this chapter, we have outlined that MBE technology can grow high-quality films by passivating conventional semiconductor substrates and using VDW epitaxial growth mode, which greatly reduces stress in the film. Second, we introduced two ways to break the TRS, so that some novel physical phenomena can be observed in magnetic TIs by means of magnetic transport, electrical transport, ARPES, and so on. Finally, we have discussed in details the various properties of the TIs with the emphases on the magnetic proximity effect, STT effects, SOT effects, and their potential applications in SOT-MRAM.

References

[1] W.Q. Zou, W. Wang, X.F. Kou, M.R. Lang, Y.B. Fan, E.S. Choi, et al., Observation of quantum Hall effect in an ultra-thin $(Bi_{0.53}Sb_{0.47})_2Te_3$ film, Appl. Phys. Lett. 110 (2017) 212401.

[2] M. König, S. Wiedmann, C. Brüne, A. Roth, H. Buhmann, L.W. Molenkamp, et al., Quantum spin Hall insulator state in HgTe quantum wells, Science 318 (2007) 766–770.

[3] C.Z. Chang, J.S. Zhang, X. Feng, J. Shen, Z.C. Zhang, M.H. Guo, et al., Experimental observation of the quantum anomalous Hall effect in a magnetic topological insulator, Science 340 (2013) 167–170.

[4] R. Yu, W. Zhang, H.J. Zhang, S.C. Zhang, X. Dai, Z. Fang, Quantized anomalous Hall effect in magnetic topological insulators, Science 329 (2010) 61–64.

[5] H.B. Zhang, H.L. Yu, D.H. Bao, S.W. Li, C.X. Wang, G.W. Yang, Magnetoresistance switch effect of a Sn-doped Bi_2Te_3 topological insulator, Adv. Mater. 24 (2012) 132–136.

[6] S.Y. Xu, M. Neupane, C. Liu, D. Zhang, A. Richardella, L.A. Wray, et al., Hedgehog spin texture and Berry's phase tuning in a magnetic topological insulator, Nat. Phys. 8 (2012) 616–622.

[7] W.K. Tse, A.H. MacDonald, Giant magneto-optical Kerr effect and universal faraday effect in thin-film topological insulators, Phys. Rev. Lett. 105 (2010) 057401.

[8] W.Q. Liu, Y.B. Xu, L. He, G. van der Laan, R. Zhang, K.L. Wang, Experimental observation of dual magnetic states in topological insulators, Sci. Adv. 5 (2019) 2088.

[9] A. Sakai, H. Kohno, Spin torques and charge transport on the surface of topological insulator, Phys. Rev. B 89 (2014) 165307.

[10] M.H. Fischer, A. Vaezi, A. Manchon, E.A. Kim, Spin-torque generation in topological insulator based heterostructures, Phys. Rev. B 93 (2016) 125303.

[11] L. He, X.F. Kou, K.L. Wang, Review of 3D topological insulator thin-film growth by molecular beam epitaxy and potential applications, Phys. Status Solidi-R. 1-2 (2013) 50–63.

[12] C.L. Tan, X.H. Cao, X.J. Wu, Q.Y. He, J. Yang, X. Zhang, et al., Recent advances in ultrathin two-dimensional nanomaterials, Chem. Rev. 117 (2017) 6225–6331.

[13] A. Kuc, T. Heine, The electronic structure calculations of two-dimensional transition-metal dichalcogenides in the presence of external electric and magnetic fields, Chem. Soc. Rev. 44 (2015) 2603.

[14] M. Chhowallal, H.S. Shin, G. Eda, L.J. Li, K.P. Loh, H. Zhang, The chemistry of two-dimensional layered transition metal dichalcogenide nanosheets, Nat. Chem. 5 (2013) 263–275.

[15] Y.F. Zhao, W. Wang, C. Li, L. He, First-principles study of nonmetal doped monolayer $MoSe_2$ for tunable electronic and photocatalytic properties, Sci. Rep. 7 (2017) 17088.

[16] F.F. Xiu, L. He, Y. Wang, L.N. Cheng, L.T. Chang, M.R. Lang, et al., Manipulating surface states in topological insulator nanoribbons, Nat. Nanotechnol. 6 (2011) 216–221.

[17] Y. Wang, F.F. Xiu, L.N. Cheng, L. He, M.R. Lang, J.S. Tang, et al., Gate-controlled surface conduction in Na-doped Bi_2Te_3 topological insulator nanoplates, Nano. Lett. 12 (3) (2012) 1170–1175.

[18] D.S. Kong, J.C. Randel, H.L. Peng, J.J. Cha, S. Meister, K.J. Lai, et al., Topological insulator nanowires and nanoribbons, Nano Lett. 10 (1) (2010) 329–333.

[19] D.S. Kong, Y.L. Chen, J.J. Cha, Q.F. Zhang, J.G. Analytis, K.J. Lai, et al., Ambipolar field effect in the ternary topological insulator $(Bi_xSb_{1-x})_2Te_3$ by composition tuning, Nat. Nanotechnol. 6 (11) (2011) 705.

[20] H.L. Peng, K.J. Lai, D.S. Kong, S. Meister, Y.L. Chen, X.L. Qi, et al., Aharonov-Bohm interference in topological insulator nanoribbons, Nat. Mater. 9 (2010) 225–229.

[21] K. Eto, Z. Ren, A.A. Taskin, K. Segawa, Y. Ando, Angular-dependent oscillations of the magnetoresistance in Bi_2Se_3 due to the three-dimensional bulk Fermi surface, Phys. Rev. B 81 (2010) 195309.

[22] Z. Ren, A.A. Taskin, S. Sasaki, K. Segawa, Y. Ando, Large bulk resistivity and surface quantum oscillations in the topological insulator Bi_2Te_2Se, Phys. Rev. B 82 (2010) 241306.

[23] Z. Ren, A.A. Taskin, S. Sasaki, K. Segawa, Y. Ando, Observations of two-dimensional quantum oscillations and ambipolar transport in the topological insulator Bi_2Se_3 achieved by Cd doping, Phys. Rev. B 84 (2011) 075316.

[24] G.H. Zhang, H.J. Qin, J. Teng, J.D. Guo, Q.L. Guo, X. Dai, et al., Quintuple-layer epitaxy of thin films of topological insulator Bi_2Se_3, Appl. Phys. Lett. 95 (2009) 053114.

[25] L. He, F.X. Xiu, Y. Wang, A.V. Fedorov, G. Huang, X.F. Kou, et al., Epitaxial growth of Bi_2Se_3 topological insulator thin films on Si (111), J. Appl. Phys. 109 (2011) 103702.

[26] M.R. Lang, L. He, F.X. Xiu, X.X. Yu, J.S. Tang, Y. Wang, et al., Revelation of topological surface states in Bi_2Se_3 thin films by in situ Al passivation, ACS Nano. 6 (1) (2012) 295–302.
[27] X.F. Kou, L. He, F.X. Xiu, M.R. Lang, Z.M. Liao, Y. Wang, et al., Epitaxial growth of high mobility Bi_2Se_3 thin films on CdS, Appl. Phys. Lett. 98 (2011) 242102.
[28] Y. Zhang, K. He, C.Z. Chang, C.L. Song, L.L. Wang, X. Chen, et al., Crossover of the three-dimensional topological insulator Bi_2Se_3 to the two-dimensional limit, Nat. Phys. 6 (2010) 584–588.
[29] L. He, F.X. Xiu, X.X. Yu, M. Teague, W.J. Jiang, Y.B. Fan, et al., Surface dominated conduction in a 6 nm thick Bi_2Se_3 thin film, Nano. Lett. 12 (2012) 1486–1490.
[30] X.F. Kou, L. He, F.X. Xiu, M.R. Lang, Z.M. Liao, Y. Wang, et al., Epitaxial growth of high mobility Bi_2Se_3 thin films on CdS, Appl. Phys. Lett. 98 (2011) 242102.
[31] A.A. Taskin, S. Sasaki, K. Segawa, Y. Ando, Manifestation of topological protection in transport properties of epitaxial Bi_2Se_3 thin films, Phys. Rev. Lett. 109 (2012) 066803.
[32] S. Borisova, J. Krumrain, M. Luysberg, G. Mussler, D. Grützmacher, Mode of growth of ultrathin topological insulator Bi_2Te_3 films on Si (111) substrates, Cryst. Growth Des. 12 (2012) 6098.
[33] Z.Y. Wang, H.D. Li, X. Guo, W.K. Ho, M.H. Xie, Growth characteristics of topological insulator Bi_2Se_3 films on different substrates, J. Cryst. Growth. 334 (2011) 96.
[34] N. Bansal, Y.S. Kim, E. Edrey, M. Brahlek, Y. Horibe, K. Iida, et al., Epitaxial growth of topological insulator Bi_2Se_3 film on Si(111) with atomically sharp interface, Thin. Solid. Films. 520 (2011) 224.
[35] R.D. Bringans, M.A. Olmstead, Bonding of Se and ZnSe to the Si(100) surface, Phys. Rev. B 39 (1989) 12985.
[36] K.H.M. Chen, H.Y. Lin, S.R. Yang, C.K. Cheng, X.Q. Zhang, C.M. Cheng, et al., Van der Waals epitaxy of topological insulator Bi_2Se_3on single layer transition metal dichalcogenide MoS_2, Appl. Phys. Lett. 111 (2017) 083106.
[37] W.Y. Ren, H.D. Li, L. Gao, Y. Li, Z.Y. Zhang, C.J. Long, et al., Epitaxial growth and thermal-conductivity limit of single-crystalline Bi_2Se_3/In_2Se_3 superlattices on mica, Nano. Res 10 (2017) 247–254.
[38] Y. Ni, Z. Zhang, C.I. Nlebedim, D.C. Jiles, Magnetotransport study of $(Sb_{1-x}Bi_x)_2Te_3$ thin films on mica substrate for ideal topological insulator, AIP Adv. 6 (2016) 055812.
[39] J.Y. Park, G.H. Lee, J. Jo, A.K. Cheng, H. Yoon, K.J. Watanabe, et al., Molecular beam epitaxial growth and electronic transport properties of high quality topological insulator Bi_2Se_3 thin films on hexagonal boron nitride, 2D Mater. 3 (2016) 035029.
[40] J.S. Dyck, P. Hájek, P. Lošťák, C. Uher, Diluted magnetic semiconductors based on $Sb_{2-x}V_xTe_3$ $(0.01 < \sim x < \sim 0.03)$, Phys. Rev. B 65 (2002) 115212.
[41] X.F. Kou, M.R. Long, Y.B. Fan, Y. Jiang, T.X. Nie, J.M. Zhang, et al., Interplay between different magnetisms in Cr-doped topological insulators, ACS Nano. 7 (2013) 9205–9212.
[42] P.P.J. Haazen, J.B. Laloe, T.J. Nummy, H.J.M. Swagten, P.J. Herrero, D. Heiman, et al., Ferromagnetism in thin-film Cr-doped topological insulator Bi_2Se_3, Appl. Phys. Lett. 100 (2012) 082404.
[43] Y.S. Hor, P. Roushan, H. Beidenkopf, J. Seo, D. Qu, J.G. Checkelsky, et al., Development of ferromagnetism in the doped topological insulator $Bi_{2-x}Mn_xTe_3$, Phys. Rev. B 81 (2010) 195203.
[44] V.A. Kulbachinskii, A.Y. Kaminskil, K. Kindo, Y. Narumi, K. Suga, P. Lostak, et al., Ferromagnetism in new diluted magnetic semiconductor $Bi_{2-x}FexTe_3$, Physica B 311 (2002) 292–297.
[45] J.S. Dyck, Č. Drasar, P. Lošt'ák, C. Uher, Low-temperature ferromagnetic properties of the diluted magnetic semiconductor $Sb_{2-x}Cr_xTe_3$, Phys. Rev. B 71 (2005) 115214.
[46] X.F. Kou, L. He, M.R. Lang, Y.B. Fan, K. Wong, Y. Jiang, et al., Manipulating surface-related ferromagnetism in modulation-doped topological insulators, Nano. Lett. 13 (2013) 4587–4593.
[47] X.F. Kou, W.J. Jiang, M.R. Lang, F.X. Xiu, L. He, Y. Wang, et al., Magnetically doped semiconducting topological insulators, J. Appl. Phys. 112 (2012) 063912–063916.
[48] Y.L. Chen, J.H. Chu, J.G. Analytis, Z.K. Liu, K. Igarashi, H.H. Kuo, et al., Massive Dirac fermion on the surface of a magnetically doped topological insulator, Science 329 (2010) 659.

[49] M.H. Liu, J.S. Zhang, C.Z. Chang, Z.C. Zhang, X. Feng, K. Li, et al., Crossover between weak antilocalization and weak localization in a magnetically doped topological insulator, Phys. Rev. Lett. 108 (2012) 036805.
[50] W.Q. Liu, D. West, L. He, Y.B. Xu, J. Liu, K.J. Wang, et al., Atomic-scale magnetism of Cr-doped Bi_2Se_3 thin film topological insulators, ACS Nano. 9 (2015) 10237–10243.
[51] M. Zhang, L.G. Liu, H. Yang, Anomalous second ferromagnetic phase transition in $Co_{0.08}Bi_{1.92}Se_3$ topological insulator, J. Alloy. Comp. 678 (2016) 463–467.
[52] C.Z. Chang, J.S. Zhang, X. Feng, J. Shen, Z.C. Zhang, M.H. Guo, et al., Experimental observation of the quantum anomalous Hall effect in a magnetic topological insulator, Science 340 (2013) 167–170.
[53] X.F. Kou, S.T. Guo, Y.B. Fan, L. Pan, M.R. Lang, Y. Jiang, et al., Scale- invariant quantum anomalous Hall effect in magnetic topological insulators beyond the two-dimensional Limit, Phys. Rev. Lett. 113 (2014) 137201.
[54] J.G. Checkelsky, J. Ye, Y. Onose, Y. Iwasa, Y. Tokura, Dirac-fermion-mediated ferromagnetism in a topological insulator, Nat. Phys. 8 (2012) 729–733.
[55] L.A. Wray, S.Y. Xu, Y.Q. Xia, D. Hsieh, A.V. Fedorov, Y.S. Hor, et al., A topological insulator surface under strong Coulomb, magnetic and disorder perturbations, Nat. Phys. 7 (2011) 32–37.
[56] J. Li, Z.Y. Wang, A. Tan, P.A. Glans, E. Arenholz, C. Hwang, et al., Magnetic dead layer at the interface between a Co film and the topological insulator Bi_2Se_3, Phys. Rev. B 86 (2012) 2–5.
[57] P.G.D. Gennes, Coupling between ferromagnets through a superconducting layer, Phys. Lett. 23 (1966) 10–11.
[58] J.S. Moodera, T.S. Santos, T. Nagahama, The phenomena of spin-filter tunnelling, J. Phys. Condens. Matter. 19 (2007) 165202.
[59] M.R. Lang, M. Montazeri, M.C. Onbasli, X.F. Kou, Y.B. Fan, P. Upadhyaya, et al., Proximity induced high-temperature magnetic order in topological insulator-ferrimagnetic insulator heterostructure, Nano Lett. 14 (2014) 3459–3465.
[60] P. Wei, F. Katmis, B.A. Assaf, H. Steinberg, P.J. Herrero, D. Heiman, et al., Exchange-coupling-induced symmetry breaking in topological insulators, Phys. Rev. Lett. 110 (2013) 186807.
[61] Y. Kubota, K. Murata, J. Miyawaki, K. Ozawa, M.C. Onbasli, T. Shirasawa, et al., Interface electronic structure at the topological insulator–ferrimagnetic insulator junction, J. Phys. Condens. Matter. 29 (2017) 055002.
[62] C.Z. Chang, P.Z. Tang, Y.L. Wang, X. Feng, K. Li, Z.C. Zhang, et al., Chemical-potential-dependent gap opening at the dirac surface states of Bi_2Se_3 induced by aggregated substitutional Cr atoms, Phys. Rev. Lett. 112 (2014) 056801.
[63] X.L. Qi, T. Hughes, S.C. Zhang, Topological field theory of time-reversal invariant insulators, Phys. Rev. B 78 (2008) 195424.
[64] J.G. Checkelsky, R. Yoshimi, A. Tsukazaki, K.S. Takahashi, Y. Kozuka, J. Falson, et al., Trajectory of the anomalous Hall effect towards the quantized state in a ferromagnetic topological insulator, Nat. Phys. 10 (2014) 731–736.
[65] M.D. Li, W.P. Cui, J. Yu, Z.Y. Dai, Z. Wang, F. Katmis, et al., Magnetic proximity effect and interlayer exchange coupling of ferromagnetic/ topological insulator/ferromagnetic trilayer, Phys. Rev. B 91 (2015) 014427.
[66] I. Vobornik, U. Manju, J. Fujii, F. Borgatti, P. Torelli, D. Krizmancic, et al., Magnetic proximity effect as a pathway to spintronic applications of topological insulators, Nano. Lett. 11 (2011) 4079–4082.
[67] W.Q. Liu, L. He, Y.B. Xu, K. Murata, M.C. Onbasli, M.R. Lang, et al., Enhancing magnetic ordering in Cr-Doped Bi_2Se_3 using high-T_C ferrimagnetic insulator, Nano Lett. 15 (2015) 764–769.
[68] M.D. Li, C.Z. Cui, B.J. Kirby, M.E. Jamer, W.P. Cui, L.J. Wu, et al., Proximity-driven enhanced magnetic order at ferromagnetic-insulator–magnetic-topological-insulator interface, Phys. Rev. Lett. 115 (2015) 087201.
[69] J.C. Slonczewski, Current-driven excitation of magnetic multilayers, J. Magn. Magn. Mater. 159 (1996) 1–7.

[70] L. Berger, Emission of spin waves by a magnetic multilayer traversed by a current, Phys. Rev. B 54 (1996) 9353.

[71] A. Brataas, A. D. Kent, H. Ohno, Current-induced torques in magnetic materials, Nat. Mater. 11 (2012) 372–381.

[72] D. Ralph, M. Stiles, Spin transfer torques, J. Magn. Magn. Mater. 320 (2008) 1190–1216.

[73] S. Bhatti, R. Sbiaa, A. Hirohata, H. Ohno, S. Fukami, S.N. Piramanayagam, Spintronics based random access memory: a review, Mater. Today. 20 (2017) 530.

[74] J. A. Katine, E. E. Fullerton, Device implications of spin-transfer torques, J. Magn. Magn. Mater. 320 (2008) 1217–1226.

[75] K. Ando, S. Takahashi, K. Harii, K. Sasage, J. Ieda, S. Maekawa, et al., Electric manipulation of spin relaxation using the spin Hall effect, Phys. Rev. Lett. 101 (2008) 036601.

[76] U.H. Pi, K.W. Kim, J.Y. Bae, S.C. Lee, Y.J. Cho, K.S. Kim, et al., Tilting of the spin orientation induced by Rashba effect in ferromagnetic metal layer, Appl. Phys. Lett. 97 (2010) 162507.

[77] L.Q. Liu, T. Moriyama, D.C. Ralph, R.A. Buhrman, Spin-torque ferromagnetic resonance induced by the spin Hall effect, Phys. Rev. Lett. 106 (2011) 036601.

[78] I.M. Miron, K. Garello, G. Gaudin, P.J. Zermatten, M.V. Costache, S. Auffret, et al., Perpendicular switching of a single ferromagnetic layer induced by in-plane current injection, Nature 476 (2011) 189–193.

[79] J. Kim, J. Sinha, M. Hayashi, M. Yamanouchi, S. Fukami, T. Suzuki, et al., Layer thickness dependence of the current-induced effective field vector in Ta|CoFeB|MgO, Nat. Mater. 12 (2013) 240–245.

[80] M.I. Dyakonov, V.I. Perel, Current-induced spin orientation of electrons in semiconductors, Phys. Lett. A 35 (1971) 459–460.

[81] J.E. Hirsch, Spin Hall effect, Phys. Rev. Lett. 83 (1999) 1834–1837.

[82] V.M. Edelstein, Spin polarization of conduction electrons induced by electric current in two-dimensional asymmetric electron systems, Solid. State. Commun. 73 (1990) 233–235.

[83] A. Chernyshov, M. Overby, X.Y. Liu, J.K. Furdyna, Y.L. Geller, L.P. Rokhinson, Evidence for reversible control of magnetization in a ferromagnetic material by means of spin–orbit magnetic field, Nat. Phys. 5 (2009) 656–659.

[84] A.A. Burkov, D.G. Hawthorn, Spin and charge transport on the surface of a topological insulator, Phys. Rev. Lett. 105 (2010) 066802.

[85] D. Culcer, E.H. Hwang, T.D. Stanescu, S. Das Sarma, Two-dimensional surface charge transport in topological insulators, Phys. Rev. B 82 (2010) 155457.

[86] D. Pesin, A.H.M. Donald, Spintronics and pseudospintronics in graphene and topological insulators, Nat. Mater. 11 (2012) 409–416.

[87] M. Z. Hasan, C. L. Kane, Colloquium: topological insulators, Rev. Mod. Phys. 82 (2010) 3045.

[88] I. Garate, M. Franz, Inverse spin-galvanic effect in the interface between a topological insulator and a ferromagnet, Phys. Rev. Lett. 104 (2010) 146802.

[89] T. Yokoyama, J. Zang, N. Nagaosa, Theoretical study of the dynamics of magnetization on the topological surface, Phys. Rev. B 81 (2010) 241410.

[90] P. Ayria, S. Tanaka, A.R.T. Nugraha, M.S. Dresselhaus, R. Saito, Phonon-assisted indirect transitions in angle-resolved photoemission spectra of graphite and graphene, Phys. Rev. B 94 (2016) 075429.

[91] Y. Ishida, T. Togashi, K. Yamamoto, M. Tanaka, T. Kiss, T. Otsu, et al., Time-resolved photoemission apparatus achieving sub-20-meV energy resolution and high stability, Rev. Sci. Instrum. 85 (2014) 123904.

[92] H. Ryu, I. Song, B. Kim, S. Cho, S. Soltani, T. Kim, et al., Photon energy dependent circular dichroism in angle-resolved photoemission from Au(111)surface states, Phys. Rev. B 95 (2017) 115144.

[93] D. Hsieh, Y. Xia, D. Qian, L. Wray, J.H. Dil, F. Meier, et al., A tunable topological insulator in the spin helical Dirac transport regime, Nature 460 (2009) 1101.

[94] S.Y. Xu, Y. Xia, L.A. Wray, S. Jia, F. Meier, J.H. Dil, et al., Topological phase transition and texture inversion in a tunable topological insulator, Science 332 (2011) 560.

[95] C.H. Li, O.M.J. Erve, J.T. Robinson, Y. Liu, L. Li, B.T. Jonker, Electrical detection of charge-current-induced spin polarization due to spin-momentum locking in Bi_2Se_3, Nat. Nanotechnol. 9 (2014) 218–224.

[96] J.S. Tang, L.T. Chang, X.F. Kou, K. Murata, E.S. Choi, M.R. Lang, et al., Electrical detection of spin-polarized surface states conduction in $(Bi_{0.53}Sb_{0.47})_2Te_3$ topological insulator, Nano Lett. 14 (2014) 5423–5429.

[97] M.H. Zhang, X.F. Wang, S. Zhang, Y. Gao, Z.H. Yu, X.Q. Zhang, et al., Unique Current-direction-dependent ON–OFFS witching in $BiSbTeSe_2$ topological insulator based spin valve transistors, IEEE. Electr. Device. L 37 (2016) 9.

[98] F. Mahfouzi, N. Nagaosa, B.K. Nikolic, Spin-orbit coupling induced spin-transfer torque and current polarization in topological-insulator/ferromagnet vertical heterostructures, Phys. Rev. Lett. 109 (2012) 166602.

[99] J.A. Hutasoit, T.D. Stanescu, Induced spin texture in semiconductor/topological insulator heterostructures, Phys. Rev. B 84 (2011) 085103.

[100] E.H. Zhao, C. Zhang, M. Lababidi, Mott scattering at the interface between a metal and a topological insulator, Phys. Rev. B 82 (2010) 205331.

[101] D. Kim, S. Cho, N.P. Butch, P. Syers, K. Kirshenbaum, S. Adam, et al., Surface conduction of topological Dirac electrons in bulk insulating Bi_2Se_3, Nat. Phys. 8 (2012) 459–463.

[102] Q.T. Zhang, K.S. Chan, J.B. Li, Spin-transfer torque generated in graphene based topological insulator heterostructures, Sci. Rep. 8 (2018) 4343.

[103] A.R. Mellnik, J.S. Lee, A. Richardella, J.L. Grab, P.J. Mintun, M.H. Fischer, et al., Spin-transfer torque generated by a topological insulator, Nature 511 (2014) 449–451.

[104] Y. Niimi, Y. Kawanishi, D.H. Wei, C. Deranlot, H.X. Yang, M. Chshiev, et al., Giant spin Hall effect Induced by skew scattering from bismuth impurities inside thin film CuBi alloys, Phys. Rev. Lett. 109 (2012) 156602.

[105] C.F. Pai, L.Q. Liu, Y. Li, H.W. Tseng, D.C. Ralph, R.A. Buhrman, Spin transfer torque devices utilizing the giant spin Hall effect of tungsten, Appl. Phys. Lett. 101 (2012) 122404.

[106] Y.B. Fan, P. Upadhyaya, X.F. Kou, M.R. Lang, S. Takei, Z.X. Wang, et al., Magnetization switching through giant spin-orbit torque in a magnetically doped topological insulator heterostructure, Nat. Mater. 13 (2014) 699–704.

[107] Y. Lu, J. Guo, Quantum simulation of topological insulator based spin transfer torque device, Appl. Phys. Lett. 102 (2013) 073106.

[108] S. Cho, D. Kim, P. Syers, N.P. Butch, J. Paglione, M.S. Fuhrer, Topological insulator quantum dot with tunable barriers, Nano. Lett. 12 (1) (2012) 469–472.

[109] S. Datta, Proceedings of the International School of Physics. Enrico Fermi, Italy (2005) 1.

[110] S. Salahuddin, S. Datta, Proceedings of the IEEE International Electron Devices Meeting (IEDM) (2006) 1–4.

[111] A.A. Yanik, G. Klimeck, S. Datta, Quantum transport with spin dephasing: a nonequlibrium Green's function approach, Phys. Rev. B 76 (2007) 045213.

[112] L.Q. Liu, C.F. Pai, Y. Li, H.W. Tseng, D.C. Ralph, R.A. Buhrman, Spin-torque switching with the giant spin Hall effect of tantalum, Science 336 (2012) 555.

[113] R. Bishnoi et al., Design Automation Conference (ASP-DAC), 2014 19th Asia and South Pacific (2014) IEEE.

CHAPTER 9

Growth and properties of magnetic two-dimensional transition-metal chalcogenides

Wen Zhang[1], Ping Kwan Johnny Wong[2], Rebekah Chua[1,3] and Andrew Thye Shen Wee[1,2]

[1]Department of Physics, National University of Singapore, 2 Science Drive 3, Singapore, Singapore, [2]Centre for Advanced 2D Materials (CA2DM) and Graphene Research Centre (GRC), National University of Singapore, 6 Science Drive 2, Singapore, Singapore, [3]NUS Graduate School for Integrative Sciences and Engineering, Centre for Life Sciences, National University of Singapore, 28 Medical Drive, Singapore, Singapore

Chapter Outline

Spintronic 2D Materials.
DOI: https://doi.org/10.1016/B978-0-08-102154-5.00007-2

9.1 Introduction

Global efforts are currently underway to develop new technologies that can integrate, improve, and even replace charge-based electronics whose continuous downscaling is expected to come to a halt soon [1]. Spintronics is one of such technologies, utilizing the electron spin in addition to charge to achieve multifunctional devices with faster operating speed and lower power consumption [2,3].

The last couple of years have witnessed an exciting synergy between spintronics (Nobel prize in 2007) and two-dimensional (2D) materials (Nobel prize in 2010) [4–8]. It started off with the successful observation of electrical spin injection and detection in graphene at low [9] and then at room temperatures [10], followed by demonstration of micrometer distance transport of spin-polarized current in graphene [11,12], and later the use of graphene as effective tunnel barriers for both lateral [13] and vertical spin devices [14–16]. What followed next was a strategic move toward other layered materials beyond graphene [17]. Notably, 2D transition-metal chalcogenides (TMCs) are promising building blocks for 2D spintronic heterostructures and devices [18–20]. This family currently consists of over 40 different species, thus extensively covering all key electronic properties from metals, semiconductors to half-metals, and from magnets to superconductors [19,20]. Besides, building spintronics around 2D-TMCs can open up unique opportunities for exploring fundamentally new physical phenomena due to the interplay between spin and other electronic degrees of freedom. The spin-valley coupling in group-VI 2D semiconducting TMCs, such as MoS_2 and WSe_2, is an appealing example, leading to a number of striking physical effects which include but are not limited to the coexistence of spin Hall and valley Hall effects [21], long-lived electron spin in the nanosecond time regime [22], and optically and electrically controlled valley polarization [23–25]. From the theoretical perspective, heterostructures that are based on 2D-TMCs also show a great variety of properties with potential applications. Examples are the giant magnetoresistance and spin-filtering phenomena in 2D spin-valves with TMCs as a nonmagnetic interlayer [26], and half-metallicity existing at some specific interfaces between 3d ferromagnets and 2D-TMCs [27]. The latest focus in this burgeoning field is on achieving robust 2D magnetism in the TMCs, either for direct exploitation of this exotic property or via magnetic proximity effects for spin-based technologies. The latter, which is conceptually similar to the previous demonstration of EuO/graphene heterostructures [28,29], permits the use of nominally nonmagnetic materials in spintronics.

High-quality growth of 2D-TMCs with well-controlled properties is clearly at the heart of the research mentioned above. This, however, represents a significant challenge, because of the intricate control required for the growth of these atomically thin layers, which is further complicated by other factors, such as the inevitable substrate effect, material stability, and reliability of characterization tools. This chapter gives an overview of the recent theoretical and experimental approaches developed to address these issues. A particular focus will be devoted to molecular beam epitaxy (MBE), a powerful growth strategy capable of achieving atomic control and layer-by-layer growth of 2D-TMCs. This chapter will also point out the challenges remaining in and ahead of this field, and provide the authors' outlook on what the future holds for MBE-grown 2D-TMCs.

The rest of this chapter is organized as follows: Section 9.2 discusses the fundamentals of MBE growth of 2D-TMCs. Section 9.3 covers the growth and properties of the magnetic 2D-TMCs reported to date. Section 9.4 lays out the opportunities and challenges of this exciting research area.

9.2 Fundamentals of molecular beam epitaxy for two-dimensional transition-metal chalcogenides

9.2.1 Uniqueness of molecular beam epitaxy

As one of the most predominant growth techniques for fabricating epitaxial films, MBE involves the creation of a molecular beam of a given source material by evaporation, and direct physical transport and condensation of the material onto a substrate in an ultrahigh vacuum (UHV) environment. These processes are usually controlled with atomic precision, and thus monolayer-by-monolayer growth of epitaxial thin films with extremely high purity and stoichiometry can be accomplished. The extreme growth controls offered by MBE have become more crucial than ever, due to the intricate and stringent growth conditions required for 2D materials that are just a few atoms thick. Relative to other more popular methods, such as physical and chemical exfoliation and chemical vapor deposition (CVD), MBE provides a unique combination of (1) wafer-size scalability, (2) atomic control over crystallinity and thickness, and (3) UHV compatibility. The third merit has become highly essential for realizing the true intrinsic properties of metallic 2D materials. Due to their higher reactive surfaces, 2D materials with a metallic character are usually more prone to adsorption of foreign species than their semiconducting counterparts, such as monolayer MoS_2. Accordingly, in situ growth and characterization by state-of-the-art surface tools are mandatory for the synthesis of high-quality 2D materials and heterostructures.

9.2.2 Concept of van der Waals epitaxy

Bulk TMCs have graphite-like layered structures, consisting of covalently bonded sheets held together by weak van der Waals forces. The clean surfaces of TMCs feature atomically flat terraces without active dangling bonds, which are ideally suitable for adsorption or epitaxial growth of other similarly bonded materials. Heteroepitaxy of such kind is referred to as van der Waals epitaxy (vdWE) and was first proposed by the group of A. Koma at the University of Tokyo [30]. Using vdWE, this group has demonstrated the epitaxial growth of $MoSe_2$ thin films on mica, despite a substantial lattice mismatch of 58% [31]. The drastic lattice relaxation observed in this work is reasonable due to the weak vdW interactions at the heterointerface, and since then has become a very unique feature of vdWE. Fig. 9.1A shows schematically the concept of vdWE on both layered and nonlayered substrates. It is essential that the surface dangling bonds of the latter are well terminated using foreign atomic species, for example, hydrogen, selenium, etc., so as to "mimic" the surface similar to a layered material. The explosive research effort into 2D materials over the past years has reignited the interests in vdWE, especially for expanding the wide collection of 2D-TMCs. Table 9.1 gives a survey of the vdWE TMC materials reported in literatures.

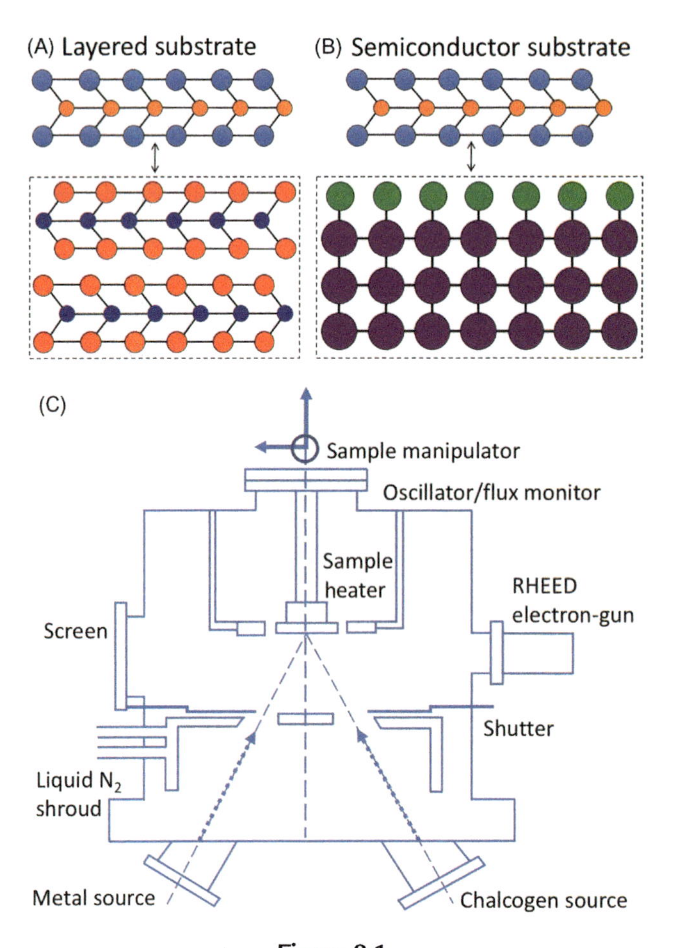

Figure 9.1
Schematic illustration for vdWE of a 2D-TMC on (A) a layered substrate and (B) surface-terminated nonlayered substrates. (C) A UHV growth chamber for 2D-TMCs. *2D-TMC*, two-dimensional transition-metal chalcogenides; *UVH*, ultrahigh vacuum; *vdWE*, van der Waals epitaxy.

Table 9.1: A survey of MBE-grown 2D-TMCs reported in the literature to date.

MBE-grown 2D-TMCs	Substrates	References
MoS_2	Au(111)	[32]
$MoSe_2$	SnS_2	[33]
	MoS_2 sulfur-terminated GaAs(111)	[34,35]
	Sapphire	[36]
	HOPG	[37]
	Graphene/SiC	[38–41]
	Bilayer graphene/SiC	[40,42]
	AlN(0001)/Si(111)	[41,43]
	$HfSe_2$/AlN(0001)/Si(111)	[44]
	$ZrSe_2$/AlN(0001)/Si(111)	[45]
	AlN(0001)/Si(111)	[46]
	Bi_2Se_3/AlN(0001)/Si(111)	[47]
$MoTe_2$	Sapphire	[37]
	MoS_2	[48]
	Bilayer graphene/SiC	[49]
WSe_2	HOPG	[50]
WTe_2	Bilayer graphene/SiC	[51,52]
FeSe	Bilayer graphene	[53]
	Nb-doped $SrTiO_3$(001)	[54]
	Graphene/SiC or $SrTiO_3$(001)	[55]
	Nb-doped $SrTiO_3$(001)	[56]
	$SrTiO_3$(001)	[57]
K-doped FeSe	$SrTiO_3$(001)	[58]
	Graphene/SiC	[59]
Fe_3Se_4	Bi_2Se_3/GaAs(111)	[60]
InSe (bulk)	InSe(0001) or GaSe(0001)	[61]
$PtSe_2$	Pt(111)	[62]
$HfSe_2$	MoS_2 or HOPG	[63,64]
	AlN(0001)/Si(111)	[44,45]
$HfTe_2$	AlN(0001)/Si(111)	[65]
$ZrSe_2$	AlN(0001)/Si(111)	[46]
$ZrTe_2$	InAs(111)	[66]
$TiSe_2$	Graphene/SiC	[67]
	Bilayer graphene/SiC	[68–70]
$TaSe_2$	Bilayer graphene/SiC	[71]
	$HfSe_2$/AlN(0001)/Si(111) or $MoSe_2$/AlN(0001)/Si(111)	[44]
GaSe	Si(111)	[72–78]
	GaAs(111)B	[79]
	H-terminated Si(111)	[80,81]
	GaAs(111)	[82,83]
	Si(111), Si(110), and Si(100)	[84,85]
	InSe(buffer layer)/InSe(0001)	[61]
SnSe	GaAs(111)B	[86]
SnTe	Si(111)	[87]
SnS	Graphene	[88]
SnS_2	Mica	[89]
CrSe	Fe/GaAs(111)B	[90]
	GaAs(100)	[91]
	NaCl(100), NaCl(110), NaCl(111), or mica	[92]
CrTe	CdTe(001)	[93]
	ZnTe/GaAs(100)	[94]
	NaCl(100), NaCl(110), NaCl(111), or mica	[92]

(*Continued*)

Table 9.1: (Continued)

MBE-grown 2D-TMCs	Substrates	References
Cr_2Te_3	Sapphire and Si(111)	[95]
	Si(111)	[96]
ZnCrTe	GaAs	[97]
CrTe	GaAs(100)	[98–101]
VSe_2	Sapphire	[102]
	MoS_2 and HOPG	[103]
Fe_3GeTe_2	GaAs(111) and sapphire	[104]
MnS	Mo substrate	[105]
MnSe	Bi_2Se_3	[106]
	GaAs(100)	[107,108]
$MnSe_2$	GaSe(0001)/GaAs(111)B and $SnSe_2$(0001)/GaAs(111)B	[109]
$SnMnSe_2$	GaAs(111)B	[86]
MnTe	SiO_2/Si(111)	[110]

It also serves to include several nonlayered chalcogenides with compatible structures, such as hexagonal NiAs, which can be interfaced with layered 2D-TMCs for construction of heterostructures. We will further comment on this aspect in Section 9.4.

9.2.3 Technical aspects of van der Waals epitaxy

Chalcogen elements, such as sulfur and selenium, are high vapor pressure materials even at room temperature, and thus would require special technical considerations. Fig. 9.1C illustrates a schematic UHV deposition chamber for vdWE of 2D-TMCs, where the chalcogen source plays as the most critical component. Previous work has employed both solid and gaseous chalcogen sources for the MBE growth of TMCs. Elemental chalcogens are by far the most popular solid sources by virtue of their high purity, availability, and ease of use [33–36,79–81,111,112]. In practice, unless during evaporation, these sources are usually isolated from a growth chamber by a gate-valve, so that chamber bake-out is not affected by the presence of those high vapor pressure materials. Stable compounds, such as SnS_2 and $SnSe_2$ [113], belong to another type of solid chalcogen sources which have several favorable characteristics to the TMC growth: (1) they are rather stable layered materials against ambient exposure, due to their low surface energies and (2) technically, they do not have to be isolated, as their decomposition temperatures are higher than typical baking temperatures of UHV systems. Gaseous sources, such as H_2S and H_2Se, have also been employed successfully for epitaxial growth of some 2D-TMCs (mainly MoS_2) [114,115]. However, due to its potential hazards of toxicity and residual hydrogen generation within a vacuum system, the application remains rather limited.

Experimentally, the actual MBE growth process for 2D-TMCs involves coevaporation of high-purity metal and chalcogen from standard Knudsen cells. An elevated substrate temperature is in most cases applied in order to promote crystallization of a deposited film. For ease of thickness control, a deposition rate of no higher than 0.1 monolayers/minutes is usually aimed for. In order to avoid chalcogen losses during growth, a high chalcogen-to-metal flux ratio of at least 10:1 is employed. Under these conditions, the growth rate will only be linearly dependent on the metal flux, and the stoichiometry of the 2D-TMCs should be self-regulating, namely, the extra chalcogen atoms cannot be incorporated into the films, since the substrate temperature is higher than the sublimation temperature of the chalcogen.

9.3 Growth and properties of magnetic two-dimensional transition-metal chalcogenides

By virtue of their exotic spin character at low-dimensionality, magnetic 2D-TMCs may potentially serve as a promising platform for both fundamental research and applications in spintronic devices. The theoretical and experimental progress is highlighted in this section, and is grouped according to extrinsic and intrinsic magnetism in 2D-TMCs.

9.3.1 Extrinsic magnetism in two-dimensional transition-metal chalcogenides

Creation of magnetism in nominally nonmagnetic 2D-TMCs has been attempted by the following routes: (1) defect engineering, (2) surface functionalization [116,117], and (3) doping control. As far as MoS_2 is concerned, magnetic signals have been experimentally identified in exfoliated nanosheets with magnetic edge states [118], proton irradiated samples with both point defects and edge states [119], and single crystals with zigzag edges at grain boundaries, respectively [120]. The created magnetism, consistent with previous theoretical calculations though [117,121–123], strongly depends on the exact defect configurations. Its enhancement is also accompanied with the creation of a considerable amount of defects, in turn leading to undesirable properties—such as reduced carrier mobility—due to formation of the scattering centers or charge-trapping sites.

For the second route, Ataca et al. have determined the minimum-energy adsorption sites for 16 different adatoms on monolayer MoS_2, based on first-principles calculations [117]. In particular, it has been predicted that the monolayer can attain local magnetic moments through the adsorption of some specific 3d transition-metal (Co, Cr, Fe, Mn, and V) atoms, as well as Si and Ge. Hydrogenation is also believed to be an effective way in inducing strong ferromagnetism in strained monolayer MoS_2 [124]. Although the hydrogenation itself does not lead to a spontaneous magnetization in MoS_2, an external tensile strain can cause a nonmagnetic-to-ferromagnetic transition with a Curie temperature of $\sim$230K. In contrast, experimental demonstrations remain rare in this respect, because the adsorption effects can only be determined unambiguously using atomically clean 2D materials. This, therefore, highlights the crucial role of MBE-grown 2D-TMCs in pushing this field forward.

Compared to defect engineering and surface functionalization, doping is a more controllable means to obtain magnetism in nonmagnetic 2D-TMCs. This route typically involves substitutions of the constituting atoms of a given TMC material, at either the metal or the chalcogen sites, by impurity atoms. Generally, the substitution of nonmetal elements (e.g., F and Cl) into the chalcogen sites would mostly cause higher formation energies than substitution at the metal sites, and therefore only a limited dopant concentration is feasible in this case [116,125]. Because of this reason, most of the studies mainly focus on doping at metal sites. For instance, Mn-doped monolayer MoS_2 has been proposed as an atomically thin dilute magnetic semiconductor [126], where the Mn substitution has been found to be energetically favorable at the Mo sites, and a dopant concentration of 5%–15% is necessary to trigger long-range ferromagnetism [127]. In reality, however, to achieve such an amount of doping remains challenging. Even though in situ doping of monolayer MoS_2 with Mn via vapor phase deposition has been demonstrated successfully, only several percent (<5%) of Mn dopants can be incorporated into the 2D material. Moreover, this amount is highly sensitive to the selection of a substrate on which the monolayer MoS_2 was exfoliated. Namely, inert substrates like graphene allow for several percentages of Mn incorporation, while substrates with reactive surface terminations, such as SiO_2 and sapphire, simply preclude the incorporation and result in defective MoS_2 [128]. On the other hand, the Ayajan group has demonstrated Re doping in monolayer $MoSe_2$ with highly tunable structural phase variations [129]. This method, based on alloying two-parent TMCs with structurally different phases, can be an excellent approach for obtaining large-area uniform alloys of 2D-$Mo_{1-x}Re_xSe_2$, accompanied with the structural phase transition from 2H to 1T′ phase as a function of Re doping. The physical mechanism behind is to provide an extra electron from Re to the Mo 4d orbitals in $MoSe_2$. In the 2H phase with trigonal prismatic coordination, this extra electron fills a higher energy level of the split Mo 4d orbitals; whereas in the 1T′ phase, this extra electron occupies a lower energy state of the Mo orbitals in the distorted octahedral coordination geometry. In both cases, the extra electron is unpaired, thus contributing to a magnetic ground state [125,130]. From zero-field cooling and field cooling measurements, both the 2H and 1T′ 2D-$Mo_{1-x}Re_xSe_2$ have shown ferromagnetism at room temperature, with the small total magnetic moment of $2.7 \times 10^{-3}\ \mu_B$ and $6.0 \times 10^{-4}\ \mu_B$ per formula, respectively, though [129].

9.3.2 Intrinsic magnetism in two-dimensional transition-metal chalcogenides

The preceding section has spanned various proposals to induce magnetism in nonmagnetic 2D-TMDs. Most of them are essentially destructive in nature, and the induced magnetism is expected to be fairly weak and localized. It is thus very timely to develop 2D-TMCs with intrinsic magnetism instead, which would make a real impact on 2D spintronics.

9.3.2.1 VX_2 (X = S, Se)

2D-VX_2 have attracted particular attention in recent years, due to a number of exciting theoretical and experimental studies on their quantum phenomena, such as charge–density waves (CDW) [131–134], and potential applications as ultrasensitive moisture sensors [135], highly conductive supercapacitors [136], 2D anode materials for lithium-ion batteries [137], and electrocatalysts for hydrogen evolution reaction [138]. Similar to a number of existing TMCs, VX_2 has two common polymorphs, that is, the 1T and 2H structures, corresponding to the octahedral and trigonal prismatic VX_6 units. Theories have predicted that both of these structures are thermodynamically stable, with their relative stability a strong function of layer thickness and temperature [139–144]. Yet, only the 1T structure has been observed so far in experiments [102,103,131,133,145–151]. Electronically, VX_2 exhibits a $3d^1$ configuration, which invokes both metallic and magnetic properties. Theoretical calculations have suggested that the magnetic moments in VX_2 mainly stem from the vanadium atoms, and their coupling is mediated by competitive effects of through-bond and through-space interactions [152]. A ferromagnetic coupling is expected to occur when the through-bond interaction dominates. Fig. 9.2 shows that, when the element X changes from S to Se, the corresponding magnetic moments per vanadium atoms increase from 0.48 μ_B to 0.68 μ_B [152]. These values, approximately one-third of those in traditional 3d ferromagnets, can be manipulated by strain.

To explore the intrinsic magnetism experimentally, 2D-VS_2 and 2D-VSe_2 have been fabricated by various methods, such as wet chemistry [131,133,145,153], exfoliation [132,153], and CVD growth [149–151]. Although some indeed observed signs of ferromagnetism [133,145], one has to take into account uncertain factors, such as contamination and/or degrade of less stable metallic TMC flakes at ambient conditions. Besides, the observed ferromagnetism was obtained by characterization tools for bulk magnetism, which failed to be correlated to the associated electronic structure of the 2D-VX_2. The above issues have subsequently motivated the development of an in situ growth and characterization strategy. For instance, combining in situ MBE and scanning tunneling microscopy, Zhang et al. have synthesized strained VSe_2 (~20 nm) thin films on sapphire and observed a ($4a \times \sqrt{3a}$) rectangular CDW structure at low temperatures for the first time [102]. Tunneling spectroscopy further shows this CDW structure transforms to an insulating phase below 500 mK, with its gap size ranging from 2 to 300 meV depending on the sample quality. In another study by Bonilla et al., room temperature ferromagnetic ordering has been observed in MBE-grown monolayer VSe_2 on highly oriented pyrolytic graphite (HOPG) and MoS_2, and the temperature-dependent magnetism shows a broad bump at around 110K–130K due to the CDW transition in the monolayers [103]. However, the observation of strong ferromagnetism in this work has been challenged by several follow-up works, indicating no signs of ferromagnetically coupled V ions. [154–156].

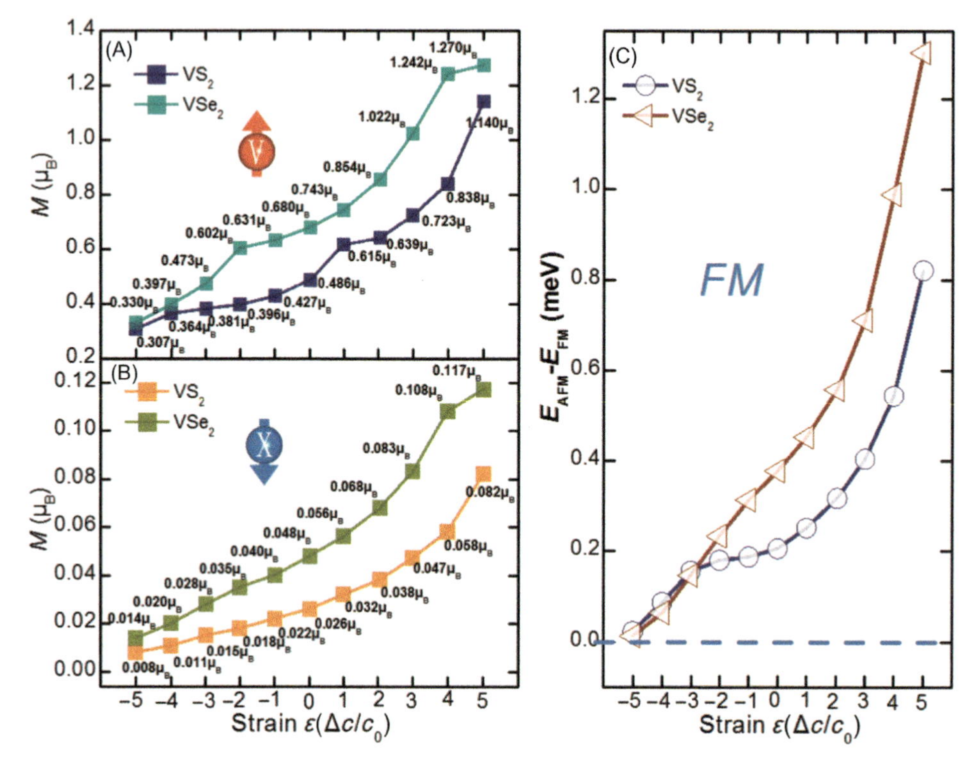

Figure 9.2

Strain-dependent magnetic moments at the V and X ionic sites in monolayer VX_2, and the corresponding energy difference between ferromagnetic and antiferromagnetic states. Source: *Reprinted with permission from Y.D. Ma, Y. Dai, M. Guo, C.W. Niu, Y.T. Zhu, B.B. Huang, Evidence of the existence of magnetism in pristine VX2 monolayers (X = S, Se) and their strain-induced tunable magnetic properties, ACS Nano 6 (2) (2012) 1695–1701. Copyright 2012 American Chemical Society.*

9.3.2.2 MnX_2 (X = S, Se)

Previous density-functional theory calculations have indicated that 2D-MnX_2 are unique among the 2D-TMCs family because they are ferromagnetic semiconductors close to room temperature [157,158]. The theoretical magnetic moments in 2D-MnX_2 are $\sim 3\ \mu_B$ per Mn atom, among which the selenide carrying a slightly higher value than sulfide by 0.16 μ_B. This difference is fundamentally attributed to the lower electronegativity of Se than S, such that the former chalcogen atoms draw less electrons from the Mn site. Consequently, this leaves more valence electrons and thus higher magnetic moments with the metal ions. Fig. 9.3 shows the MBE growth of monolayer $MnSe_2$ on GaSe-seeded GaAs [109], which exhibits a large saturation magnetization of $\sim 4\ \mu_B$ per Mn atom. With the thickness increasing to about 40 nm, the saturation magnetization was enhanced due to the emergence of α-MnSe, and this insulating MnSe layer is attractive for serving as a gate dielectric for spin field-effect structures.

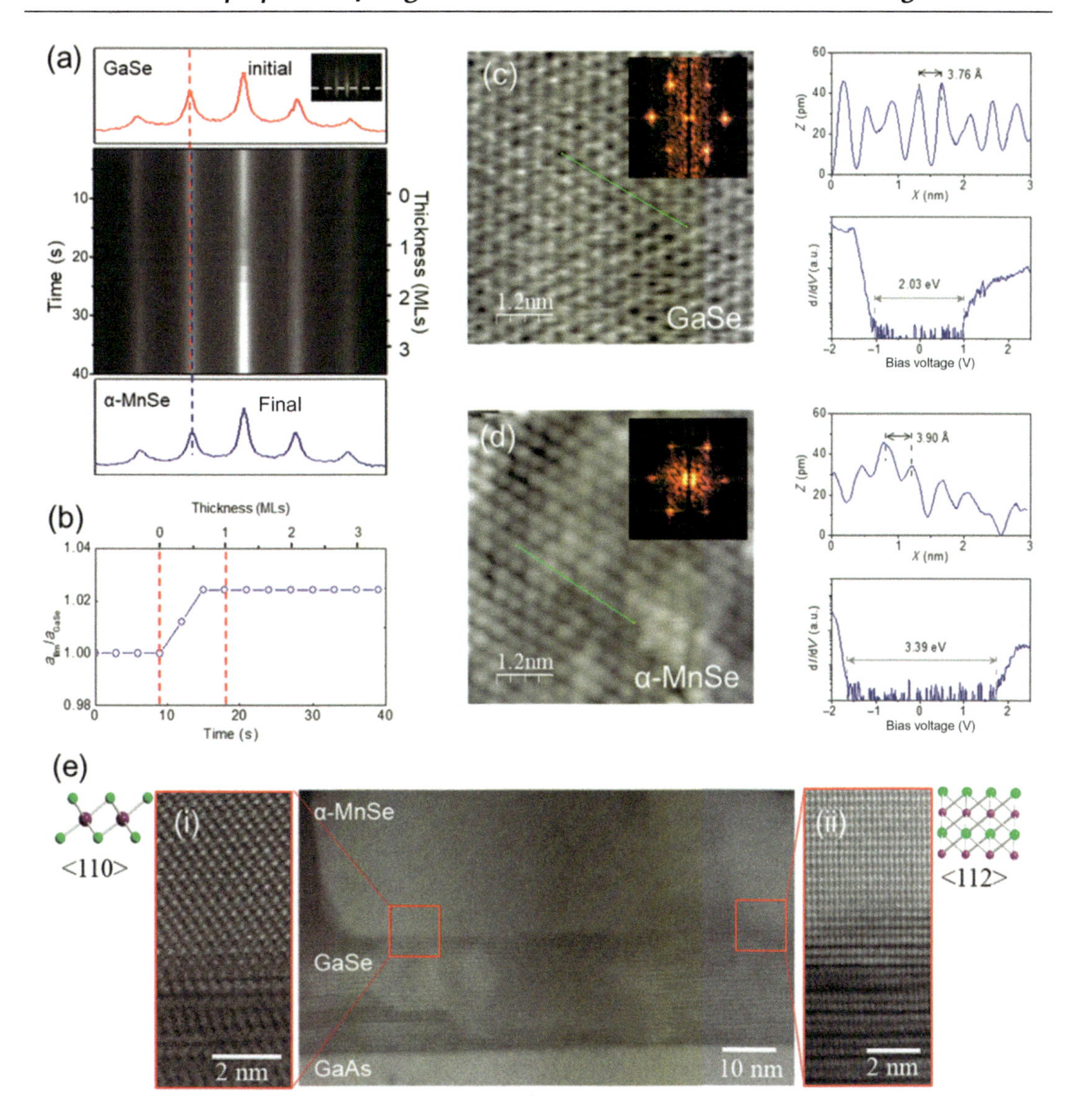

Figure 9.3

Molecular beam epitaxy growth evolution from GaSe seed layer to ~3 layers of $MnSe_x$. STS has been used to identify the GaSe and MnSe surfaces, as they exhibit different bandgap value. High resolution STEM shows the atomically sharp $MnSe_x$/GaSe interface. *MBE*, molecular beam epitaxy; *STS*, scanning tunnelling spectroscopy; *STEM*, scanning transmission electron microscopy. Source: *Reprinted with permission from D.J. O'Hara, T. Zhu, A.H. Trout, A.S. Ahmed, Y.K. Luo, C.H. Lee, et al., Room temperature intrinsic ferromagnetism in epitaxial manganese selenide films in the monolayer limit, Nano. Lett. 18 (5) (2018) 3125–3131. Copyright 2018 American Chemical Society.*

9.3.2.3 $CrXTe_3$ (X = Si, Ge, Sn)

$CrXTe_3$, with the centrosymmetric structure, has attracted significant attention as well. Besides MnX_2, they also belong to a rare group of 2D-layered ferromagnetic semiconductors [159–166]. $CrSiTe_3$ is an indirect gap semiconductor with indirect and direct band gaps at 0.4 and 1.2 eV, respectively [167]. Theories have suggested that 2D-$CrSiTe_3$ is dynamically stable and its Curie temperature may shift to a higher temperature when the thickness is reduced to monolayer. This seems to be confirmed by initial experimental results signifying a change in resistivity at 80K–120K when the $CrSiTe_3$ crystals are a few layers thick [168]. More recent studies confirm that exfoliated $CrGeTe_3$ exhibits a Curie temperature of 61K and a uniaxial magnetic anisotropy, which can be tuned by an external magnetic field. Even down to the monolayer regime, $CrGeTe_3$ remains ferromagnetic below 30K [159]. The authors also observed a nonconstant Curie temperature of the 2D nanosheets, which is dependent on the magnitude of the external magnetic field and explained within the framework of spin–wave excitations [159]. This effect is absent in the bulk counterpart, where the Curie temperature is predominantly determined by exchange interactions. In theory, due to the enhanced ionicity of the Sn–Te bond and stronger superexchange coupling between the magnetic Cr atoms, monolayer $CrSnTe_3$ should exhibit the highest Curie temperature (170K) among the $CrXTe_3$ series, while an experimental study in this regard is still lacking [166]. Fig. 9.4 shows a recent demonstration of channel resistance modulation in an exfoliated 2D-$CrGeTe_3$ device by both back gating [169] and ionic liquid gating [170]. The latter method has been found to be more effective in influencing the channel resistance than the former does, which is similar to previous reports on the dual gating responses in other 2D devices [171,172].

9.3.2.4 Fe_3GeTe_2

Fe_3GeTe_2, whose layered structure is built up with Fe_3Ge heterometallic slabs sandwiched between two Te layers, has recently emerged as an alternative of $CrXTe_3$ for 2D spintronic applications. It is a quasi-2D itinerant ferromagnet with higher Curie temperature ($\sim$230K) and higher stability against oxidation than $CrXTe_3$ [173]. Its perpendicular magnetic anisotropy energy ($\sim 10^7$ erg/cm^3) is almost two orders of magnitude higher than that of $CrGeTe_3$ (10^5 erg/cm^3), making it a better candidate for current-induced switching in magnetic tunnel junctions and magnetic random access memory applications [174].

Using X-ray magnetic circular dichroism and corresponding sum rule analysis, Zhu et al. have extracted the spin and orbital magnetic moments of bulk Fe_3GeTe_2 crystal as 1.58 μ_B and 0.10 μ_B per Fe atom, respectively [175]. These values are quite close to those calculated by the spin-polarized generalized gradient approximation [176]. Fig. 9.5 illustrates a recent demonstration of wafer-scale growth of ultrathin Fe_3GeTe_2 on sapphire and GaAs by MBE [104]. Through anomalous Hall effect and conventional magnetic

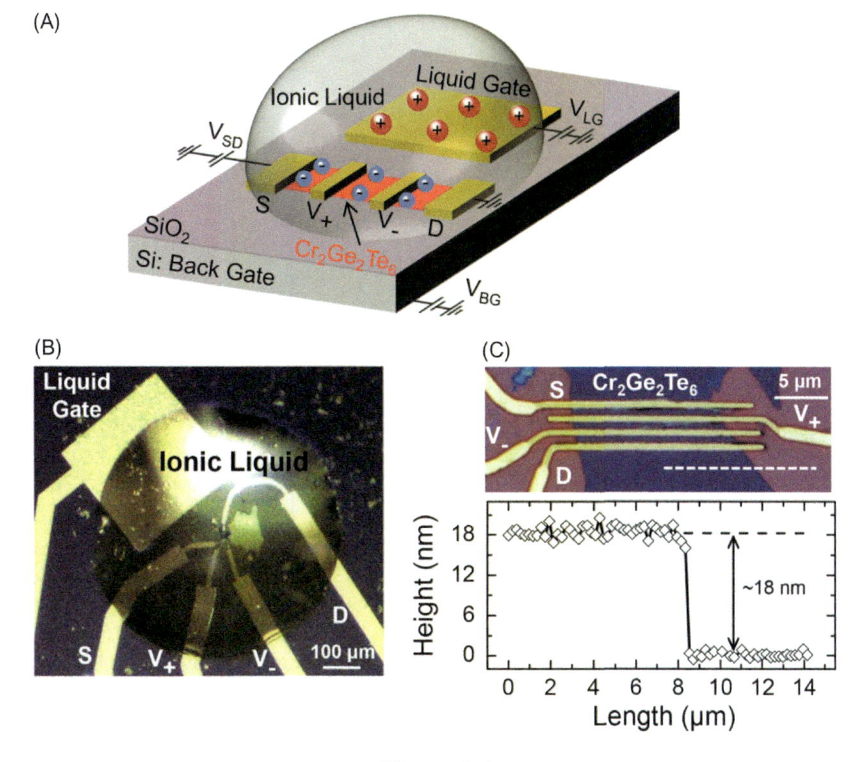

Figure 9.4

Demonstration of ionic liquid (HMIM-TFSI) gating on an 18 nm thick $CrGeTe_3$ device fabricated by mechanical exfoliation. *HMIM-TFSI*, 1-Hexyl-3-methylimidazolium bis(trifluormethylsulfonyl) imide. Source: *Reprinted with permission from Y.Y. Chen, W.Y. Xing, X.R. Wang, B.W. Shen, W. Yuan, T. Su, et al., Role of oxygen in ionic liquid gating on two-dimensional Cr2Ge2Te6: a non-oxide material, ACS Appl. Mater. Interfaces, 10 (1) (2018) 1383–1388. Copyright 2018 American Chemical Society.*

measurements, an easy-magnetization axis out of the film plane and a Curie temperature of ~200K have been discovered in these high-quality 2D films.

9.4 Opportunities and challenges

The blooming of 2D-TMCs in the last decade has extended the synergy between spintronics and 2D materials, starting from graphene. As we have pointed out in previous sections, this field is currently transforming into a critical phase in which increasingly stringent requirements are necessary for realizing exotic 2D-TMCs and their heterostructures. In this respect, MBE offers a unique advantage by combinations of wafer-size scalability, crystallinity and thickness controls (atomically precise growth), and UHV compatibility. These key features are foreseen to make high-quality 2D-TMCs and relevant heterostructures well accessible in practice, which in turn opens up new opportunities and

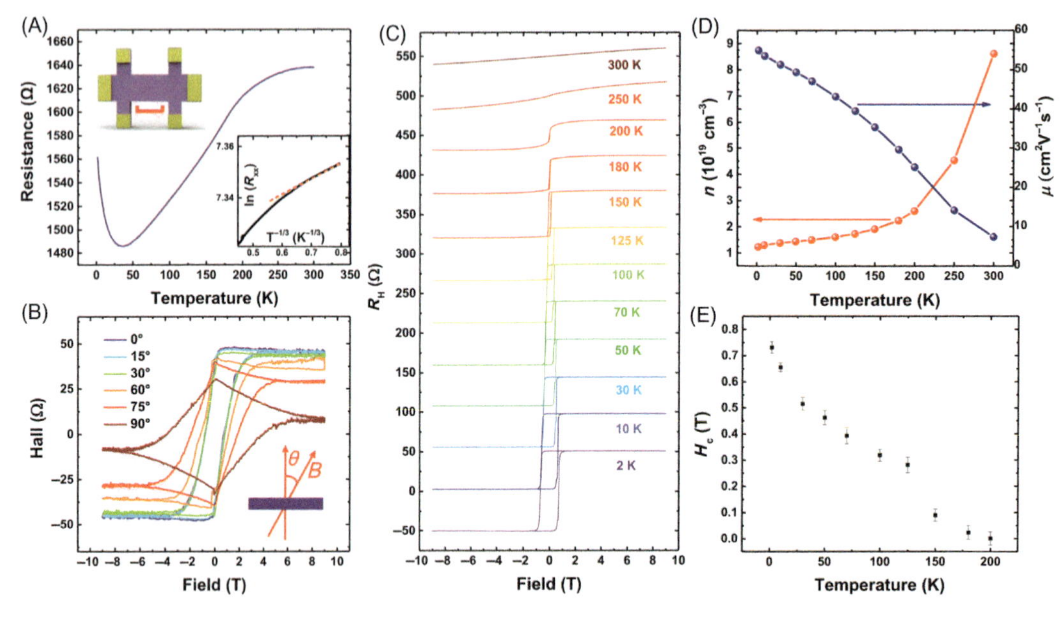

Figure 9.5

Temperature-dependent anomalous Hall effect of 8 nm Fe3GeTe_2 grown by MBE, showing an out-of-plane magnetic anisotropy and a Curie temperature of ~200K. *MBE*, molecular beam epitaxy. Source: *Reprinted from S.S. Liu, X. Yuan, Y.C. Zhou, Y. Sheng, C. Huang, E.Z. Zhang, et al., Wafer-scale two-dimensional ferromagnetic Fe3GeTe2 thin films were grown by molecular beam epitaxy, NPJ 2D Mater. Appl., 1 (2017) 1–7. This figure is licensed under the Creative Commons Attribution 4.0 International License. To view a copy of this license, visit http://creativecommons.org/licenses/by/4.0/ or send a letter to Creative Commons, PO Box 1866, Mountain View, CA 94042, USA.*

challenges on both fundamental and applied levels. Some important aspects will be summarized in this section.

9.4.1 Fundamental issues of van der Waals epitaxy

The concept of vdWE provides a powerful strategy for fabricating high-quality 2D-TMCs and their heterostructures. The primary advantages of vdWE over other ex situ approaches, such as micromechanical exfoliation, include minimal ambient contamination, atomically sharp interfaces, and less constraint in lattice mismatch. On substrates without active dangling bonds, the requirement for lattice mismatch is less stringent, given by the weak vdW or Coulombic interactions along with the normal directions to the 2D plane. As already shown in the previous sections and in Table 9.1, vdWE has been successfully employed for obtaining a number of TMCs down to the 2D regime. However, one major challenge is to achieve controlled monolayer growth, which is hampered by our limited understanding of the mechanisms governing this type of epitaxy. One particular example is the substrate-induced effect. Compared to the bulk counterparts, 2D-TMCs are generally characterized by their lower density of states, which leads to two major consequences: (1) reduced screening effect of the deposited 2D materials and (2) significant interface charge redistribution. In the case of heterostructures, additional factors such as polarity and polarizability between different materials have to be taken into account as well. Therefore, understanding the role of each interaction and their interplay is clearly important, which is in turn useful for tailoring the monolayer growth of TMCs by vdWE.

9.4.2 Molecular doping

The interfacial interactions between 2D-TMCs and foreign molecules provide another dimension to modify the properties of the 2D materials. This concept is not new and has been well demonstrated for graphene [177]. The underlying rationale is to take advantage of the vdW surfaces of 2D-TMCs, which, in combination with organic molecules, becomes a nondestructive approach of charge doping. The degree of charge doping is precisely controlled by not only the alignment of 2D material–molecule energy level, but also the surface coverage of molecules. In the context of magnetic 2D-TMCs, this approach may further allow spin doping [178–183] and manipulation [184–188] via adsorption of organic molecules, both of which are entirely new in the field at present. These studies can be facilitated by use of the MBE-grown 2D-TMCs, where the complications due to structural defects and ambient contamination that could lead to significant inhomogeneity of surface electronic properties can be largely avoided.

9.4.3 Hybrid three-dimensional/two-dimensional structures

vdW heterostructures allow for the maximal exploitation of 2D-TMCs through combining the wide spectrum of electronic characters of these materials. Here, we want to point out that this notion is not limited to layered materials, but can be extended to many other materials, without layered in structure. In fact, this possibility has been partly confirmed by previous works on the growth of the three-dimensional (3D) semiconductors with zinc blende and wurtzite structures, for example, GaAs/graphene [189,190], GaN/graphene [191–193], and ZnO/h–BN [194]. In these cases, the 2D materials serve as a buffer layer for the vdWE. Similarly, NiAs-type TMCs have also been considered to be highly relevant to the construction of vdW heterostructures: when leaving out every other cation layer, the NiAs structure can be transformed into the CdI_2 structure, which is typical for many layered materials and thus establishes structural compatibility between the two lattices. Such flexibility is very exciting for the exploration and exploitation of spin-based interaction across hybrid 3D–2D interfaces. An existing example is Fe_3GeTe_2/MnTe, where the epitaxial growth is enabled by the compatible NiAs structure of MnTe and its compatible lattice parameters with those of Fe_3GeTe_2 [104]. Another example of a hybrid 3D/2D structure is the interfaces between 3d ferromagnets and 2D materials, which have been predicted to host a variety of interesting spin-dependent phenomena [26,27,195,196]. Despite this, experimental observations of these physical effects are largely hindered by the low surface energies of 2D materials, which are unfavorable for the formation of highly ordered interfaces [197]. To tackle this issue, a strategy has been put forth, which is based on solid-phase epitaxy of a ferromagnetic amorphous alloy and can be generalized for the 2D materials family potentially [198].

9.5 Summary

This chapter provides an intensive review on the current progress of the growth and properties of magnetic 2D-TMCs and discusses how these materials can be fitted in the field of spintronics. MBE, in this regard, sets itself apart from other existing growth approaches, offering unique capabilities for achieving wafer-size scalability, high crystallinity, well-controlled thickness, and full compatibility to state-of-the-art vacuum technology. These abilities are expected to bring about further unprecedented opportunities for spintronics: (1) The exploration of well-defined 2D-TMCs with intrinsic magnetic functionality persisting up to room temperature; and, perhaps even more exciting and (2) novel 2D-based heterostructures interfacing with organic molecules and nonlayered inorganic materials.

Acknowledgment

This work was financially supported by the Singapore Ministry of Education (Tier 2 grant MOE2016-T2-2-110), A*STAR Pharos (R-144-000-359-305), the National Research Foundation (Medium Sized Centre Programme R-723-000-001-281), and the National Key Research and Development Program of China (2017YFA0204800).

References

[1] G.E. Moore, Cramming more components onto integrated circuitsReprinted from Electronics, pg 114-117, April 19, 1965 Proc. IEEE 86 (1) (1998) 82–85.
[2] S.A. Wolf, D.D. Awschalom, R.A. Buhrman, J.M. Daughton, S. von Molnar, M.L. Roukes, et al., Spintronics: a spin-based electronics vision for the future, Science 294 (5546) (2001) 1488–1495.
[3] Y.B. Xu, S.M. Thompson, Spintronic Materials and Technology., Taylor and Francis Group, London, 2006.
[4] M.N. Baibich, J.M. Broto, A. Fert, F.N. Vandau, F. Petroff, P. Eitenne, et al., Giant magnetoresistance of (001)Fe/(001) Cr magnetic superlattices, Phys. Rev. Lett. 61 (21) (1988) 2472–2475.
[5] G. Binasch, P. Grunberg, F. Saurenbach, W. Zinn, Enhanced magnetoresistance in layered magnetic-structures with antiferromagnetic interlayer exchange, Phys. Rev. B 39 (7) (1989) 4828–4830.
[6] K.S. Novoselov, A.K. Geim, S.V. Morozov, D. Jiang, Y. Zhang, S.V. Dubonos, et al., Electric field effect in atomically thin carbon films, Science 306 (5696) (2004) 666–669.
[7] K.S. Novoselov, D. Jiang, F. Schedin, T.J. Booth, V.V. Khotkevich, S.V. Morozov, et al., Two-dimensional atomic crystals, Proc. Natl. Acad. Sci. USA. 102 (30) (2005) 10451–10453.
[8] A.K. Geim, K.S. Novoselov, The rise of graphene, Nat. Mater. 6 (3) (2007) 183–191.

[9] N. Tombros, C. Jozsa, M. Popinciuc, H.T. Jonkman, B.J. van Wees, Electronic spin transport and spin precession in single graphene layers at room temperature, Nature 448 (7153) (2007) 571–574.
[10] W. Han, K. Pi, K.M. McCreary, Y. Li, J.J.I. Wong, A.G. Swartz, et al., Tunneling spin injection into single layer graphene, Phys. Rev. Lett. 105 (16) (2010) 167202.
[11] M.V. Kamalakar, C. Groenveld, A. Dankert, S.P. Dash, Long distance spin communication in chemical vapour deposited graphene, Nat. Commun. 6 (2015) 6766.
[12] B. Dlubak, M.B. Martin, C. Deranlot, B. Servet, S. Xavier, R. Mattana, et al., Highly efficient spin transport in epitaxial graphene on SiC, Nat. Phys. 8 (7) (2012) 557–561.
[13] O.M.J. van 't Erve, A.L. Friedman, E. Cobas, C.H. Li, J.T. Robinson, B.T. Jonker, Low-resistance spin injection into silicon using graphene tunnel barriers, Nat. Nanotechnol 7 (11) (2012) 737–742.
[14] B. Dlubak, M.B. Martin, R.S. Weatherup, H. Yang, C. Deranlot, R. Blume, et al., Graphene-passivated nickel as an oxidation-resistant electrode for spintronics, ACS Nano. 6 (12) (2012) 10930–10934.
[15] E. Cobas, A.L. Friedman, O.M.J. van't Erve, J.T. Robinson, B.T. Jonker, Graphene as a tunnel barrier: graphene-based magnetic tunnel junctions, Nano. Lett. 12 (6) (2012) 3000–3004.
[16] M.B. Martin, B. Dlubak, R.S. Weatherup, M. Piquemal-Banci, H. Yang, R. Blume, et al., Protecting nickel with graphene spin-filtering membranes: a single layer is enough, Appl. Phys. Lett. 107 (1) (2015) 012408.
[17] G.R. Bhimanapati, Z. Lin, V. Meunier, Y. Jung, J. Cha, S. Das, et al., Recent advances in two-dimensional materials beyond graphene, ACS Nano. 9 (12) (2015) 11509–11539.
[18] Q.H. Wang, K. Kalantar-Zadeh, A. Kis, J.N. Coleman, M.S. Strano, Electronics and optoelectronics of two-dimensional transition metal dichalcogenides, Nat. Nanotechnol. 7 (11) (2012) 699–712.
[19] S.Z. Butler, S.M. Hollen, L.Y. Cao, Y. Cui, J.A. Gupta, H.R. Gutierrez, et al., Progress, challenges, and opportunities in two-dimensional materials beyond graphene, ACS Nano. 7 (4) (2013) 2898–2926.
[20] M. Chhowalla, H.S. Shin, G. Eda, L.J. Li, K.P. Loh, H. Zhang, The chemistry of two-dimensional layered transition metal dichalcogenide nanosheets, Nat. Chem. 5 (4) (2013) 263–275.
[21] D. Xiao, G.B. Liu, W.X. Feng, X.D. Xu, W. Yao, Coupled spin and valley physics in monolayers of MoS2 and other group-VI dichalcogenides, Phys. Rev. Lett. 108 (19) (2012) 196802.
[22] L.Y. Yang, N.A. Sinitsyn, W.B. Chen, J.T. Yuan, J. Zhang, J. Lou, et al., Long-lived nanosecond spin relaxation and spin coherence of electrons in monolayer MoS2 and WS2, Nat. Phys 11 (10) (2015) 830–834.
[23] K.F. Mak, K.L. He, J. Shan, T.F. Heinz, Control of valley polarization in monolayer MoS2 by optical helicity, Nat. Nanotechnol. 7 (8) (2012) 494–498.
[24] H.L. Zeng, J.F. Dai, W. Yao, D. Xiao, X.D. Cui, Valley polarization in MoS2 monolayers by optical pumping, Nat. Nanotechnol. 7 (8) (2012) 490–493.
[25] Y. Ye, J. Xiao, H.L. Wang, Z.L. Ye, H.Y. Zhu, M. Zhao, et al., Electrical generation and control of the valley carriers in a monolayer transition metal dichalcogenide, Nat. Nanotechnol. 11 (7) (2016) 597–602.
[26] K. Dolui, A. Narayan, I. Rungger, S. Sanvito, Efficient spin injection and giant magnetoresistance in Fe/MoS2/Fe junctions, Phys. Rev. B 90 (4) (2014) 041401.
[27] Q. Chen, Y.X. Ouyang, S.J. Yuan, R.Z. Li, J.L. Wang, Uniformly wetting deposition of Co atoms on MoS2 monolayer: a promising two-dimensional robust half-metallic ferromagnet, ACS Appl. Mater. Interfaces 6 (19) (2014) 16835–16840.
[28] A.G. Swartz, P.M. Odenthal, Y.F. Hao, R.S. Ruoff, R.K. Kawakami, Integration of the ferromagnetic insulator EuO onto graphene, ACS Nano. 6 (11) (2012) 10063–10069.
[29] H.X. Yang, A. Hallal, D. Terrade, X. Waintal, S. Roche, M. Chshiev, Proximity effects induced in graphene by magnetic insulators: first-principles calculations on spin filtering and exchange-splitting gaps, Phys. Rev. Lett. 110 (4) (2013) 046603.
[30] A. Koma, Van der Waals epitaxy—a new epitaxial-growth method for a highly lattice-mismatched system, Thin Solid Films 216 (1) (1992) 72–76.
[31] K. Ueno, K. Saiki, T. Shimada, A. Koma, Epitaxial-growth of transition-metal dichalcogenides on cleaved faces of mica, J. Vac. Sci. Technol. A-Vac. Surf. Films 8 (1) (1990) 68–72.

[32] S.G. Sorensen, H.G. Fuechtbauer, A.K. Tuxen, A.S. Walton, J.V. Lauritsen, Structure and electronic properties of in situ synthesized single-layer MoS2 on a gold surface, ACS Nano. 8 (7) (2014) 6788–6796.

[33] F.S. Ohuchi, B.A. Parkinson, K. Ueno, A. Koma, van der Waals epitaxial-growth and characterization of MoSe2 thin-films on SNS2, J. Appl. Phys. 68 (5) (1990) 2168–2175.

[34] H. Murata, A. Koma, Modulated STM images of ultrathin MoSe2 films grown on MoS2(0001) studied by STM/STS, Phys. Rev. B 59 (15) (1999) 10327–10334.

[35] T. Mori, H. Abe, K. Saiki, A. Koma, Characterization of epitaxial-films of layered materials using moire images of scanning tunneling microscope, Jpn. J. Appl. Phys Part 1-Regul. Papers Short Notes Rev. Papers 32 (6B) (1993) 2945–2949.

[36] K. Ueno, T. Shimada, K. Saiki, A. Koma, Heteroepitaxial growth of layered transition-metal dichalcogenides on sulfur-terminated GAAS (111) surfaces, Appl. Phys. Lett. 56 (4) (1990) 327–329.

[37] A. Roy, H.C.P. Movva, B. Satpati, K. Kim, R. Dey, A. Rai, et al., Structural and electrical properties of MoTe2 and MoSe2 grown by molecular beam epitaxy, ACS Appl. Mater. Interfaces 8 (11) (2016) 7396–7402.

[38] H. Liu, L. Jiao, F. Yang, Y. Cai, X. Wu, W. Ho, et al., Dense network of one-dimensional midgap metallic modes in monolayer MoSe2 and their spatial undulations, Phys. Rev. Lett. 113 (6) (2014) 066105.

[39] H. Liu, H. Zheng, F. Yang, L. Jiao, J. Chen, W. Ho, et al., Line and point defects in MoSe2 bilayer studied by scanning tunneling microscopy and spectroscopy, ACS Nano. 9 (6) (2015) 6619–6625.

[40] L. Jiao, H.J. Liu, J.L. Chen, Y. Yi, W.G. Chen, Y. Cai, et al., Molecular-beam epitaxy of monolayer MoSe2: growth characteristics and domain boundary formation, New J. Phys. 17 (2015) 053023.

[41] M.M. Ugeda, A.J. Bradley, S.-F. Shi, F.H. da Jornada, Y. Zhang, D.Y. Qiu, et al., Giant bandgap renormalization and excitonic effects in a monolayer transition metal dichalcogenide semiconductor, Nat. Mater. 13 (12) (2014) 1091–1095.

[42] Y. Zhang, T.-R. Chang, B. Zhou, Y.-T. Cui, H. Yan, Z. Liu, et al., Direct observation of the transition from indirect to direct bandgap in atomically thin epitaxial MoSe2, Nat. Nanotechnol. 9 (2) (2014) 111–115.

[43] S. Barja, S. Wickenburg, Z.-F. Liu, Y. Zhang, H. Ryu, M.M. Ugeda, et al., Charge density wave order in 1D mirror twin boundaries of single-layer MoSe2, Nat. Phys. 12 (8) (2016) 751–756.

[44] D. Tsoutsou, K.E. Aretouli, P. Tsipas, J. Marquez-Velasco, E. Xenogiannopoulou, N. Kelaidis, et al., Epitaxial 2D MoSe2 (HfSe2) semiconductor/2D TaSe2 metal van der Waals heterostructures, ACS Appl. Mater. Interfaces 8 (3) (2016) 1836–1841.

[45] K.E. Aretouli, P. Tsipas, D. Tsoutsou, J. Marquez-Velasco, E. Xenogiannopoulou, S.A. Giamini, et al., Two-dimensional semiconductor HfSe2 and MoSe2/HfSe2 van der Waals heterostructures by molecular beam epitaxy, Appl. Phys. Lett. 106 (14) (2015) 143105.

[46] P. Tsipas, D. Tsoutsou, J. Marquez-Velasco, K.E. Aretouli, E. Xenogiannopoulou, E. Vassalou, et al., Epitaxial ZrSe2/MoSe2 semiconductor v.d. Waals heterostructures on wide band gap AlN substrates, Microelectron. Eng. 147 (2015) 269–272.

[47] E. Xenogiannopoulou, P. Tsipas, K.E. Aretouli, D. Tsoutsou, S.A. Giamini, C. Bazioti, et al., High-quality, large-area MoSe2 and MoSe2/Bi2Se3 heterostructures on AlN(0001)/Si(111) substrates by molecular beam epitaxy, Nanoscale 7 (17) (2015) 7896–7905.

[48] H.C. Diaz, R. Chaghi, Y. Ma, M. Batzill, Molecular beam epitaxy of the van der Waals heterostructure MoTe2 on MoS2: phase, thermal, and chemical stability, 2D Mater. 2 (4) (2015) 044010.

[49] Y. Yu, G. Wang, S. Qin, N. Wu, Z. Wang, K. He, et al., Molecular beam epitaxy growth of atomically ultrathin MoTe2 lateral heterophase homojunctions on graphene substrates, Carbon. N. Y. 115 (2017) 526–531.

[50] J.H. Park, S. Vishwanath, X. Liu, H. Zhou, S.M. Eichfeld, S.K. Fullerton-Shirey, et al., Scanning tunneling microscopy and spectroscopy of air exposure effects on molecular beam epitaxy grown WSe2 monolayers and bilayers, ACS Nano. 10 (4) (2016) 4258–4267.

[51] S. Tang, C. Zhang, D. Wong, Z. Pedramrazi, H.-Z. Tsai, C. Jia, et al., Quantum spin Hall state in monolayer 1T'-WTe2, Nat. Phys. 13 (7) (2017) 683.
[52] Z.-Y. Jia, Y.-H. Song, X.-B. Li, K. Ran, P. Lu, H.-J. Zheng, et al., Direct visualization of a two-dimensional topological insulator in the single-layer 1T '-WTe2, Phys. Rev. B 96 (4) (2017) 041108(R).
[53] C.-L. Song, Y.-L. Wang, Y.-P. Jiang, Z. Li, L. Wang, K. He, et al., Molecular-beam epitaxy and robust superconductivity of stoichiometric FeSe crystalline films on bilayer graphene, Phys. Rev. B 84 (2) (2011) 020503(R).
[54] J.-F. Ge, Z.-L. Liu, C. Liu, C.-L. Gao, D. Qian, Q.-K. Xue, et al., Superconductivity above 100 K in single-layer FeSe films on doped SrTiO3, Nat. Mater. 14 (3) (2015) 285–289.
[55] L.-L. Wang, X.-C. Ma, X. Chen, Q.-K. Xue, Molecular beam epitaxy and superconductivity of stoichiometric FeSe and KxFe2-ySe2 crystalline films, Chin. Phys. B 22 (8) (2013) 086801.
[56] Q.-Y. Wang, Z. Li, W.-H. Zhang, Z.-C. Zhang, J.-S. Zhang, W. Li, et al., Interface-induced high-temperature superconductivity in single unit-cell FeSe films on $SrTiO_3$, Chin. Phys. Lett. 29 (3) (2012) 037402.
[57] W.-H. Zhang, Y. Sun, J.-S. Zhang, F.-S. Li, M.-H. Guo, Y.-F. Zhao, et al., Direct observation of high-temperature superconductivity in one-unit-cell FeSe films, Chin. Phys. Lett. 31 (1) (2014) 017401.
[58] W. Li, H. Ding, Z. Li, P. Deng, K. Chang, K. He, et al., KFe2Se2 is the parent compound of K-doped iron selenide superconductors, Phys. Rev. Lett. 109 (5) (2012) 057003.
[59] W. Li, H. Ding, P. Deng, K. Chang, C. Song, K. He, et al., Phase separation and magnetic order in K-doped iron selenide superconductor, Nat. Phys. 8 (2) (2012) 126–130.
[60] H.M. do Nascimento Vasconcelos, M. Eddrief, Y. Zheng, D. Demaille, S. Hidki, E. Fonda, et al., Magnetically hard Fe_3Se_4 embedded in Bi_2Se_3 topological insulator thin films grown by molecular beam epitaxy, ACS Nano. 10 (1) (2016) 1132–1138.
[61] J.Y. Emery, L. Brahimostmane, C. Hirlimann, A. Chevy, Reflection high-energy electron-diffraction studies of InSe and gase layered compounds grown by molecular-beam epitaxy, J. Appl. Phys. 71 (7) (1992) 3256–3259.
[62] Y. Wang, L. Li, W. Yao, S. Song, J.T. Sun, J. Pan, et al., Monolayer PtSe2, a new semiconducting transition-metal-dichalcogenide, epitaxially grown by direct selenization of Pt, Nano. Lett. 15 (6) (2015) 4013–4018.
[63] A.T. Barton, R. Yue, S. Anwar, H. Zhu, X. Peng, S. McDonnell, et al., Transition metal dichalcogenide and hexagonal boron nitride heterostructures grown by molecular beam epitaxy, Microelectron. Eng. 147 (2015) 306–309.
[64] R. Yue, A.T. Barton, H. Zhu, A. Azcatl, L.F. Pena, J. Wang, et al., HfSe2 thin films: 2D transition metal dichalcogenides grown by molecular beam epitaxy, ACS Nano. 9 (1) (2015) 474–480.
[65] S. Aminalragia-Giamini, J. Marquez-Velasco, P. Tsipas, D. Tsoutsou, G. Renaud, A. Dimoulas, Molecular beam epitaxy of thin HfTe2 semimetal films, 2D Mater. 4 (1) (2017) 015001.
[66] P. Tsipas, D. Tsoutsou, S. Fragkos, R. Sant, C. Alvarez, H. Okuno, et al., Massless dirac fermions in ZrTe2 semimetal grown on InAs(111) by van der Waals epitaxy, ACS Nano. 12 (2) (2018) 1696–1703.
[67] J.-P. Peng, J.-Q. Guan, H.-M. Zhang, C.-L. Song, L. Wang, K. He, et al., Molecular beam epitaxy growth and scanning tunneling microscopy study of TiSe2 ultrathin films, Phys. Rev. B 91 (12) (2015) 121113(R).
[68] K. Sugawara, Y. Nakata, R. Shimizu, P. Han, T. Hitosugi, T. Sato, et al., Unconventional charge-density-wave transition in monolayer 1T-TiSe2, ACS Nano. 10 (1) (2016) 1341–1345.
[69] P. Chen, Y.H. Chan, X.Y. Fang, Y. Zhang, M.Y. Chou, S.K. Mo, et al., Charge density wave transition in single-layer titanium diselenide, Nat. Commun. 6 (2015) 8943.
[70] P. Chen, Y.H. Chan, M.H. Won, X.Y. Fang, M.Y. Chou, S.K. Mo, et al., Dimensional effects on the charge density waves in ultrathin films of TiSe2, Nano. Lett. 16 (10) (2016) 6331–6336.
[71] H. Ryu, Y. Chen, H. Kim, H.-Z. Tsai, S. Tang, J. Jiang, et al., Persistent charge-density-wave order in single-layer TaSe2, Nano. Lett. 18 (2) (2018) 689–694.

[72] A. Koebel, Y. Zheng, J.F. Petroff, M. Eddrief, L.T. Vinh, C. Sebenne, et al., A transmission electron-microscopy structural-analysis of gase thin-films grown on Si(111) substrates, J. Cryst. Growth 154 (3–4) (1995) 269–274.
[73] H. Reqqass, J.P. Lacharme, M. Eddrief, C.A. Sebenne, V. LeThanh, Y.L. Zheng, et al., Influence of surface reconstruction on MBE growth of layered GaSe on Si(111) substrates, Appl. Surf. Sci. 104 (1996) 557–562.
[74] H. Reqqass, J.P. Lacharme, C. Sebenne, M. Eddrief, V. Lethanh, Silicon(111) surface-properties upon uhv thermal-dissociation of a gase epitaxial layer, Surf. Sci. 331 (1995) 464–467.
[75] R. Fritsche, E. Wisotzki, A. Islam, A. Thissen, A. Klein, W. Jaegermann, et al., Electronic passivation of Si(111) by Ga-Se half-sheet termination, Appl. Phys. Lett. 80 (8) (2002) 1388–1390.
[76] R. Fritsche, E. Wisotzki, A. Thissen, A. Islam, A. Klein, W. Jaegermann, et al., Preparation of a Si(111): GaSe van der Waals surface termination by selenization of a monolayer Ga on Si(111), Surf. Sci. 515 (2–3) (2002) 296–304.
[77] R. Rudolph, C. Pettenkofer, A.A. Bostwick, J.A. Adams, F. Ohuchi, M.A. Olmstead, et al., Electronic structure of the Si(111): GaSe van der Waals-like surface termination, New J. Phys. 7 (2005) 108.
[78] R. Rudolph, C. Pettenkofer, A. Klein, W. Jaegermann, Chemical passivation of Si(111) capped by a thin GaSe layer, Appl. Surf. Sci. 167 (1–2) (2000) 122–124.
[79] K. Ueno, H. Abe, K. Saiki, A. Koma, Heteroepitaxy of layered semiconductor GaSe on a GaAs(111)B surface, Jpn. J. Appl. Phys. Part 2-Letters & Express Letters 30 (8A) (1991) L1352–L1354.
[80] K. Ueno, M. Sakurai, A. Koma, Van der Waals epitaxy on hydrogen-terminated Si(111) surfaces and investigation of its growth-mechanism by atomic-force microscope, J. Cryst. Growth 150 (1–4) (1995) 1180–1185.
[81] K.Y. Liu, K. Ueno, Y. Fujikawa, K. Saiki, A. Koma, Heteroepitaxial growth of layered semiconductor gase on a hydrogen-terminated Si(111) surface, Jpn. J. Appl. Phys. Part 2-Letters 32 (3B) (1993) L434–L437.
[82] L.E. Rumaner, J.L. Gray, F.S. Ohuchi, Nucleation and growth of GaSe on GaAs by Van der Waal epitaxy, J. Cryst. Growth 177 (1–2) (1997) 17–27.
[83] L.E. Rumaner, M.A. Olmstead, F.S. Ohuchi, Interaction of GaSe with GaAs(111): formation of heterostructures with large lattice mismatch, J. Vac. Sci. Technol. B 16 (3) (1998) 977–988.
[84] R. Rudolph, C. Pettenkofer, A. Klein, W. Jaegermann, Van der Waals-xenotaxy: growth of GaSe(0001) on low index silicon surfaces, Appl. Surf. Sci. 166 (1–4) (2000) 437–441.
[85] W. Jaegermann, R. Rudolph, A. Klein, C. Pettenkofer, Perspectives of the concept of van der Waals epitaxy: growth of lattice mismatched GaSe (0001) films on Si(111), Si(110) and Si(100), g Solid Films 380 (1–2) (2000) 276–281.
[86] S. Dong, X. Liu, X. Li, V. Kanzyuba, T. Yoo, S. Rouvimov, et al., Room temperature weak ferromagnetism in $Sn_{1-x}Mn_xSe_2$ 2D films grown by molecular beam epitaxy, APL Mater. 4 (3) (2016) 032601.
[87] C.-H. Yan, H. Guo, J. Wen, Z.-D. Zhang, L.-L. Wang, K. He, et al., Growth of topological crystalline insulator SnTe thin films on Si(111). substrate by molecular beam epitaxy, Surf. Sci. 621 (2014) 104–108.
[88] W. Wang, K.K. Leung, W.K. Fong, S.F. Wang, Y.Y. Hui, S.P. Lau, et al., Molecular beam epitaxy growth of high quality p-doped SnS van der Waals epitaxy on a graphene buffer layer, J. Appl. Phys. 111 (9) (2012) 0935220.
[89] K.W. Nebesny, G.E. Collins, P.A. Lee, L.K. Chau, J. Danziger, E. Osburn, et al., Organic inorganic molecular-beam epitaxy—formation of an ordered phthalocyanine/SnS2 heterojunction, Chem. Mater. 3 (5) (1991) 829–838.
[90] C. Wang, B. Zhang, B. You, S.K. Lok, S.K. Chan, X.X. Zhang, et al., Competitive antiferromagnetic and ferromagnetic coupling in a CrSe/Fe/GaAs(111) B structure, J. Appl. Phys. 104 (2) (2008) 023916.
[91] C. Wang, B. Zhang, B. You, S.K. Lok, S.K. Chan, X.X. Zhang, et al., Molecular-beam-epitaxy-grown CrSe/Fe bilayer on GaAs(100) substrate, J. Appl. Phys. 102 (8) (2007) 083901.

[92] A. Goswami, P.S. Nikam, Study of vacuum-deposited films of CrTe and CrSe on single-crystals, Thin Solid Films 11 (2) (1972) 353.

[93] K. Kanazawa, K. Yamawaki, N. Sekita, Y. Nishio, S. Kuroda, M. Mitome, et al., Structural and magnetic properties of hexagonal Cr1-delta Te films grown on CdTe(001) by molecular beam epitaxy, J. Cryst. Growth 415 (2015) 31–35.

[94] M.G. Sreenivasan, X.J. Hou, K.L. Teo, M.B.A. Jalil, T. Liew, T.C. Chong, Growth of CrTe thin films by molecular-beam epitaxy, Thin Solid Films 505 (1–2) (2006) 133–136.

[95] A. Roy, S. Guchhait, R. Dey, T. Pramanik, C.-C. Hsieh, A. Rai, et al., Perpendicular magnetic anisotropy and spin glass-like behavior in molecular beam epitaxy grown chromium telluride thin films, ACS Nano 9 (4) (2015) 3772–3779.

[96] T. Pramanik, A. Roy, R. Dey, A. Rai, S. Guchhait, H.C.P. Movva, et al., Angular dependence of magnetization reversal in epitaxial chromium telluride thin films with perpendicular magnetic anisotropy, J. Magn. Magn. Mater. 437 (2017) 72–77.

[97] H. Saito, W. Zaets, S. Yamagata, Y. Suzuki, K. Ando, Ferromagnetism in II-VI diluted magnetic semiconductor $Zn_{1-x}Cr_xTe$, J. Appl. Phys. 91 (10) (2002) 8085–8087.

[98] J.F. Bi, H. Lu, M.G. Sreenivasan, K.L. Teo, Exchange bias in zinc-blende CrTe-MnTe bilayer, Appl. Phys. Lett. 94 (25) (2009) 252504.

[99] J.F. Bi, M.G. Sreenivasan, K.L. Teo, T. Liew, Observation of strong magnetic anisotropy in zinc-blende CrTe thin films, J. Phys. D: Appl. Phys. 41 (4) (2008) 045002.

[100] M.G. Sreenivasan, K.L. Teo, M.B.A. Jalil, T. Liew, T.C. Chong, A.Y. Du, Zinc-blende structure of CrTe epilayers grown on GaAs, IEEE. Trans. Magn. 42 (10) (2006) 2691–2693.

[101] M.G. Sreenivasan, J.F. Bi, K.L. Teo, T. Liew, Systematic investigation of structural and magnetic properties in molecular beam epitaxial growth of metastable zinc-blende CrTe toward half-metallicity, J. Appl. Phys. 103 (4) (2008) 043908.

[102] D. Zhang, J. Ha, H. Baek, Y.-H. Chan, F.D. Natterer, A.F. Myers, et al., Strain engineering a 4a x root 3a charge-density-wave phase in transition-metal dichalcogenide 1T-VSe2, Phys. Rev. Mater. 1 (2) (2017) 024005.

[103] M. Bonilla, S. Kolekar, Y. Ma, H.C. Diaz, V. Kalappattil, R. Das, et al., Strong room-temperature ferromagnetism in VSe2 monolayers on van der Waals substrates, Nat. Nanotechnol. 13 (4) (2018) 289.

[104] S.S. Liu, X. Yuan, Y.C. Zhou, Y. Sheng, C. Huang, E.Z. Zhang, et al., Wafer-scale two-dimensional ferromagnetic Fe3GeTe2 thin films were grown by molecular beam epitaxy, NPJ 2D Mater. Appl. 1 (2017) 1–7.

[105] H. Sato, T. Mihara, A. Furuta, M. Tamura, K. Mimura, N. Happo, et al., Chemical trend of occupied and unoccupied Mn 3d states in MnY (Y = S, Se, Te), Phys. Rev. B 56 (12) (1997) 7222–7231.

[106] A.V. Matetskiy, I.A. Kibirev, T. Hirahara, S. Hasegawa, A.V. Zotov, A.A. Saranin, Direct observation of a gap opening in topological interface states of MnSe/Bi2Se3 heterostructure, Appl. Phys. Lett. 107 (9) (2015) 091604.

[107] I. Ishibe, Y. Nabetani, T. Kato, T. Matsumoto, MBE growth and RHEED characterization of MnSe/ZnSe superlattices on GaAs (100) substrates, J. Cryst. Growth 214 (2000) 172–177.

[108] N. Samarth, P. Klosowski, H. Luo, T.M. Giebultowicz, J.K. Furdyna, J.J. Rhyne, et al., Antiferromagnetism in ZnSe/MnSe strained-layer superlattices, Phys. Rev. B 44 (9) (1991) 4701–4704.

[109] D.J. O'Hara, T. Zhu, A.H. Trout, A.S. Ahmed, Y.K. Luo, C.H. Lee, et al., Room temperature intrinsic ferromagnetism in epitaxial manganese selenide films in the monolayer limit, Nano. Lett. 18 (9) (2018) 3125–3131.

[110] Z.H. Wang, D.Y. Geng, W.J. Gong, J. Li, Y.B. Li, Z.D. Zhang, Effect of adding Cr on magnetic properties and metallic behavior in MnTe film, Thin Solid Films 522 (2012) 175–179.

[111] Y. Nakata, K. Sugawara, R. Shimizu, Y. Okada, P. Han, T. Hitosugi, et al., Monolayer 1 T-NbSe2 as a Mott insulator, NPG Asia Mater. 8 (2016) e321.

[112] H. Yamamoto, K. Yoshii, K. Saiki, A. Koma, Improved heteroepitaxial growth of layered NbSe2 on GaAs (111)B, J. Vac. Sci. Technol. A: Vac. Surf. Films 12 (1) (1994) 125–129.

[113] T. Shimada, F.S. Ohuchi, B.A. Parkinson, Thermal-decomposition of SnS2 and SnSE2: novel molecular-beam epitaxy sources for sulfur and selenium, J. Vac. Sci. Technol. A: Vac. Surf. Films 10 (3) (1992) 539–542.
[114] S.S. Gronborg, S. Ulstrup, M. Bianchi, M. Dendzik, C.E. Sanders, J.V. Lauritsen, et al., Synthesis of epitaxial single-layer MoS2 on Au(111), Langmuir 31 (35) (2015) 9700–9706.
[115] J.A. Miwa, M. Dendzik, S.S. Gronborg, M. Bianchi, J.V. Lauritsen, P. Hofmann, et al., Van der Waals epitaxy of two-dimensional MoS2-graphene heterostructures in ultrahigh vacuum, ACS Nano. 9 (6) (2015) 6502–6510.
[116] A.M. Hu, L.L. Wang, W.Z. Xiao, G. Xiao, Q.Y. Rong, Electronic structures and magnetic properties in nonmetallic element substituted MoS2 monolayer, Comput. Mater. Sci. 107 (2015) 72–78.
[117] C. Ataca, S. Ciraci, Functionalization of single-layer MoS2 honeycomb structures, J. Phys. Chem. C 115 (27) (2011) 13303–13311.
[118] J. Zhang, J.M. Soon, K.P. Loh, J.H. Yin, J. Ding, M.B. Sullivian, et al., Magnetic molybdenum disulfide nanosheet films, Nano. Lett. 7 (8) (2007) 2370–2376.
[119] S. Mathew, K. Gopinadhan, T.K. Chan, X.J. Yu, D. Zhan, L. Cao, et al., Magnetism in MoS2 induced by proton irradiation, Appl. Phys. Lett. 101 (10) (2012) 102103.
[120] S. Tongay, S.S. Varnoosfaderani, B.R. Appleton, J.Q. Wu, A.F. Hebard, Magnetic properties of MoS2: existence of ferromagnetism, Appl. Phys. Lett. 101 (12) (2012) 123105.
[121] A.R. Botello-Mendez, F. Lopez-Urias, M. Terrones, H. Terrones, Metallic and ferromagnetic edges in molybdenum disulfide nanoribbons, Nanotechnology. 20 (32) (2009) 325703.
[122] Y.F. Li, Z. Zhou, S.B. Zhang, Z.F. Chen, MoS2 nanoribbons: high stability and unusual electronic and magnetic properties, J. Am. Chem. Soc. 130 (49) (2008) 16739–16744.
[123] A. Vojvodic, B. Hinnemann, J.K. Norskov, Magnetic edge states in MoS2 characterized using density-functional theory, Phys. Rev. B 80 (12) (2009) 125416.
[124] H.L. Shi, H. Pan, Y.W. Zhang, B.I. Yakobson, Strong ferromagnetism in hydrogenated monolayer MoS2 tuned by strain, Phys. Rev. B 88 (20) (2013) 205305.
[125] K. Dolui, I. Rungger, C. Das Pemmaraju, S. Sanvito, Possible doping strategies for MoS2 monolayers: an ab initio study, Phys. Rev. B 88 (7) (2013) 075420.
[126] A. Ramasubramaniam, D. Naveh, Mn-doped monolayer MoS2: an atomically thin dilute magnetic semiconductor, Phys. Rev. B 87 (19) (2013) 195201.
[127] R. Mishra, W. Zhou, S.J. Pennycook, S.T. Pantelides, J.C. Idrobo, Long-range ferromagnetic ordering in manganese-doped two-dimensional dichalcogenides, Phys. Rev. B 88 (14) (2013) 144409.
[128] K.H. Zhang, S.M. Feng, J.J. Wang, A. Azcatl, N. Lu, R. Addou, et al., Manganese doping of monolayer MoS2: the substrate is criticalvol. 15, pg 6586, 2015 Nano. Lett. 16 (3) (2016). 2125-2125.
[129] V. Kochat, A. Apte, J.A. Hachtel, H. Kumazoe, A. Krishnamoorthy, S. Susarla, et al., Re doping in 2D transition metal dichalcogenides as a new route to tailor structural phases and induced magnetism, Adv. Mater. 29 (43) (2017) 1703754.
[130] P.J. Zhao, J.M. Zheng, P. Guo, Z.Y. Jiang, L.K. Cao, Y. Wan, Electronic and magnetic properties of Re-doped single-layer MoS2: a DFT study, Comput. Mater. Sci. 128 (2017) 287–293.
[131] X. Sun, T. Yao, Z.P. Hu, Y.Q. Guo, Q.H. Liu, S.Q. Wei, et al., In situ unravelling structural modulation across the charge-density-wave transition in vanadium disulfide, Phys. Chem. Chem. Phys. 17 (20) (2015) 13333–13339.
[132] J.Y. Yang, W.K. Wang, Y. Liu, H.F. Du, W. Ning, G.L. Zheng, et al., Thickness dependence of the charge-density-wave transition temperature in VSe2, Appl. Phys. Lett. 105 (6) (2014) 063109.
[133] K. Xu, P.Z. Chen, X.L. Li, C.Z. Wu, Y.Q. Guo, J.Y. Zhao, et al., Ultrathin nanosheets of vanadium diselenide: a metallic two-dimensional material with ferromagnetic charge-density-wave behavior, Angew. Chem.-Int. Ed. 52 (40) (2013) 10477–10481.
[134] M. Mulazzi, A. Chainani, N. Katayama, R. Eguchi, M. Matsunami, H. Ohashi, et al., Absence of nesting in the charge-density-wave system 1T-VS2 as seen by photoelectron spectroscopy, Phys. Rev. B 82 (7) (2010) 075130.
[135] J. Feng, L.L. Peng, C.Z. Wu, X. Sun, S.L. Hu, C.W. Lin, et al., Giant moisture responsiveness of VS2 ultrathin nanosheets for novel touchless positioning interface, Adv. Mater. 24 (15) (2012) 1969–1974.

[136] J. Feng, X. Sun, C.Z. Wu, L.L. Peng, C.W. Lin, S.L. Hu, et al., Metallic few-layered VS2 ultrathin nanosheets: high two-dimensional conductivity for in-plane supercapacitors, J. Am. Chem. Soc. 133 (44) (2011) 17832–17838.

[137] Y. Jing, Z. Zhou, C.R. Cabrera, Z.F. Chen, Metallic VS2 monolayer: a promising 2D anode material for lithium ion batteries, J. Phys. Chem. C 117 (48) (2013) 25409–25413.

[138] W.W. Zhao, B.H. Dong, Z.L. Guo, G. Su, R.J. Gao, W. Wang, et al., Colloidal synthesis of VSe2 single-layer nanosheets as novel electrocatalysts for the hydrogen evolution reaction, Chem. Commun. 52 (59) (2016) 9228–9231.

[139] M. Kan, B. Wang, Y.H. Lee, Q. Sun, A density functional theory study of the tunable structure, magnetism and metal-insulator phase transition in VS2 monolayers induced by in-plane biaxial strain, Nano Res. 8 (4) (2015) 1348–1356.

[140] H. Zhang, L.M. Liu, W.M. Lau, Dimension-dependent phase transition and magnetic properties of VS2, J. Mater. Chem. A 1 (36) (2013) 10821–10828.

[141] H.L. Zhuang, R.G. Hennig, Stability and magnetism of strongly correlated single-layer VS2, Phys. Rev. B 93 (5) (2016) 054429.

[142] H.R. Fuh, C.R. Chang, Y.K. Wang, R.F.L. Evans, R.W. Chantrell, H.T. Jeng, Newtype single-layer magnetic semiconductor in transition-metal dichalcogenides VX2 (X = S, Se and Te), Sci. Rep. 6 (2016) 32625.

[143] B. Liu, L.J. Wu, Y.Q. Zhao, L.Z. Wang, M.Q. Cai, A first-principles study of magnetic variation via doping vacancy in monolayer VS2, J. Magn. Magn. Mater. 420 (2016) 218–224.

[144] H. Pan, Electronic and magnetic properties of vanadium dichalcogenides monolayers tuned by hydrogenation, J. Phys. Chem. C 118 (24) (2014) 13248–13253.

[145] D.Q. Gao, Q.X. Xue, X.Z. Mao, W.X. Wang, Q. Xu, D.S. Xue, Ferromagnetism in ultrathin VS2 nanosheets, J. Mater. Chem. C 1 (37) (2013) 5909–5916.

[146] A. Thompson, B. Silbernagel, Magnetic properties of VSe2: inferences from the study of metal-rich V1 + δSe2 compounds, J. Appl. Phys. 49 (3) (1978) 1477–1479.

[147] A.H. Thompson, B.G. Silbernagel, Correlated magnetic and transport properties in the charge-density-wave states of VSe2, Phys. Rev. B 19 (7) (1979) 3420–3426.

[148] C.F. Vanbruggen, C. Haas, Magnetic-susceptibility and electrical-properties of VSe2 single-crystals, Solid State Commun. 20 (3) (1976) 251–254.

[149] Q.Q. Ji, C. Li, J.L. Wang, J.J. Niu, Y. Gong, Z.P. Zhang, et al., Metallic vanadium disulfide nanosheets as a platform material for multifunctional electrode applications, Nano. Lett. 17 (8) (2017) 4908–4916.

[150] J.T. Yuan, J.J. Wu, W.J. Hardy, P. Loya, M. Lou, Y.C. Yang, et al., Facile synthesis of single crystal vanadium disulfide nanosheets by chemical vapor deposition for efficient hydrogen evolution reaction, Adv. Mater. 27 (37) (2015) 5605–5609.

[151] Z.P. Zhang, J.J. Niu, P.F. Yang, Y. Gong, Q.Q. Ji, J.P. Shi, et al., Van der Waals epitaxial growth of 2D metallic vanadium diselenide single crystals and their extra-high electrical conductivity, Adv. Mater. 29 (37) (2017) 1702359.

[152] Y.D. Ma, Y. Dai, M. Guo, C.W. Niu, Y.T. Zhu, B.B. Huang, Evidence of the existence of magnetism in pristine VX2 monolayers (X = S, Se) and their strain-induced tunable magnetic properties, ACS Nano. 6 (2) (2012) 1695–1701.

[153] Y. Wang, Z. Sofer, J. Luxa, M. Pumera, Lithium exfoliated vanadium dichalcogenides (VS2, VSe2, VTe2) exhibit dramatically different properties from their bulk counterparts, Adv. Mater. Interfaces 3 (23) (2016) 1600433.

[154] J. Feng, D. Biswas, A. Rajan, M.D. Watson, F. Mazzola, O.J. Clark, et al., Electronic structure and enhanced charge-density wave order of monolayer VSe_2, Nano Lett. 18 (2018) 4493–4499.

[155] P.K.J. Wong, W. Zhang, F. Bussolotti, X. Yin, T.S. Herng, L. Zhang, et al., Evidence of spin frustration in a vanadium diselenide monolayer magnet, Adv. Mater. 31 (2019) 1901185.

[156] W. Zhang, L. Zhang, P.K.J. Wong, J. Yuan, G. Vinai, P. Torelli, et al., Magnetic transition in monolayer vse2via interface hybridization, ACS Nano 13 (2019) 8997–9004.

[157] M. Kan, S. Adhikari, Q. Sun, Ferromagnetism in MnX2 (X = S, Se) monolayers, Phys. Chem. Chem. Phys. 16 (10) (2014) 4990–4994.

[158] C. Ataca, H. Sahin, S. Ciraci, Stable, single-layer MX2 transition-metal oxides and dichalcogenides in a honeycomb-like structure, J. Phys. Chem. C 116 (16) (2012) 8983–8999.

[159] C. Gong, L. Li, Z.L. Li, H.W. Ji, A. Stern, Y. Xia, et al., Discovery of intrinsic ferromagnetism in two-dimensional van der Waals crystals, Nature 546 (7657) (2017) 265.

[160] N. Sivadas, M.W. Daniels, R.H. Swendsen, S. Okamoto, D. Xiao, Magnetic ground state of semiconducting transition-metal trichalcogenide monolayers, Phys. Rev. B 91 (23) (2015) 235425.

[161] V. Carteaux, F. Moussa, M. Spiesser, 2D Ising-like ferromagnetic behavior for the lamellar Cr2Si2Te6 compound: a neutron-scattering investigation, Europhys. Lett. 29 (3) (1995) 251–256.

[162] S. Lebegue, T. Bjorkman, M. Klintenberg, R.M. Nieminen, O. Eriksson, Two-Dimensional materials from data filtering and Ab initio calculations, Phys. Rev. X 3 (3) (2013) 031002.

[163] B.J. Liu, Y.M. Zou, S.M. Zhou, L. Zhang, Z. Wang, H.X. Li, et al., Critical behavior of the van der Waals bonded high T-C ferromagnet Fe3GeTe2, Sci. Rep. 7 (2017) 6184.

[164] Y. Tian, M.J. Gray, H.W. Ji, R.J. Cava, K.S. Burch, Magneto-elastic coupling in a potential ferromagnetic 2D atomic crystal, 2D Mater. 3 (2) (2016) 025035.

[165] T.J. Williams, A.A. Aczel, M.D. Lumsden, S.E. Nagler, M.B. Stone, J.Q. Yan, et al., Magnetic correlations in the quasi-two-dimensional semiconducting ferromagnet CrSiTe3, Phys. Rev. B 92 (14) (2015) 144404.

[166] H.L.L. Zhuang, Y. Xie, P.R.C. Kent, P. Ganesh, Computational discovery of ferromagnetic semiconducting single-layer CrSnTe3, Phys. Rev. B 92 (3) (2015) 035407.

[167] L.D. Casto, A.J. Clune, M.O. Yokosuk, J.L. Musfeldt, T.J. Williams, H.L. Zhuang, et al., Strong spin-lattice coupling in CrSiTe3, APL Mater. 3 (4) (2015) 041515.

[168] M.W. Lin, H.L.L. Zhuang, J.Q. Yan, T.Z. Ward, A.A. Puretzky, C.M. Rouleau, et al., Ultrathin nanosheets of CrSiTe3: a semiconducting two-dimensional ferromagnetic material, J. Mater. Chem. C 4 (2) (2016) 315–322.

[169] W. Xing, Y. Chen, P.M. Odenthal, X. Zhang, W. Yuan, T. Su, et al., Electric field effect in multilayer Cr2Ge2Te6: a ferromagnetic 2D material, 2D Mater. 4 (2) (2017) 024009.

[170] Y.Y. Chen, W.Y. Xing, X.R. Wang, B.W. Shen, W. Yuan, T. Su, et al., Role of oxygen in ionic liquid gating on two-dimensional Cr2Ge2Te6: a non-oxide material, ACS Appl. Mater. Interfaces 10 (1) (2018) 1383–1388.

[171] J.T. Ye, M.F. Craciun, M. Koshino, S. Russo, S. Inoue, H.T. Yuan, et al., Accessing the transport properties of graphene and its multilayers at high carrier density, Proc. Natl. Acad. Sci. USA 108 (32) (2011) 13002–13006.

[172] J.T. Ye, Y.J. Zhang, R. Akashi, M.S. Bahramy, R. Arita, Y. Iwasa, Superconducting dome in a gate-tuned band insulator, Science 338 (6111) (2012) 1193–1196.

[173] H.J. Deiseroth, K. Aleksandrov, C. Reiner, L. Kienle, R.K. Kremer, Fe3GeTe2 and Ni3GeTe2—two new layered transition-metal compounds: crystal structures, HRTEM investigations, and magnetic and electrical properties, Eur. J. Inorg. Chem. (8)(2006) 1561–1567.

[174] W. Zhang, P.K.J. Wong, P. Yan, J. Wu, S.A. Morton, X.R. Wang, et al., Observation of current-driven oscillatory domain wall motion in Ni80Fe20/Co bilayer nanowire, Appl. Phys. Lett. 103 (4) (2013) 042403.

[175] J.X. Zhu, M. Janoschek, D.S. Chaves, J.C. Cezar, T. Durakiewicz, F. Ronning, et al., Electronic correlation and magnetism in the ferromagnetic metal Fe3GeTe2, Phys. Rev. B 93 (14) (2016) 144404.

[176] V.Y. Verchenko, A.A. Tsirlin, A.V. Sobolev, I.A. Presniakov, A.V. Shevelkov, Ferromagnetic order, strong magnetocrystalline anisotropy, and magnetocaloric effect in the layered telluride Fe3-delta GeTe2, Inorg. Chem. 54 (17) (2015) 8598–8607.

[177] W. Chen, D.C. Qi, X.Y. Gao, A.T.S. Wee, Surface transfer doping of semiconductors, Prog. Surf. Sci. 84 (9–10) (2009) 279–321.

[178] C. Wackerlin, F. Donati, A. Singha, R. Baltic, A.C. Uldry, B. Delley, et al., Strong antiferromagnetic exchange between manganese phthalocyanine and ferromagnetic europium oxide, Chem. Commun. 51 (65) (2015) 12958–12961.

[179] C. Wackerlin, P. Maldonado, L. Arnold, A. Shchyrba, J. Girovsky, J. Nowakowski, et al., Magnetic exchange coupling of a synthetic Co(II)-complex to a ferromagnetic Ni substrate, Chem. Commun. 49 (91) (2013) 10736–10738.
[180] C. Wackerlin, K. Tarafder, J. Girovsky, J. Nowakowski, T. Hahlen, A. Shchyrba, et al., Ammonia coordination introducing a magnetic moment in an on-surface low-spin porphyrin, Angew. Chem. -Int. Ed. 52 (17) (2013) 4568–4571.
[181] T. Gang, M.D. Yilmaz, D. Atac, S.K. Bose, E. Strambini, A.H. Velders, et al., Tunable doping of a metal with molecular spins, Nat. Nanotechnol. 7 (4) (2012) 232–236.
[182] J. Girovsky, K. Tarafder, C. Waeckerlin, J. Nowakowski, D. Siewert, T. Haehlen, et al., Antiferromagnetic coupling of Cr-porphyrin to a bare Co substrate, Phys. Rev. B 90 (22) (2014) 220404(R).
[183] A. Atxabal, M. Ribeiro, S. Parui, L. Urreta, E. Sagasta, X. Sun, et al., Spin doping using transition metal phthalocyanine molecules, Nat. Commun. 7 (2016) 13751.
[184] T.L.A. Tran, D. Cakir, P.K.J. Wong, A.B. Preobrajenski, G. Brocks, W.G. van der Wiel, et al., Magnetic properties of bcc-Fe(001)/C-60 interfaces for organic spintronics, ACS Appl. Mater. Interfaces 5 (3) (2013) 837–841.
[185] T.L.A. Tran, P.K.J. Wong, M.P. de Jong, W.G. van der Wiel, Y.Q. Zhan, M. Fahlman, Hybridization-induced oscillatory magnetic polarization of C-60 orbitals at the C-60/Fe(001) interface, Appl. Phys. Lett. 98 (22) (2011) 222505.
[186] P.K.J. Wong, T.L.A. Tran, P. Brinks, W.G. van der Wiel, M. Huijben, M.P. de Jong, Highly ordered C-60 films on epitaxial Fe/MgO(001) surfaces for organic spintronics, Org. Electron. 14 (2) (2013) 451–456.
[187] P.K.J. Wong, W. Zhang, G. van der Laan, M.P. de Jong, Hybridization-induced charge rebalancing at the weakly interactive C-60/Fe3O4(001) spinterface, Org. Electron. 29 (2016) 39–43.
[188] P.K.J. Wong, W. Zhang, K. Wang, G. van der Laan, Y.B. Xu, W.G. van der Wiel, et al., Electronic and magnetic structure of C-60/Fe3O4(001): a hybrid interface for organic spintronics, J. Mater. Chem. C 1 (6) (2013) 1197–1202.
[189] Y. Alaskar, S. Arafin, Q.Y. Lin, D. Wickramaratne, J. McKay, A.G. Norman, et al., Theoretical and experimental study of highly textured GaAs on silicon using a graphene buffer layer, J. Cryst. Growth 425 (2015) 268–273.
[190] Y. Kim, S.S. Cruz, K. Lee, B.O. Alawode, C. Choi, Y. Song, et al., Remote epitaxy through graphene enables two-dimensional material-based layer transfer, Nature 544 (7650) (2017) 340.
[191] K. Chung, C.H. Lee, G.C. Yi, Transferable GaN layers grown on ZnO-coated graphene layers for optoelectronic devices, Science 330 (6004) (2010) 655–657.
[192] K. Chung, S.I. Park, H. Baek, J.S. Chung, G.C. Yi, High-quality GaN films grown on chemical vapor-deposited graphene films, NPG Asia Mater. 4 (2012) e24.
[193] J. Kim, C. Bayram, H. Park, C.W. Cheng, C. Dimitrakopoulos, J.A. Ott, et al., Principle of direct van der Waals epitaxy of single-crystalline films on epitaxial graphene, Nat. Commun. (2014) 5, 4836.
[194] H. Oh, Y.J. Hong, K.S. Kim, S. Yoon, H. Baek, S.H. Kang, et al., Architectured van der Waals epitaxy of ZnO nanostructures on hexagonal BN, NPG Asia Mater. 6 (2014) e145.
[195] V.M. Karpan, G. Giovannetti, P.A. Khomyakov, M. Talanana, A.A. Starikov, M. Zwierzycki, et al., Graphite and graphene as perfect spin filters, Phys. Rev. Lett. 99 (17) (2007) 176602.
[196] V.M. Karpan, P.A. Khomyakov, G. Giovannetti, A.A. Starikov, P.J. Kelly, Ni(111)/graphene/h-BN junctions as ideal spin injectors, Phys. Rev. B 84 (15) (2011) 153406.
[197] P.K.J. Wong, M.P. de Jong, L. Leonardus, M.H. Siekman, W.G. van der Wiel, Growth mechanism and interface magnetic properties of Co nanostructures on graphite, Phys. Rev. B 84 (5) (2011) 054420.
[198] P.K.J. Wong, E. van Geijn, W. Zhang, A.A. Starikov, T.L.A. Tran, J.G.M. Sanderink, et al., Crystalline CoFeB/graphite interfaces for carbon spintronics fabricated by solid phase epitaxy, Adv. Funct. Mater. 23 (39) (2013) 4933–4940.

CHAPTER 10

Spin-valve effect of 2D-materials based magnetic junctions

Muhammad Zahir Iqbal

Nanotechnology Research Laboratory, Faculty of Engineering Sciences, GIK Institute of Engineering Sciences and Technology, Topi, Pakistan

Chapter Outline

10.1 Introduction

Many bulk materials exist in the form of stacking layers that are bonded through weak interlayer interactions, permitting exfoliation into separate thin layers [1]. Such class of atomically thick materials is classified as two-dimensional (2D) layered materials. Among these, graphene is the first ever discovered 2D material, which is the single layer of graphite. The linear dispersion relation and the massless nature of its charge carriers known as Dirac fermions provide new physics for scientists. Graphene is an exceptional thinnest electrical and thermal conductor with high carrier mobility [2,3].

A new class of 2D crystals is also discovered including transition metal dichalcogenides (TMDs), the analog of graphenes, such as hexagonal boron nitride (hBN) and transition metal oxides like perovskite- and titania-based oxides [4,5]. Particularly, TMDs have revealed a wide range of mechanical, chemical, optical, thermal, electronic, and spintronic properties [6–11]. The 2D version of TMDs possesses such remarkable properties, which

Spintronic 2D Materials.
DOI: https://doi.org/10.1016/B978-0-08-102154-5.00009-6

are completely distinct from that of graphene. Graphene offers extremely high mobility of $\sim 10^6$ $cm^2/V/s$ at lower temperature [12]. Since graphene does not have a bandgap, however graphene-based field-effect transistors (FETs) offers low on/off current ratios and cannot be effectively switched on/off. Bandgaps are generated in graphene through chemical functionalization, applying high electric field and nanostructuring, but such approaches add complexity and degrade the carrier mobility [13–17]. In contrast, TMDs have sizable bandgaps (1–2 eV [6]), which promises fascinating FETs and optoelectronic devices. Besides, the spin-related properties are highly influenced by the interface formed between the ferromagnetic (FM) and 2D materials in a spin-valve device, and due to the strong spin–orbit coupling, 2D materials could be ideal candidates for spintronic devices.

The general formula for TMDs is MX_2, where "M" represents transition metal (Ti, Zr, etc.) from group IV and group V (such as V, Nb, and so on) otherwise group VI (including W and Mo), while "X" stands for chalcogens such as S, Se, etc. These materials constitute flat structures in the form of X–M–X; the chalcogen (X) atoms form two hexagonal planes which are separated with a layer of metal atoms, as demonstrated in Fig. 10.1A. TMDs

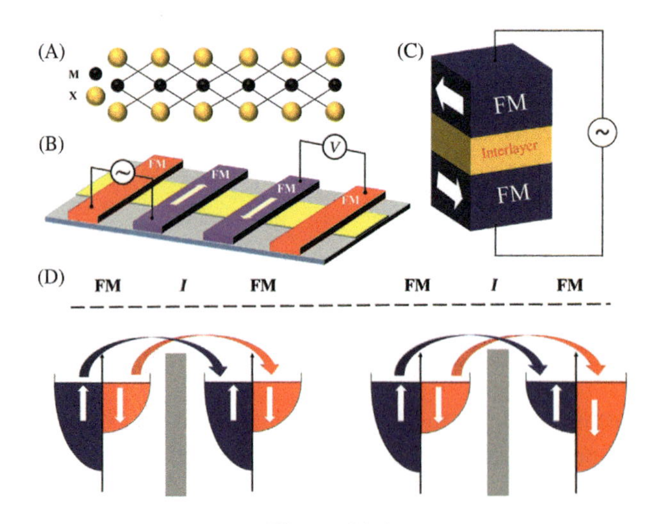

Figure 10.1

(A) The schematic structure of TMD, wherein a plane of transition metal atoms (M) separate chalcogen atoms (X). (B) Schematics showing lateral spin-valve structure constituting CIP geometry. (C) Graphical representation of vertical spin-valve structure (CPP geometry); the two FMs are separated with a nonmagnetic spacer. (D) Illustration of spin-DOS of FM electrodes. The left and right panels show the spin parallel and antiparallel situations, respectively. The higher the spin-DOS the higher is the MR ratio. *CIP*, Current in-plane; *CPP*, current perpendicular to plane; *DOS*, density of states; *FM*, ferromagnetic; *MR*, magnetoresistance; *TMD*, transition metal dichalcogenide. Source: *From M.Z. Iqbal, et al., Electron spin dynamics in vertical magnetic junctions incorporating two-dimensional layered materials, J. Mater. Chem. C 5(43) (2017) 11174–11184.*

display rhombohedral or hexagonal symmetry, wherein the metal atoms possess trigonal or octahedral prismatic coordination. The electrical properties of these crystalline materials vary from semiconducting to metallic, but they also demonstrate some exotic behaviors, for instance, superconductivity and charge-density waves [18–20], however, these are beyond the context of this chapter.

TMDs have recently received a great deal of interest owing to their layer-dependent properties. For example, the indirect bandgap in several bulk TMDs [molybdenum disulfide (MoS_2) and tungsten disulfide (WS_2)] transforms into direct bandgap when peeled in monolayer form. This direct bandgap enables photoluminescence in single-layer form, which makes these materials useful for optoelectronic applications [21]. The unique electronic structure allows MoS_2 to be utilized in valley polarization that is not observed in bilayer form [22–24]. Pristine MoS_2 behaves as nonmagnetic (NM); the NM nature is retained even after creating vacancies for S or Mo alone. However, the substitution of Co at Mo site induces spin-polarized state [25]. Also, Zn, Fe, Mn, and Co doping in single-layer MoS_2 results in 2D-diluted magnetic semiconductors [26]. Further, the spintronic attributes are highly affected by the interface states between the FM and 2D crystal. For instance, the interface formed in Fe_4N/MoS_2 superlattice between Fe and S induces spin polarization in MoS_2 due to strong interfacial hybridization [27]. In general, 2D materials including graphene, TMDs, hBN, and various heterostructures of these materials have revealed remarkable properties, and therefore have found outstanding applications in the very emerging fields of science and technology.

Spintronics, dealing with the study of control and optimization of the spin degree of electrons has found revolutionary applications in information storage and programmable logic devices [28–37]. In order to detect the spin signal, the spin-valve device is realized that comprises at least two FM electrodes and a NM layer [38]. The resistance of a spin-valve junction when measured displays two distinct resistance states owing to the relative orientations of magnetization vectors of FM films. These devices are categorized into (1) lateral spin valve and (2) vertical spin valve. In the lateral configuration, FM electrodes are deposited over the top of the NM film, wherein the spin-polarized current flows along the plane of the device forming current in-plane configuration (Fig. 10.1B). Whereas in vertical spin-valve geometry, the NM layer is incorporated between the FM electrodes and the spin-polarized current flows vertically making current perpendicular to plane (CPP) geometry (Fig. 10.1C) [36]. Here, the accumulated results on vertical spin-valve structures obtained for different intervening layers are presented. The spin-valve signal in a vertical spin device (CPP geometry) is determined by measuring the magnetoresistance (MR) ratio as a function of $\vec{B}$. The schematic illustration of spin filtering for FM/NM/FM vertical junction is also presented in Fig. 10.2A. Fig. 10.2B depicts the optical micrograph for vertical spin-valve structure incorporating Gr as the intervening layer. The mechanism used for the spin-valve signal measurement is illustrated in

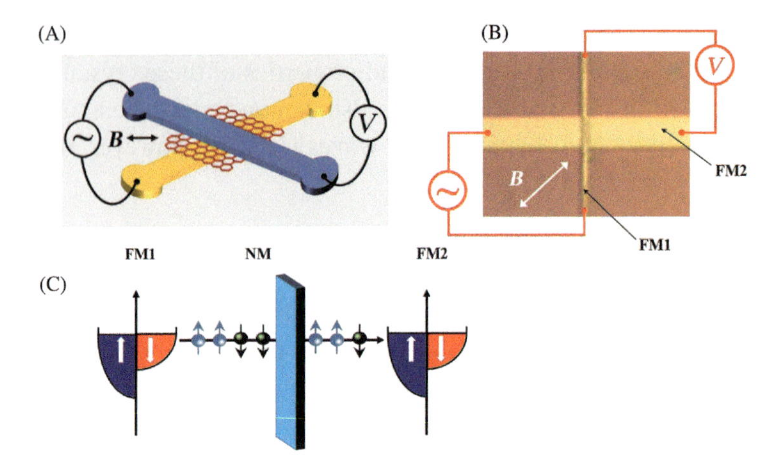

Figure 10.2
(A) Schematic illustration of vertical spin valve comprising of two FM electrodes with graphene interlayer. (B) Optical image of NiFe/graphene/NiFe junction showing the mechanism of transport measurement and the direction of the applied magnetic field. (C) Schematic illustration of spin filtering for FM/NM/FM vertical junction. *FM*, Ferromagnetic; *NM*, nonmagnetic. Source: *From M.Z. Iqbal, et al., Spin valve effect of 2D-materials based magneticjunctions, Adv. Eng. Mater. 20 (2018) 1700692.*

Fig. 10.2C, where an in-plane magnetic field (B) is swept to achieve parallel and antiparallel magnetization states.

In this chapter, the spin-valve effect in CPP devices integrating TMDCs as the intervening layers is discussed. For the sake of comparison, other 2D materials such as graphene, graphene–graphene (Gr–Gr), and graphene–hBN (Gr–hBN) incorporated spin-valve devices are also discussed. Further, the inverted MR signal observed in Gr–hBN device is explained through Julliere model. The current–voltage (I–V) curves are particularly discussed to determine the metallic or insulating nature of magnetic junctions.

10.2 Theoretical model

Theoretically, graphene layers are anticipated to be a perfect cause of spin filtering in FM/graphene/FM magnetic junctions [39,40]. A perfect spin filter is one that allows all carriers of same spin orientation to pass through and block all other carriers having different spins. The spin filtering of graphene at the FM/graphene interface can be attributed to its unique structural properties. The in-plane lattice constants of graphite (vertically stacked layers of graphene) match almost perfectly with the lattice constants of fcc-Co and hcp-Ni structures (i.e., only 1.3% lattice mismatch for graphene/Ni and 1.9% for graphene/Co structures), which leads to a perfect spin filtering in these junctions [39,40]. In addition, graphene has occupied states only around the vicinity of K point at the Fermi level in the first Brillouin

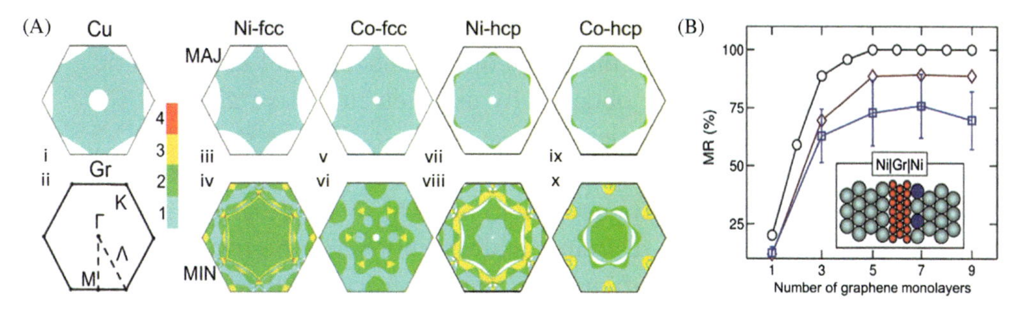

Figure 10.3

Fermi surface projections on close-packed planes for (A) (i) fcc-Cu. (ii) Gr K point, (iii), (v), (vii), and (ix) representing the majority spins for fcc-Ni, fcc-Co, hcp-Ni, and hcp-Co, respectively. (iv), (vi), (viii), and (x) minority spins of fcc-Ni, fcc-Co, hcp-Ni, and hcp-Co, respectively. (B) MR ratio as a function of the number of graphene layers for ideal junction, junction with alloy disorder, and junction with rough top layer of electrode representing by black circles, brown diamonds, and blue squares, respectively. The spread in MR is demonstrated by error bars, while alloy disorder at Ni/graphene interface is shown by the schematic (inset). *MR*, Magnetoresistance. Source: *From V. Karpan, et al., Graphite and graphene as perfect spin filters, Phys. Rev. Lett. 99 (17) (2007) 176602.*

zone. Fig. 10.3A shows the absence of majority spin states for Ni and Co in the vicinity of K point. However, it reveals the presence of minority spin states almost everywhere in the locality of K point. Thus only minority spins can transmit from Ni or Co surface into the graphene. The graphene states close to K point are responsible for the transport of spins from one FM electrode to the other. Therefore it is predicted that if the optimum number of graphene layers will be intervened between FM electrodes, conductance due to majority spins will be snuffed out and only minority spins will have a continuous transport channel. Hence, graphene is supposed to be a perfect spin filter in FM/graphene/FM junctions.

The spin filtering mechanism in FM/graphene/FM junction depends on various factors like the number of graphene layers, the geometry of ferromagnet/graphene interface, lattice mismatch, and the roughness or disorder of the interface [39,40]. Fig. 10.3B illustrates the dependence of MR ratio on number of layers, where MR ratio increases with number of graphene layers and for ideal junctions, it becomes 100% (black circles with no disorders). Junctions with alloy disorders are represented with brown curve while the blue curve depicts the MR ratio for junction exhibiting a rough top layer of one of the two FM electrodes. This type of disordered or roughed junctions is responsible for the reduction of MR's magnitude to 70%–90%. The inset of Fig. 10.3B represents the Ni/graphene/Ni junction with blue Cu spheres as alloy disorder and bluish green Ni spheres, whereas the small red spheres denote the locality of C atoms of graphene.

The measurement of resistance under the influence of the magnetic field for such a junction reveals high- and low-resistance states due to the difference in coercivities of

FMs. This corresponds to the spin-valve signal. FM electrodes of different widths can be used to achieve different switching fields of magnetization of the magnetic materials [41], which can also be obtained by using electrodes of different FM materials having different coercivities. In case of low-resistance state, the magnetization vectors of FM electrodes are oriented parallel to each other and the spin-polarized electrons can easily move from one FM to the other giving a large current [42]. However, the high-resistance state occurs when the two FMs are magnetized in the antiparallel configuration and thus very small spin current flows through the junction [43]. The sign of spin signal or spin polarization depends on the type of spins transmitting through the NM spacer. As predicted by many theoretical studies, the electron-spin dynamics is not only determined from the spin density of states (DOS), but also have a strong dependence on the mobility of particular type of spins. The study was initially highlighted by Gadzuk [44], when he reported positive spin polarization for Ni. Although the DOS of minority "d" band electrons is much larger compared with DOS of majority "s" electrons. However, the high mobile "s" electrons become dominant in tunneling across the barrier as compared with localized "d" electrons. Meservey and Tedrow performed experiments regarding spin-polarized tunneling, wherein they demonstrated that tunneling current is spin-dependent [45]. This motivated Julliere to put forward a model explaining the transport mechanism in FM junctions, which is now named as Julliere model [46]. The model beautifully relates the MR of the magnetic junction to spin polarization of FM layers. It defines the MR as:

$$\mathrm{MR} = \frac{R_{\mathrm{AP}} - R_{\mathrm{P}}}{R_{\mathrm{P}}} = \frac{2P_1P_2}{1 - P_1P_2} \tag{10.1}$$

Based on the preliminary predictions, Julliere presumed explicitly that spin polarization can only be defined from FM metals and is independent of interface or barrier. Intuitively, the MR ratio increases by increasing the spin polarization of the magnetic films. Fig. 10.1D illustrates the schemes for electron-spin transport in spin-valve structure. The spin parallel situation is depicted in the left panel while the spin antiparallel configuration is shown in the right panel of the figure. The available spin-DOS of FM layers determines the magnitude of MR and is increased by using FMs with higher spin polarization [47]. Also, the MR can be increased when the spin filtering efficiency of the NM spacer is increased. It may be achieved by increasing the thickness of the intervening layer or exchange splitting.

10.3 Growth of two-dimensional materials and device fabrication

Several methods are being used to synthesize 2D materials (graphene, TMDs, hBN, etc.) such as micromechanical cleavage, physical vapor deposition (PVD), and chemical vapor deposition (CVD) techniques. In micromechanical cleavage, single and few layers of 2D materials are obtained through exfoliation of the bulk crystal. However, it is difficult to control the number of layers and the size of the 2D flakes for large-scale production. In PVD, a heated atomic source is used that controllably deposits the material on the desired substrate [48]. This method requires an ultrahigh vacuum and well-calibrated condition, highly pure atomic sources. The samples obtained under these conditions are then studied *in-situ* using surface sensitive techniques. While the CVD process involves the growth of 2D crystals by the decomposition or reaction of the liquid, solid and gas precursors under controlled environment [49,50]. This method not only allows the growth of single-layer crystals but also facilitates the large-scale synthesis of 2D materials.

In device fabrication, the bottom FM electrode of a particular thickness is deposited on Si substrate via e-beam lithography. Then, a 2D layer is transferred on the top surface of FM and finally, the top FM electrode is deposited using e-beam lithography. The unwanted 2D layer excluding the junction area is removed through O_2 plasma etching. Lock-in technique at a frequency of 11.7 Hz and with root mean square current amplitude of 45 μA is employed to carry out the transport measurements along with the obliquely applied magnetic field. The schematic illustration of transferring graphene layer and device fabrication is given in Fig. 10.4.

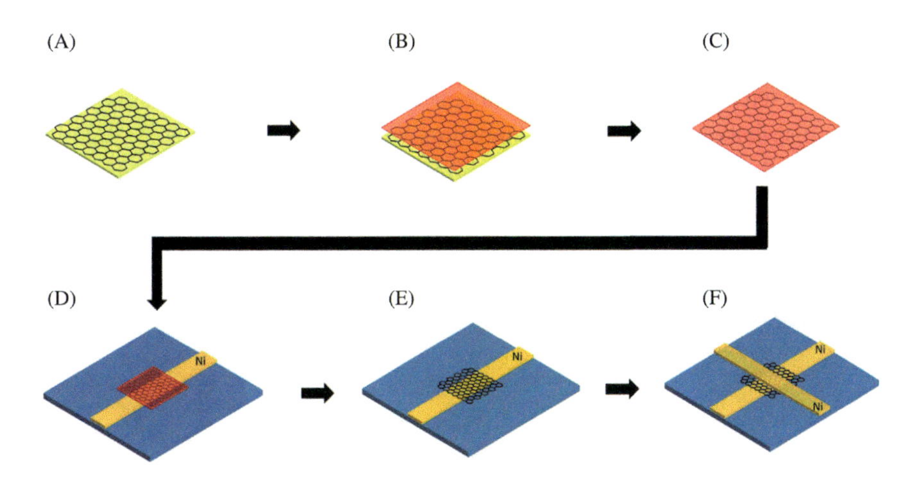

Figure 10.4

Schematic representation for the transfer process of graphene layer and device fabrication. (A) Graphene grown on Cu substrate (B) Poly(methyl methacrylate) (PMMA) spin coated on graphene to etch the bottom Cu foil. (C) PMMA/graphene membrane after etching of the Cu foil APS solution. (D) Bottom NiFe electrode patterned on SiO_2/Si substrate and the transferred PMMA/graphene membrane. (E) PMMA removed via acetone solution. (F) Top Ni electrode is patterned to form the final device. *APS*, Ammonium persulfate. Source*: From M.Z. Iqbal, et al., Spin valve effect of 2D-materials based magneticjunctions, Adv. Eng. Mater. 20 (2018) 1700692.*

10.4 Tungsten disulfide–based spin-valve device

To construct a spin-valve device, two FM layers are contacted through a NM spin transport layer. The electrical resistance of the device is controlled by the spin states of FM layers leads by applying an external magnetic field. The field of spin electronics is revolutionized by discoveries, such as giant MR and tunneling MR (TMR). The atomically flat 2D crystals are used as spacers in vertical magnetic structures for studying various spintronic applications [8,11,51–54]. The optical image of WS_2-based spin-valve structure (Co/WS_2/NiFe) is shown in Fig. 10.5A, the WS_2 layer is sandwiched between Co (top) and NiFe (bottom) electrodes. In tungsten disulfide, because of its strong spin–orbit coupling and other excellent properties is applicable and can be considered as an ideal system in spintronics. An in-plane magnetic field (B) polarizes the magnetizations of FM films either semiparallel or semiantiparallel by sweeping B. This gives rise to two distinct states of resistance for the magnetic junction, the so-called spin-valve effect. Fig. 10.5B illustrates the relative MR for WS_2-based spin-valve structure (Co/WS_2/NiFe) that clearly depicts the low and high states of resistance [8]. The relative MR ratio was calculated by using the above mentioned equation 9.1 of polarization (Eq. 10.1), wherein R_{AP} represents the high-resistance state signifying antiparallel alignment of spins of FM layers, while R_P corresponds to low-resistance state during which magnetization vectors are in the parallel configuration. The magnitude for MR was estimated to be approximately 0.18% at 300K, which increased to 0.47% when the measurement was performed at 4.2K. The temperature dependence of MR is illustrated in Fig. 10.9C(ii), where MR ratio monotonically increases from 0.17% at 300K to 0.50% at 10K. Actually, when the temperature is raised, the magnetic impurity scattering increases and the electron energy distribution functions start smearing, which decreases the MR value [55]. WS_2 incorporated magnetic junction was further analyzed by measuring I–V characteristics at

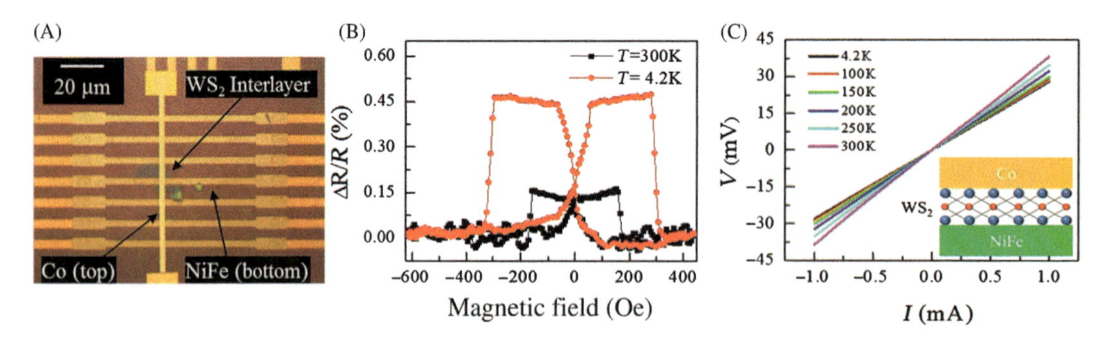

Figure 10.5
(A) Optical image of WS_2-based spin-valve structure (Co/WS_2/NiFe). (B) MR traces for WS_2-based spin-valve structure obtained at room temperature (300K) and 4.2K, respectively. (C) I–V characteristics of the junction indicating metallic behavior for WS_2. *MR*, Magnetoresistance. Source: *From M.Z. Iqbal, et al., Room temperature spin valve effect in NiFe/WS2/Co junctions, Sci. Rep. 6 (2016) 21038.*

progressive temperatures (Fig. 10.5C). Unexpectedly linear curves were observed suggesting that WS_2 act as a conducting film rather than behaving as a semiconducting barrier inserted between the two FM leads. This metallic behavior of semiconducting WS_2 arose due to strong hybridization of S atoms with FM atoms at the interface [9].

10.5 Molybdenum disulfide–based spin-valve device

Sometimes the exposure of the bottom electrode to the ambient environment causes oxidation at the surface, which may reduce the magnitude of MR [56,57]. For instance, the MR calculated for NiFe/MoS_2/NiFe/Co magnetic junction is approximately 0.4%; the low value of MR can be attributed to the oxidation of NiFe electrode during the transfer process of MoS_2. To avoid the possibility of oxidation, ~2 nm thick film of Au was sputtered over NiFe film constituting NiFe/MoS_2/Au/NiFe/Co architecture. This resulted in an increase of MR to ~0.73% due to the absence of oxide layer on NiFe electrode (as depicted in Fig. 10.6A) [58]. Similar results were obtained when Ni electrode was covered with a thin layer of Au before transferring graphene [59]. Besides, theoretical calculations predicted reasonable MR (~9%) for MoS_2 incorporated vertical magnetic structure [60]. Fig. 10.6A shows the experimentally observed spin-valve signals for NiFe/MoS_2/Au/NiFe/Co device at various temperatures [58]. The decrease in MR at higher temperatures is attributed to the smearing of electron energy distribution and impurity scattering as stated earlier [61]. The temperature dependence of resistance of NiFe/MoS_2/Au/NiFe/Co junction is demonstrated in Fig. 10.6B; the resistance increases monotonically with increasing temperature revealing

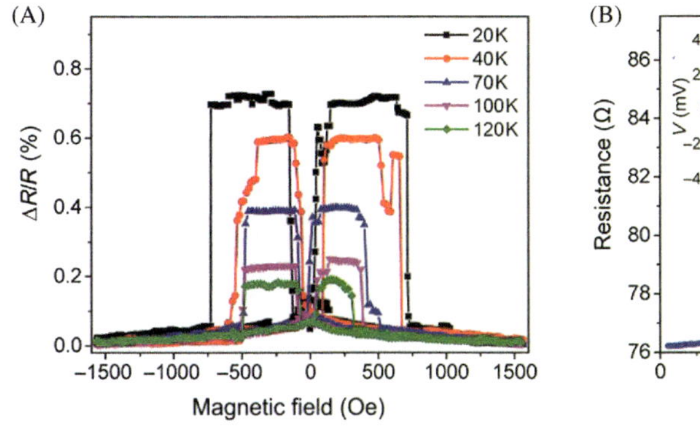

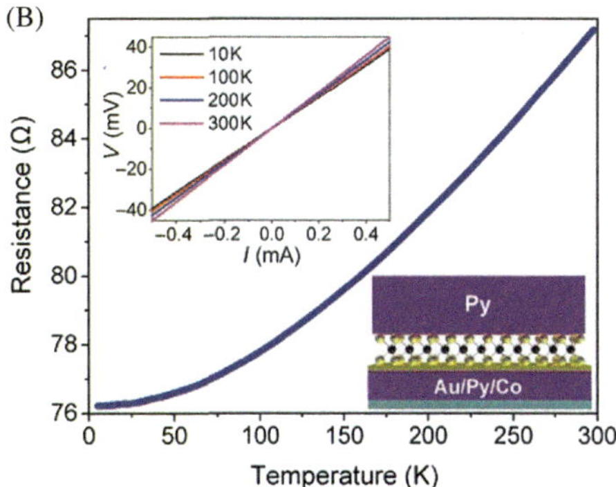

Figure 10.6
(A) Spin-valve effect in NiFe/MoS_2/Au/NiFe/Co magnetic junction at different temperatures. (B) The temperature dependent resistance of the junction resistance of NiFe/MoS_2/Au/NiFe/Co structure, (inset: I-V curves at different temperatures and schematic of the device). Despite semiconducting nature, MoS_2 show metallic behavior in vertical magnetic junctions as indicated by the linearity of I-V curves. Source: *From W. Wang, et al., Spin-valve effect in NiFe/MoS2/NiFe junctions, Nano Lett. 5 (8) (2015) 5261–5267.*

metallic behavior for MoS_2. The ohmic behavior is further confirmed through I–V curves (the inset). This metallic behavior is ascribed to strong hybridization between S atoms and Ni/Fe atoms of FM layer. The distance between the FM surface atoms and the contiguous S atoms is extremely small which enables overlap of wave functions between S and FM atoms, leading to metallic behavior of the junction [60].

Wang et al. provide further details of the physical mechanism causing the MR by first-principle density function theory–based transport calculations for the Py/MoS_2/Py junctions (Fig. 10.7A). The results depict the metallic junction behavior of single-layer MoS_2 exhibiting a large conductance of 1.2×10^{14} S/m^2. Transmittance (T) as a function of energy (with respect to Fermi energy) is plotted in Fig. 10.7B. For parallel configuration of junctions, T is larger for up spin ($T\uparrow$) as compared with down spin ($T\downarrow$), for energy values around E_F. Thus Py/MoS_2 junction works as a spin filter by transmitting spin-up electrons.

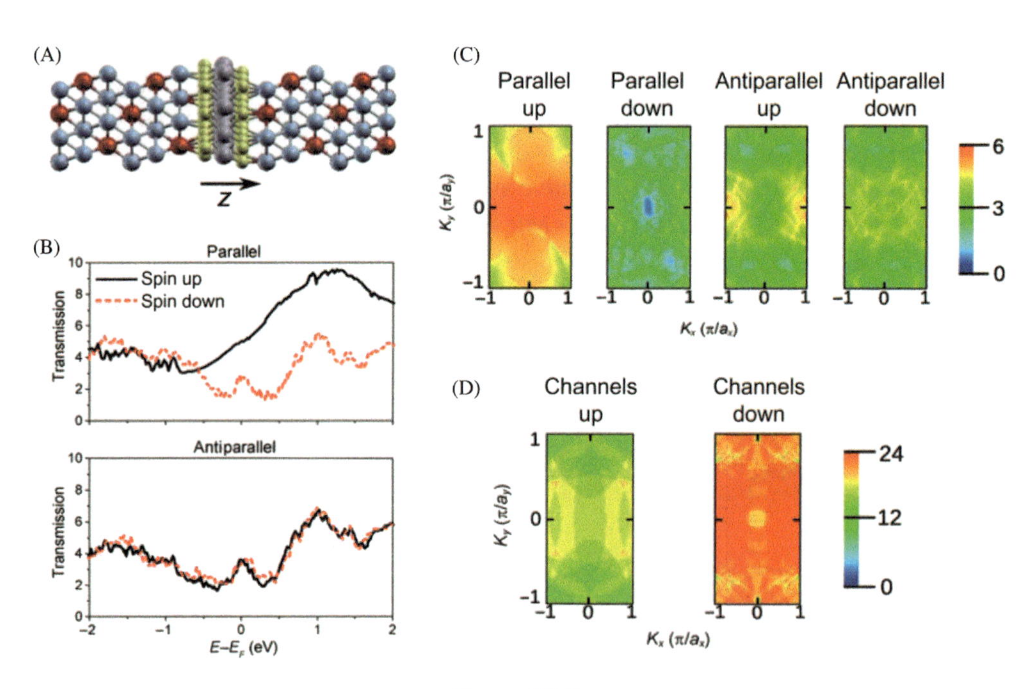

Figure 10.7

Transport calculations for Py/MoS_2/Py junction. (A) Scattering region for the ab initio calculations, with the transport along the z-direction. Self-energies corresponding to semiinfinite Py electrodes are attached to the six left- and right-hand Py layers inside the scattering region. (B) Transmission as a function of energy for spin-up and spin-down channels, for parallel and antiparallel setup of the junction. (C) $K_x - K_y$ resolved plot of the transmission at $E = E_F$, for the spin-up and spin-down directions for parallel and antiparallel configurations. (D) Spin-up and spin-down open channels for the electrodes at $E = E_F$, plotted across the $K_x - K_y$ plane. *Source: From W. Wang, et al., Spin-valve effect in NiFe/MoS2/NiFe junctions, Nano Lett. 15 (8) (2015) 5261–5267.*

In addition, $T\uparrow$ is a smooth function of E for $E - E_F > -1$ eV, whereas $T\downarrow$ is a less smooth function. The fact behind this smoothness is s-type character of Py for these energies while the extended spin-down d-states of the Py over a wide energy range around E_F leads to comparatively less smooth $T\downarrow$ function. The $T\uparrow$ and $T\downarrow$ transmissions in the antiparallel setups are almost identical [9].

10.6 Spin-valve effect in $Fe_3O_4/MoS_2/Fe_3O_4$

Studies have shown that the spin-valve effect diminishes and ultimately vanishes with increasing temperature. In this regard, Curie temperature (T_c) of the FM electrodes can play a critical role in observing room temperature spin-valve effect. Fe_3O_4, known as magnetite, is a half-metal with high Curie temperature and has extraordinary potential for magnetic tunnel junctions (MTJs). It has a very well-defined spin-polarized electrons band near the Fermi level (i.e., half-metallic character) and a T_c of approximately 858K, which make it an ideal candidate in room temperature spintronic applications [62–64]. It has revealed outstanding spin transport properties such as spin filter effect, large transversal MR, spin-Seebeck effect, and electrical field–induced phase transition [65–70]. Fig. 10.8A shows TMR for the $Fe_3O_4/MoS_2/Fe_3O_4$ magnetic junction [11]. This study highlighted the role of magnetic layers in spin-valve architectures; featuring transport of spins through tunneling mechanism. The TMR ratio was determined as a function of the applied in-plane magnetic field, indicating an abrupt increase of TMR owing to the magnetization reversal from parallel to antiparallel configuration. Its magnitude is calculated using Eq. (10.1), which comes out to be 0.2%. The small value of TMR was attributed to the low-quality interface

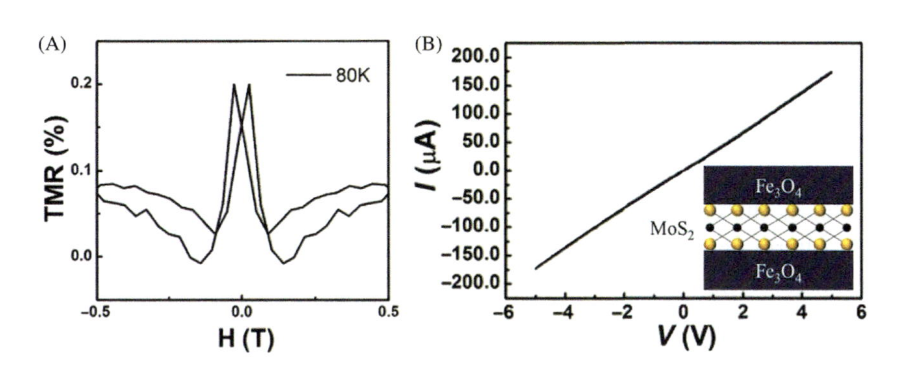

Figure 10.8
(A) TMR for $Fe_3O_4/MoS_2/Fe_3O_4$ vertical magnetic junction measured at 80K. (B) I–V measurement, demonstrating nonlinear I–V characteristics for the structure that defines the formation of a tunnel junction (inset: schematic of the $Fe_3O_4/MoS_2/Fe_3O_4$ structure). *TMR*, Tunneling magnetoresistance. Source: *From H.-C. Wu, et al., Spin-dependent transport properties of Fe3O4/MoS2/Fe3O4 junctions, Sci. Rep. 5 (2015) 15984.*

formed between the top Fe_3O_4 and MoS_2 intervening layers [11]. The tunneling characteristics of magnetotransport was verified through I–V measurements, indicating nonlinear behavior which follows that $Fe_3O_4/MoS_2/Fe_3O_4$ junction is in the tunneling regime (Fig. 10.8B). In Section 10.5, the FM layers were not oxidized and thus made the I–V curves linear, facilitating the formation of ohmic junction.

10.7 Comparison between transition metal dichalcogenide's with different two-dimensional materials-based spin valves

Besides TMDs, other 2D materials such as graphene, hBN, and their heterostructures have been exploited as interlayers in the spin-valve devices. Fig. 10.9A(i) elucidates the results of MR ratio for the graphene spin valve at two different temperatures. The relative MR ratio is observed to increase abruptly as a result of the magnetization switch from parallel to antiparallel alignment. The coercivities of magnetic layers are dependent on the dimensions of FM films and are different for different FM metals [52]. The magnitude of MR is calculated to be approximately ~0.085% at room temperature that is raised to ~0.14% at 10K. To understand the transport mechanism through the graphene magnetic junction, I–V measurements were performed at two different temperatures, as illustrated in Fig. 10.9A (iii). The obtained linear I–V characteristic curves were in contrast to previously reported results [71,72], which indicate ohmic behavior of graphene in perpendicular magnetic junction. Trend of temperature-dependent MR is demonstrated in Fig. 10.9(ii), showing an increase in MR as the temperature is lowered to 10K.

In the second attempt, double-layered graphene obtained through a simple transfer process, that is, Gr–Gr was incorporated in vertical spin-valve structure as the NM spacer. The device resistance was measured via a four-probe measurement setup, thereby sweeping the magnetic field from negative values to positive values. The MR traces for this double-layered graphene magnetic device are demonstrated in Fig. 10.9B(i), giving the magnitude of ~0.27% at 300K and ~0.47% at 10K for the junction. In contrast to Gr incorporated magnetic junction, the Gr–Gr magnetic device is found to offer higher MR. This suggests the high spin filtering effect for Gr–Gr homostructure when sandwiched between the FMs. Therefore to increase the spin filtering effect and making the spin properties more effective, the number of graphene layers is increased as claimed by some theoretical and experimental results [73–75]. According to I–V characteristics of the junction showing a linear trend (Fig. 10.9B(iii)), the Gr–Gr homostructure behaves as metallic rather than as an insulating barrier in the vertical spin-valve configuration. The MR ratio versus temperature is also shown in Fig. 10.9B(ii), which decreases with rising temperature.

Fig. 10.9C(i) illustrates the magnetic field dependence of MR ratio for hBN MTJ at 300K and 10K, thereby varying the magnetic field from negative to positive direction. The

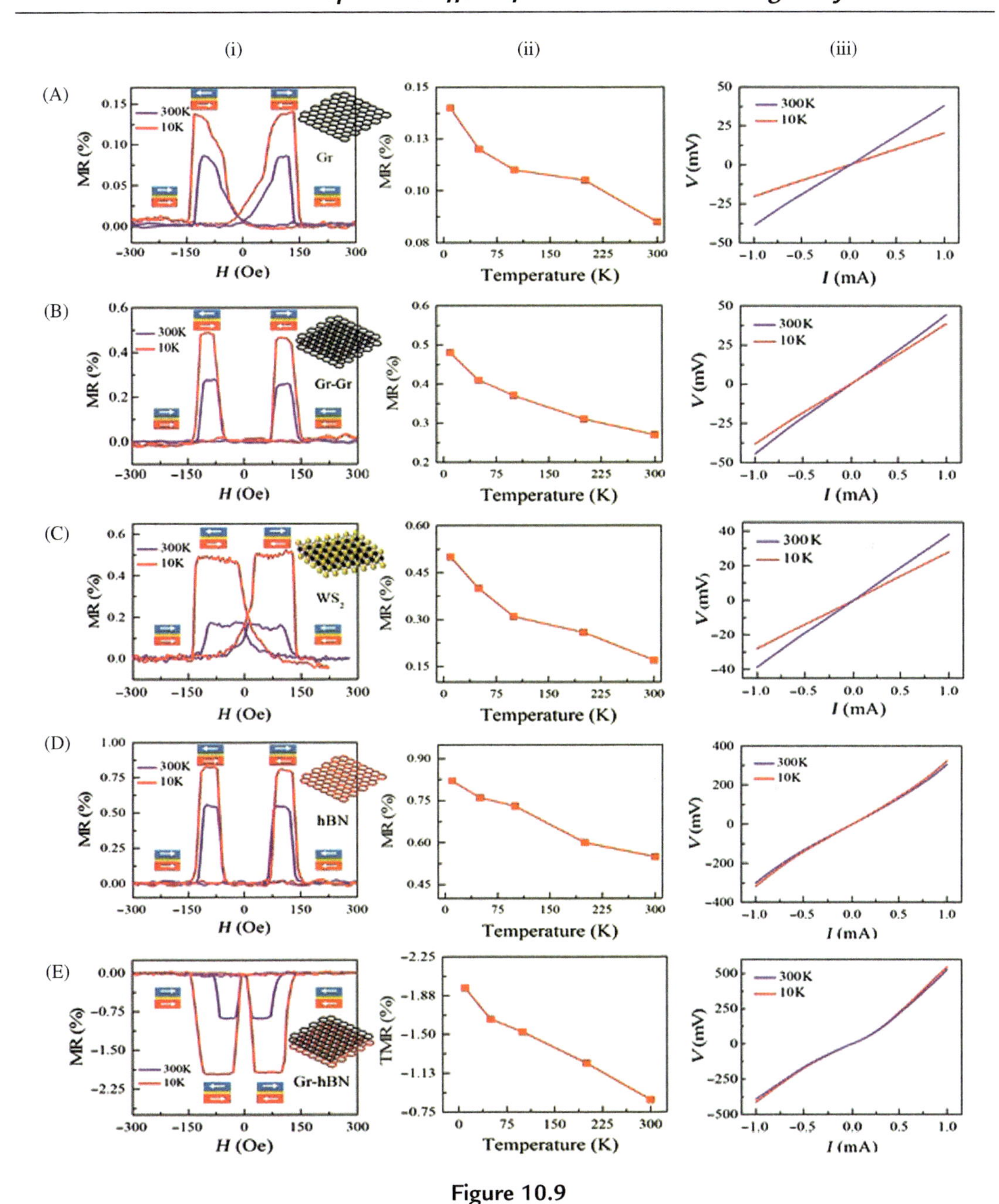

Figure 10.9

MR curves, temperature dependent MR behavior and I–V characteristics spin valve (A) Gr (B) Gr-Gr (C) WS_2 (D) hBN and (E) Gr-hBN. Sources: *From M.Z. Iqbal, et al., Spin valve effect of NiFe/graphene/NiFe junctions, Nano. Res. 6 (5) (2013) 373–380; M.Z. Iqbal, et al., Room temperature spin valve effect in NiFe/WS₂/Co junctions, Sci. Rep. 6 (2016) 21038; M.Z. Iqbal, et al., Room temperature spin valve effect in the NiFe/Gr–hBN/Co magnetic tunnel junction, J. Mater. Chem. C 4 (37) (2016) 8711–8715.*

relative MR is obtained to be 0.55% at 300K, which is systematically enhanced to 0.82% at 10K. The temperature dependence of MR ratio for hBN MTJ is also presented in Fig. 10.9C (ii), whose MR monotonically progresses as the temperature is lower down. This increase in MR ratio could be due to high magnetizations of FM electrodes with decreasing temperature. Moreover, I–V measurements are performed to realize the transport mechanism; which revealed nonlinear curves indicating the tunneling behavior as expected (Fig. 10.9C(iii)). Besides this, the curves are almost identical at both 300K and 10K and do not vary with temperature. Such temperature independence reveals the good quality of the hBN intervening layer, which is free from pinholes and interface states [76].

Finally, Gr–hBN heterostructured magnetic device was constructed to study the magnetotransport properties of this multilayered spin-valve configuration. The MR signal is negative in polarity offering an inverted signal, as depicted in Fig. 10.9E(i). Instead, the spin-valve signal for Gr–Gr homostructure was observed to be positive as revealed in Fig. 10.9B(i). This is apparently suggesting that if identical layers (homostructure) are inserted in the spin-valve structure, it will have no effect on the polarity of signal; however, insertion of different layers in the form of hybrid system will definitely change the sign of MR signal. The explanation of the sign of spin-valve signal involves many causes, one of the reasons is the formation of same interfaces between the FM films and intervening layer when double-layered graphene (Gr–Gr) is used, while the incorporation of Gr–hBN heterostructure in the magnetic junction results in different interfaces formed between FMs and interlayer [77]. The polarity of the signal can further be justified using $MR = 2P_1 \times P_2/(1 - P_1 \times P_2)$; the Julliere model [46]. Here, the terms P_1 and P_2 are used to represent spin polarizations of the interfaces formed between the FM layers and the NM interlayer. From the model, one can see that the product $P_1 \times P_2$ is serving in evaluating the polarity or sign of spin-valve signal. To apply the model here, assume that P_1 and P_2 represent spin polarization vectors at Gr/FM and hBN/FM interfaces, respectively. In literature, negative spin polarization is reported for Gr/FM interface [78], which implies P_1 is negative. This is intuitively suggesting positive spin polarization (P_2) for hBN/FM interface, and thus the product becomes negative resulting in inverted MR signal. By following a similar procedure for double-layered Gr–Gr structure positive MR can be obtained. The amplitudes of MR are calculated to be −0.85% and −1.95% at room temperature and 10K, respectively. A similar reversal of the signal is reported in Gr-Al_2O_3 vertical MTJ [79,80]. Also, the decrease of the magnitude by increasing temperature is ascribed to magnetic impurity scattering or thermal smearing of electron energy distribution in FM leads [41,81,82]. Temperature dependency of MR ratio is also shown in Fig. 10.9E(ii). Its magnitude decreases with increasing temperature, which is attributed to the magnetic impurity scattering and thermal smearing of electron energy distribution in FMs as mentioned earlier. The I–V characteristics of the junction revealed nonlinear trend as elucidated in Fig. 10.9E (iii), indicating that Gr–hBN heterostructure acts as a tunneling barrier for the charge transport. When the temperature was lowered to 10K, the I–V curve is hardly affected and remained almost invariant indicating direct tunneling current through the Gr–hBN magnetic junction [83,84].

Fig. 10.10A represents the 2D materials utilized as an interlayer for spin-valve devices. All spin-valve devices were fabricated with the same dimensions of FMs whose junction area is

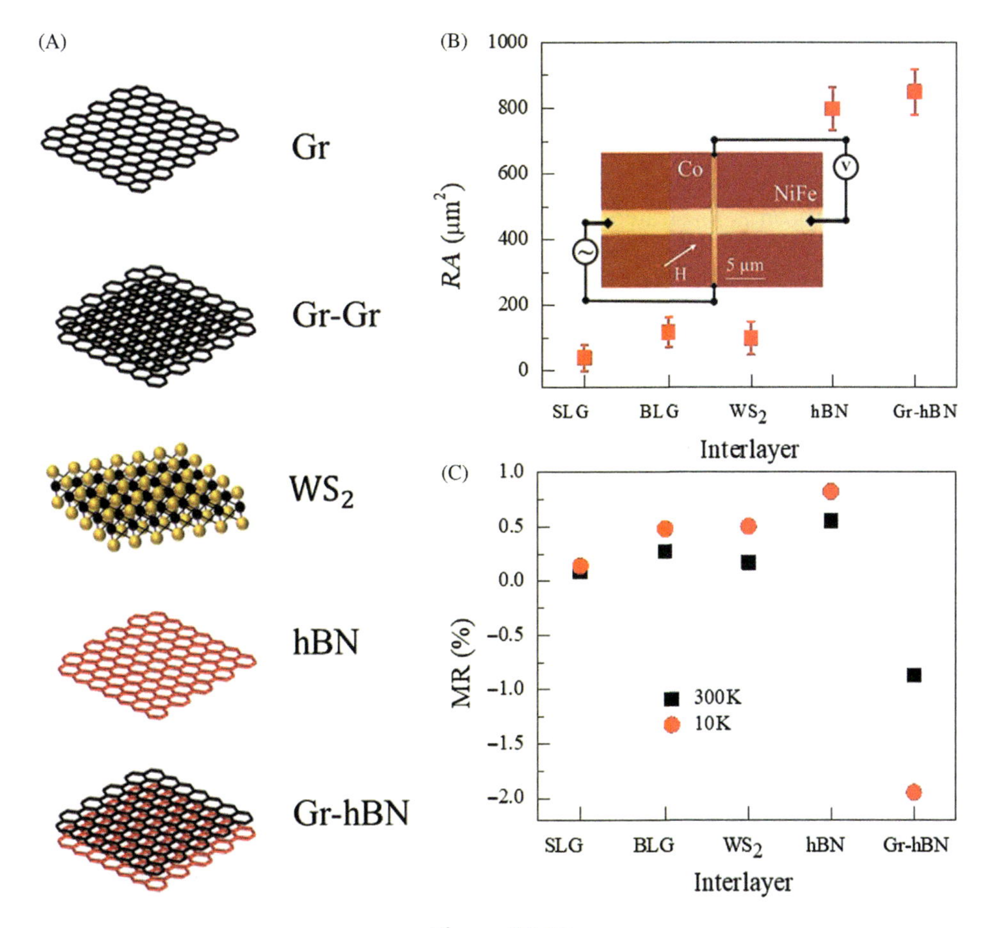

Figure 10.10

(A) Schematic illustration of 2D layers used as an interlayer for spin-valve devices. (B) Junction resistance for vertical magnetic junctions incorporating various intervening layers (Gr, Gr–Gr, WS_2, hBN, and Gr–hBN). The uncertainty in *RA* product values is represented by error bars. Optical image of one of the devices is demonstrated in the inset. (C) The magnitude of MR ratio for different 2D material-based spin-valve devices measured at 300K and 10K. *MR*, Magnetoresistance; *RA*, resistance × area. Sources: *From M.Z. Iqbal, et al., Spin valve effect of NiFe/graphene/NiFe junctions, Nano. Res. 6 (5) (2013) 373–380; M.Z. Iqbal, et al., Room temperature spin valve effect in NiFe/WS_2/Co junctions, Sci. Rep. 6 (2016) 21038; M.Z. Iqbal, et al., Room temperature spin valve effect in the NiFe/Gr–hBN/Co magnetic tunnel junction, J. Mater. Chem. C 4 (37) (2016) 8711–8715.*

calculated as 3.3 μm × 0.78 μm for each magnetic junction. A comparative analysis for resistance × area (*RA*) products of all spin-valve structures is carried out against each sandwiching layer is illustrated in Fig. 10.10B. *RA* is noticed to be lowest for single layer graphene (SLG) vertical magnetic structure and highest for Gr–hBN device. The error bars reveal standard deviation of *RA* products for each spin-valve junctions as a function of the intervening layer. Fig. 10.10C presents the relative MR ratios as a function of the interlayer measured at 300K and 10K, respectively. The magnitudes of MR become high as the temperature is lower down. The MR ratio is found to be highest for Gr–hBN device. The inset of Fig. 10.10B shows an optical image and measurement configuration for one of the devices. This confirms high spin filtering action of Gr–hBN as compared to rest of the interlayers. Furthermore, MR of all 2D materials is discussed comprehensively in the chapter and comparison is illustrated in Fig. 10.9.

10.8 Summary

In summary, the spin-valve effect for different vertical magnetic architectures incorporating different 2D layers is comprehensively demonstrated. The *MR* ratio is calculated at several temperatures ranging from 10K to 300K, where a monotonic decrease in MR is observed as the temperature rises to 300K. This temperature dependency reflects how the spin polarization of FM electrodes decreases as the temperature increases. The important facts have been disclosed regarding the role of TMDCs as an intervening layer in magnetoresistive devices and further corroborated by comparing with other 2D materials. The overview of the spin-valve effect in 2D materials offers comprehensive knowledge pertinent to spin reversal and filtering mechanism.

References

[1] K. Novoselov, D. Jiang, F. Schedin, T. Booth, V. Khotkevich, S. Morozov, et al., Two-dimensional atomic crystals, Proc. Natl. Acad. Sci. U.S.A. 102 (30) (2005) 10451–10453.

[2] A.A. Balandin, S. Ghosh, W. Bao, I. Calizo, D. Teweldebrhan, F. Miao, et al., Superior thermal conductivity of single-layer graphene, Nano Lett. 8 (3) (2008) 902–907.

[3] A.S. Mayorov, R.V. Gorbachev, S.V. Morozov, L. Britnell, R. Jalil, L.A. Ponomarenko, et al., Micrometer-scale ballistic transport in encapsulated graphene at room temperature, Nano Lett. 11 (6) (2011) 2396–2399.

[4] M. Osada, T. Sasaki, Two-dimensional dielectric nanosheets: novel nanoelectronics from nanocrystal building blocks, Adv. Mater. 24 (2) (2012) 210–228.

[5] A. Ayari, E. Cobas, O. Ogundadegbe, M.S. Fuhrer, Realization and electrical characterization of ultrathin crystals of layered transition-metal dichalcogenides, J. Appl. Phys. 101 (1) (2007) 014507.

[6] J. Wilson, A. Yoffe, The transition metal dichalcogenides discussion and interpretation of the observed optical, electrical and structural properties, Adv. Phys. 18 (73) (1969) 193–335.

[7] A.D. Yoffe, Low-dimensional systems: quantum size effects and electronic properties of semiconductor microcrystallites (zero-dimensional systems) and some quasi-two-dimensional systems, Adv. Phys. 42 (2) (1993) 173–262.

[8] M.Z. Iqbal, M.W. Iqbal, S. Siddique, M.F. Khan, S.M. Ramay, Room temperature spin valve effect in NiFe/WS_2/Co junctions, Sci. Rep. 6 (2016) 21038.

[9] W. Wang, A. Narayan, L. Tang, K. Dolui, Y. Liu, X. Yuan, et al., Spin-valve effect in NiFe/MoS_2/NiFe junctions, Nano Lett. 15 (8) (2015) 5261–5267.

[10] A.T. Neal, H. Liu, J. Gu, P.D. Ye, Magneto-transport in MoS_2: phase coherence, spin–orbit scattering, and the hall factor, ACS Nano. 7 (8) (2013) 7077–7082.

[11] H.-C. Wu, C.Ó. Coileáin, M. Abid, O. Mauit, A. Syrlybekov, A. Khalid, et al., Spin-dependent transport properties of Fe_3O_4/MoS_2/Fe_3O_4 junctions, Sci. Rep. 5 (2015) 15984.

[12] D. Elias, R. Gorbachev, A. Mayorov, S. Morozov, A. Zhukov, P. Blake, et al., Dirac cones reshaped by interaction effects in suspended graphene, Nat. Phys. 7 (9) (2011) 701–704.

[13] M.-W. Lin, C. Ling, Y. Zhang, H.J. Yoon, M.M.-C. Cheng, L.A. Agapito, et al., Room-temperature high on/off ratio in suspended graphene nanoribbon field-effect transistors, Nanotechnology 22 (26) (2011) 265201.

[14] X. Li, X. Wang, L. Zhang, S. Lee, H. Dai, Chemically derived, ultrasmooth graphene nanoribbon semiconductors, Science 319 (5867) (2008) 1229–1232.

[15] R. Balog, B. Jørgensen, L. Nilsson, M. Andersen, E. Rienks, M. Bianchi, et al., Bandgap opening in graphene induced by patterned hydrogen adsorption, Nat. Mater. 9 (4) (2010) 315–319.

[16] M.Y. Han, B. Özyilmaz, Y. Zhang, P. Kim, Energy band-gap engineering of graphene nanoribbons, Phys. Rev. Lett. 98 (20) (2007) 206805.

[17] T.-T. Tang, Y. Zhang, C.-H. Park, B. Geng, C. Girit, Z. Hao, et al., A tunable phonon–exciton Fano system in bilayer graphene, Nat. Nanotechnol. 5 (1) (2010) 32–36.

[18] B. Sipos, A.F. Kusmartseva, A. Akrap, H. Berger, L. Forró, E. Tutiš, From Mott state to superconductivity in 1T-TaS2, Nat. Mater. 7 (12) (2008) 960.

[19] J.A. Wilson, F. Di Salvo, S. Mahajan, Charge-density waves and superlattices in the metallic layered transition metal dichalcogenides, Adv. Phys. 24 (2) (1975) 117–201.

[20] A.C. Neto, Charge density wave, superconductivity, and anomalous metallic behavior in 2D transition metal dichalcogenides, Phys. Rev. Lett. 86 (19) (2001) 4382.

[21] A. Kuc, N. Zibouche, T. Heine, Influence of quantum confinement on the electronic structure of the transition metal sulfide TS_2, Phys. Rev. B 83 (24) (2011) 245213.

[22] K.F. Mak, K. He, J. Shan, T.F. Heinz, Control of valley polarization in monolayer MoS_2 by optical helicity, Nat. Nanotechnol. 7 (8) (2012) 494–498.

[23] H. Zeng, J. Dai, W. Yao, D. Xiao, X. Cui, Valley polarization in MoS_2 monolayers by optical pumping, Nat. Nanotechnol. 7 (8) (2012) 490–493.

[24] T. Cao, G. Wang, W. Han, H. Ye, C. Zhu, J. Shi, et al., Valley-selective circular dichroism of monolayer molybdenum disulphide, Nat. Commun. 3 (2012) 887.

[25] X. Zhang, W. Mi, X. Wang, Y. Cheng, U. Schwingenschlögl, The interface between Gd and monolayer MoS_2: a first-principles study, Sci. Rep. 4 (2014) 7368.

[26] Y. Cheng, Z. Zhu, W. Mi, Z. Guo, U. Schwingenschlögl, Prediction of two-dimensional diluted magnetic semiconductors: doped monolayer MoS_2 systems, Phys. Rev. B 87 (10) (2013) 100401.

[27] N. Feng, W. Mi, Y. Cheng, Z. Guo, U. Schwingenschlögl, H. Bai, Magnetism by interfacial hybridization and p-type doping of MoS_2 in Fe_4N/MoS_2 superlattices: a first-principles study, ACS Appl. Mater. Interfaces 6 (6) (2014) 4587-–4594.

[28] I. Žutić, J. Fabian, S.D. Sarma, Spintronics: fundamentals and applications, Rev. Mod. Phys. 76 (2) (2004) 323.

[29] C. Chappert, A. Fert, F.N. Van Dau, The emergence of spin electronics in data storage, Nat. Mater. 6 (11) (2007) 813.

[30] H. Dery, P. Dalal, L. Sham, Spin-based logic in semiconductors for reconfigurable large-scale circuits, Nature 447 (7144) (2007) 573–576.

[31] J.-G.J. Zhu, C. Park, Magnetic tunnel junctions, Mater. Today 9 (11) (2006) 36–45.

[32] J.R. Childress, R.E. Fontana, Magnetic recording read head sensor technology, Comptes Rendus Physique 6 (9) (2005) 997–1012.

[33] E. Chen, D. Apalkov, Z. Diao, A. Driskill-Smith, D. Druist, D. Lottis, et al., Advances and future prospects of spin-transfer torque random access memory, IEEE. Trans. Magn. 46 (6) (2010) 1873–1878.
[34] M.Z. Iqbal, G. Hussain, Electron spin dynamics in vertical magnetic junctions incorporating two-dimensional layered materials, J. Mater. Chem. C 5 (43) (2017) 11174–11184.
[35] M.Z. Iqbal, G. Hussain, S. Siddique, M.W. Iqbal, Graphene spin valve: an angle sensor, J. Magn. Magn. Mater. 432 (2017) 135–139.
[36] M.Z. Iqbal, N.A. Qureshi, G. Hussain, Recent advancements in 2D-materials interface based magnetic junctions for spintronics, J. Magn. Magn. Mater. 457 (2018) 110–125.
[37] M.Z. Iqbal, G. Hussain, S. Siddique, T. Hussain, M.J. Iqbal, Influence of DC-biasing on the performance of graphene spin valve, Solid State Commun. 272 (2018) 33–36.
[38] M.Z. Iqbal, G. Hussain, S. Siddique, M.W. Iqbal, G. Murtaza, S.M. Ramay, Interlayer quality dependent graphene spin valve, J. Magn. Magn. Mater. 422 (2017) 322–327.
[39] V. Karpan, G. Giovannetti, P. Khomyakov, M. Talanana, A. Starikov, M. Zwierzycki, et al., Graphite and graphene as perfect spin filters, Phys. Rev. Lett. 99 (17) (2007) 176602.
[40] V. Karpan, P. Khomyakov, A. Starikov, G. Giovannetti, M. Zwierzycki, M. Talanana, et al., Theoretical prediction of perfect spin filtering at interfaces between close-packed surfaces of Ni or Co and graphite or graphene, Phys. Rev. B 78 (19) (2008) 195419.
[41] M.Z. Iqbal, M.W. Iqbal, J.H. Lee, Y.S. Kim, S.-H. Chun, J. Eom, Spin valve effect of NiFe/graphene/NiFe junctions, Nano. Res. 6 (5) (2013) 373–380.
[42] R. Meservey, P. Tedrow, J. Moodera, Electron spin polarized tunneling study of ferromagnetic thin films, J. Magn. Magn. Mater. 35 (1-3) (1983) 1–6.
[43] R. Meservey, Tunnelling in a magnetic field with spin-polarized electrons, Phys. Scr. 38 (2) (1988) 272.
[44] J. Gadzuk, Band-structure effects in the field-induced tunneling of electrons from metals, Phys. Rev. 182 (2) (1969) 416.
[45] P.M. Tedrow, R. Meservey, Spin-dependent tunneling into ferromagnetic nickel, Phys. Rev. Lett. 26 (4) (1971) 192.
[46] M. Julliere, Tunneling between ferromagnetic films, Phys. Lett. A 54 (3) (1975) 225–226.
[47] G.-X. Miao, M. Münzenberg, J.S. Moodera, Tunneling path toward spintronics, Rep. Prog. Phys. 74 (3) (2011) 036501.
[48] A. Rockett, The Materials Science of Semiconductors, Springer Science & Business Media, 2007.
[49] X. Li, W. Cai, J. An, S. Kim, J. Nah, D. Yang, et al., Large-area synthesis of high-quality and uniform graphene films on copper foils, Science 324 (5932) (2009) 1312–1314.
[50] J.-H. Lee, E.K. Lee, W.-J. Joo, Y. Jang, B.-S. Kim, J.Y. Lim, et al., Wafer-scale growth of single-crystal monolayer graphene on reusable hydrogen-terminated germanium, Science 344 (6181) (2014) 286–289.
[51] J. Meng, J.-J. Chen, Y. Yan, D.-P. Yu, Z.-M. Liao, Vertical graphene spin valve with Ohmic contacts, Nanoscale 5 (19) (2013) 8894–8898.
[52] A. Dankert, M.V. Kamalakar, A. Wajid, R. Patel, S.P. Dash, Tunnel magnetoresistance with atomically thin two-dimensional hexagonal boron nitride barriers, Nano. Res. 8 (4) (2015) 1357–1364.
[53] O.V. Yazyev, A. Pasquarello, Magnetoresistive junctions based on epitaxial graphene and hexagonal boron nitride, Phys. Rev. B 80 (3) (2009) 035408.
[54] M.Z. Iqbal, S. Siddique, G. Hussain, Spin valve effect of 2D-materials based magneticjunctions, J. Adv. Eng. Mater. 20 (2018) 1700692.
[55] I.V. Roshchin, I.K. Schuller, J.J. Akerman, R. Whig Dave, J. Slaughter, Origin of temperature dependence in tunneling magnetoresistance, APS Meeting Abstracts, 2002, p. 15006.
[56] S. Entani, H. Naramoto, S. Sakai, Magnetotransport properties of a few-layer graphene-ferromagnetic metal junctions in vertical spin valve devices, J. Appl. Phys. 117 (17) (2015) 17A334.
[57] M.Z. Iqbal, G. Hussain, S. Siddique, M.W. Iqbal, Enhanced magnetoresistance in graphene spin valve, J. Magn. Magn. Mater. 429 (2017) 330–333.
[58] W. Wang, A. Narayan, L. Tang, K. Dolui, Y. Liu, X. Yuan, et al., Spin-valve effect in NiFe/MoS_2/NiFe junctions, Nano Lett. 5 (8) (2015) 5261–5267.

[59] M.Z. Iqbal, G. Hussain, S. Siddique, M.W. Iqbal, Interlayer reliant magnetotransport in graphene spin valve, J. Magn. Magn. Mater. 441 (2017) 39–42.
[60] K. Dolui, A. Narayan, I. Rungger, S. Sanvito, Efficient spin injection and giant magnetoresistance in Fe/MoS_2/Fe junctions, Phys. Rev. B 90 (4) (2014) 041401.
[61] J. Åkerman, I. Roshchin, J. Slaughter, R. Dave, I. Schuller, Origin of temperature dependence in tunneling magnetoresistance, Europhys. Lett. 63 (1) (2003) 104.
[62] F. Walz, The Verwey transition—a topical review, J. Phys.: Condens. Matter 14 (12) (2002) R285.
[63] J. Brabers, F. Walz, H. Kronmüller, The influence of a finite bandwidth on the Verwey transition in magnetite, J. Phys.: Condens. Matter 12 (25) (2000) 5437.
[64] M. Ziese, Extrinsic magnetotransport phenomena in ferromagnetic oxides, Rep. Prog. Phys. 65 (2) (2002) 143.
[65] Z.-M. Liao, Y.-D. Li, J. Xu, J.-M. Zhang, K. Xia, D.-P. Yu, Spin-filter effect in magnetite nanowire, Nano Lett. 6 (6) (2006) 1087–1091.
[66] A. Fernández-Pacheco, J. De Teresa, J. Orna, L. Morellon, P. Algarabel, J. Pardo, et al., Giant planar hall effect in epitaxial Fe_3O_4 thin films and its temperature dependence, Phys. Rev. B 78 (21) (2008) 212402.
[67] H.-C. Wu, R. Ramos, R. Sofin, Z.-M. Liao, M. Abid, I. Shvets, Transversal magneto-resistance in epitaxial Fe_3O_4 and Fe_3O_4/NiO exchange biased system, Appl. Phys. Lett. 101 (5) (2012) 052402.
[68] R. Ramos, T. Kikkawa, K. Uchida, H. Adachi, I. Lucas, M. Aguirre, et al., Observation of the spin Seebeck effect in epitaxial Fe_3O_4 thin films, Appl. Phys. Lett. 102 (7) (2013) 072413.
[69] J. Gooth, R. Zierold, J.G. Gluschke, T. Boehnert, S. Edinger, S. Barth, et al., Gate voltage induced phase transition in magnetite nanowires, Appl. Phys. Lett. 102 (7) (2013) 073112.
[70] S. Lee, A. Fursina, J.T. Mayo, C.T. Yavuz, V.L. Colvin, R.S. Sofin, et al., Electrically driven phase transition in magnetite nanostructures, Nat. Mater. 7 (2) (2008) 130.
[71] E. Cobas, A.L. Friedman, O.M. van't Erve, J.T. Robinson, B.T. Jonker, Graphene as a tunnel barrier: graphene-based magnetic tunnel junctions, Nano Lett. 12 (6) (2012) 3000–3004.
[72] W. Li, L. Xue, H. Abruna, D. Ralph, Magnetic tunnel junctions with single-layer-graphene tunnel barriers, Phys. Rev. B 89 (18) (2014) 184418.
[73] V. Karpan, G. Giovannetti, P. Khomyakov, M. Talanana, A. Starikov, M. Zwierzycki, et al., Graphite and graphene as perfect spin filters, Phys. Rev. Lett. 99 (17) (2007) 176602.
[74] V. Karpan, P. Khomyakov, A. Starikov, G. Giovannetti, M. Zwierzycki, M. Talanana, et al., Theoretical prediction of perfect spin filtering at interfaces between close-packed surfaces of Ni or Co and graphite or graphene, Phys. Rev. B 78 (19) (2008) 195419.
[75] J.-H. Park, H.-J. Lee, Out-of-plane magnetoresistance in ferromagnet/graphene/ferromagnet spin-valve junctions, Phys. Rev. B 89 (16) (2014) 165417.
[76] M.V. Kamalakar, A. Dankert, J. Bergsten, T. Ive, S.P. Dash, Enhanced tunnel spin injection into graphene using chemical vapor deposited hexagonal boron nitride, Sci. Rep. 4 (2014).
[77] M.Z. Iqbal, S. Siddique, G. Hussain, M.W. Iqbal, Room temperature spin valve effect in the NiFe/Gr–hBN/Co magnetic tunnel junction, J. Mater. Chem. C 4 (37) (2016) 8711–8715.
[78] B. Dlubak, M.-B. Martin, R.S. Weatherup, H. Yang, C. Deranlot, R. Blume, et al., Graphene-passivated nickel as an oxidation-resistant electrode for spintronics, ACS Nano. 6 (12) (2012) 10930–10934.
[79] M.Z. Iqbal, M.W. Iqbal, X. Jin, C. Hwang, J. Eom, Interlayer dependent polarity of magnetoresistance in graphene spin valves, J. Mater. Chem. C 3 (2) (2015) 298–302.
[80] M.-B. Martin, B. Dlubak, R.S. Weatherup, H. Yang, C. Deranlot, K. Bouzehouane, et al., Sub-nanometer atomic layer deposition for spintronics in magnetic tunnel junctions based on graphene spin-filtering membranes, ACS Nano 8 (8) (2014) 7890–7895.
[81] I.V. Roshchin, I.K. Schuller, J.J. Akerman, R. Whig Dave, J. Slaughter, Origin of temperature dependence in tunneling magnetoresistance, APS Meeting Abstracts, 2002.
[82] C.H. Shang, J. Nowak, R. Jansen, J.S. Moodera, Temperature dependence of magnetoresistance and surface magnetization in ferromagnetic tunnel junctions, Phys. Rev. B 58 (6) (1998) R2917.

[83] T. Roy, L. Liu, S. De La Barrera, B. Chakrabarti, Z. Hesabi, C. Joiner, et al., Tunneling characteristics in chemical vapor deposited graphene–hexagonal boron nitride–graphene junctions, Appl. Phys. Lett. 104 (12) (2014) 123506.
[84] M.Z. Iqbal, M.M. Faisal, Fowler-Nordheim tunneling characteristics of graphene/hBN/metal heterojunctions, J. Appl. Phys. 125 (2019) 084902.
[85] M.Z. Iqbal, S. Siddique, M.F. Khan, A.U. Rehman, A. Rehman, J. Eom, Gate-Dependent Tunnelling Current Modulation of Graphene/hBN Vertical Heterostructures, Adv. Eng. Mater. 20 (2018) 1800159.

CHAPTER 11

Layered topological semimetals for spintronics

Xuefeng Wang and Minhao Zhang

School of Electronic Science and Engineering, and Collaborative Innovation Center of Advanced Microstructures, Nanjing University, Nanjing, P.R. China

Chapter Outline

11.1 Introduction

11.1.1 Introduction to Spintronics

Spintronics is proposed to encode and transmit information using the electron spins instead of the electron charges [1,2]. A prominent example from the field of spintronics is the giant magnetoresistance (MR) effect discovered by Baibich et al. [3] and Binasch et al. [4], which leads to a considerable increase in storage densities of hard disk drives. Besides the applications in magnetic disk storages, spintronics is now expected as a possible way to solve the hard limit in scaling today's Si field-effect transistor (FET). There are two states of the spin, that is, spin up and spin down, which are analogous to the ON and OFF states

Spintronic 2D Materials.
DOI: https://doi.org/10.1016/B978-0-08-102154-5.00011-4

of a conventional FET. To realize the logical function, the spintronics needs to achieve all-electrical spin injection, transport, manipulation, and detection of spin currents. Datta and Das first proposed the structure of a spin-FET [5], which comprises these essential components, spin transport, spin manipulation, and spin injector/detector. After the proposal, workable solutions have been implemented to search for an ideal spin-FET.

11.1.1.1 Spin transport in graphene

The spin transport properties have been widely studied on graphene [6–8] and silicon [9] in previous works. Tombros et al. [6] have observed the spin transport over micrometer-scale distances in graphene in 2007. They fabricated the spin-valve devices based on graphene with at least four ferromagnetic (FM) electrodes (Fig. 11.1A). Three resistance levels are observed (Fig. 11.1A) in the nonlocal measurements as the magnetic field is swept from positive (negative) to negative (positive) values. The different resistance levels are associated with the configurations of magnetization directions of the four FM electrodes, which indicate that the polarized spins are injected by the FM electrodes in a magnetic field. Two resistance levels are observed in a local two-terminal geometry (Fig. 11.1B). The spin transport in graphene shows a weak dependence on temperature (Fig. 11.1C). Owing to the weak intrinsic spin–orbit coupling (SOC), a very long spin coherence length has been observed up to room temperature, which makes graphene an ideal candidate for nonmagnetic spin transport channels in spin-FETs [8].

11.1.1.2 Spin current in topological insulators

FM metals have long been considered as a spin injector/detector to inject polarized spins in a magnetic field. However, complications have been described in electrical spin injection from a FM metal into a semiconductor, such as impedance mismatch problem at the metal/semiconductor interfaces [10]. Besides, the stray field effects have been observed in the downscaling process, which would disturb the configurations of magnetization directions of the FM electrodes. Therefore all-electrical technology (without FM materials) in spintronics is encouraged, especially for the applications in logical spin-FET.

Materials with strong SOC are able to generate the spin-polarized currents electrically without any magnetic field [11]. Three-dimensional (3D) topological insulators (TIs), with an insulating bulk, have two-dimensional (2D) gapless conducting states on their surfaces [12–15]. The property of spin–momentum locking naturally induces self-polarized spins as long as there is a charge current on surface states (SSs) [16–26]. People have demonstrated spin polarization on the surfaces of TIs by an all-electrical method without magnetic fields [16–26]. In TIs the amplitude and direction of the net spin polarization are directly governed by the charge current on the surface effectively.

The all-electrical spin injection was realized in a TI (Bi_2Te_2Se) – graphene lateral van der Waals heterostructure without any magnetic fields. First, Vaklinova et al. measured the

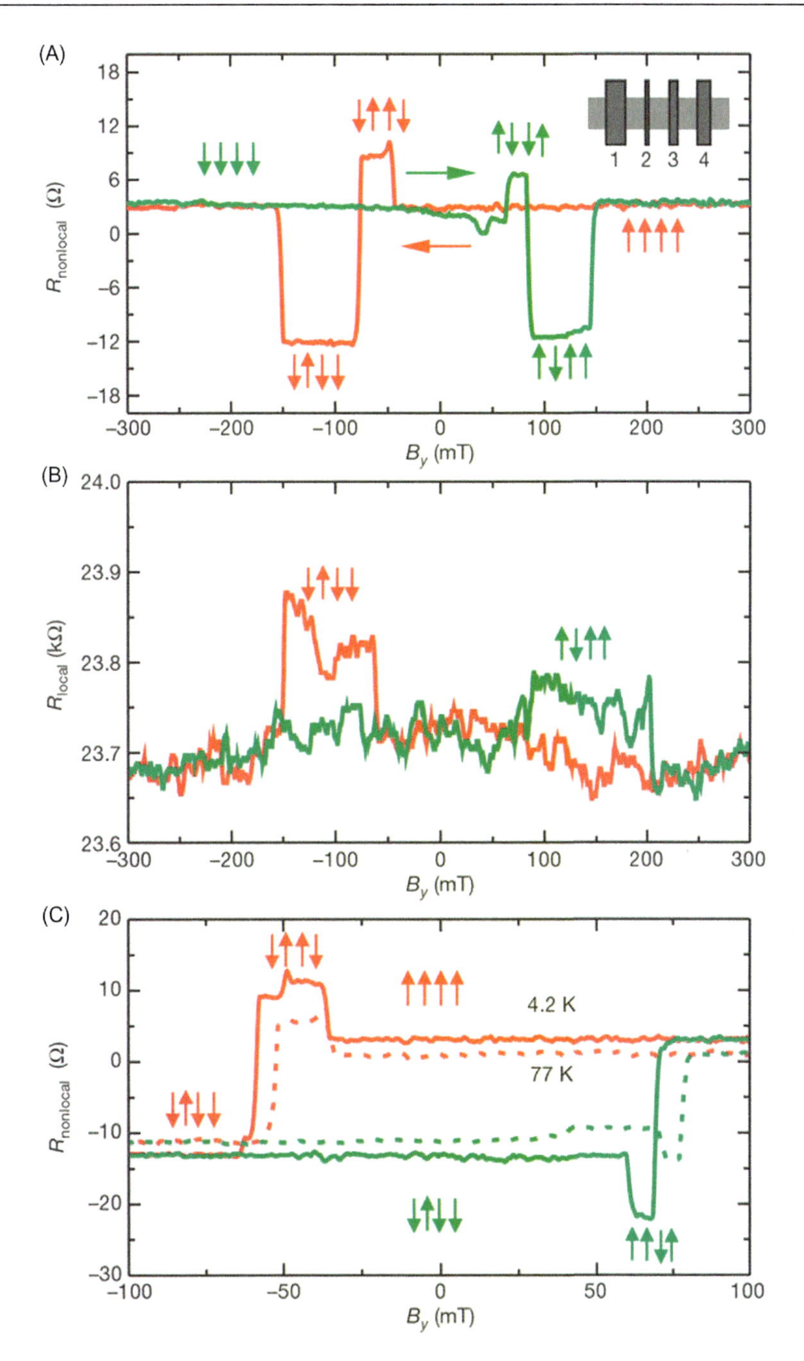

Figure 11.1

Spin transport in spin-valve devices based on graphene at 4.2K and 77K, respectively [6]. (A) Nonlocal measurements of spin-valve devices at 4.2K. (B) Local measurements of spin-valve devices at 4.2K. (C) Nonlocal measurements of spin-valve devices at 4.2K and 77K, respectively. The magnetic configurations of the FM electrodes are illustrated as the red and green arrows in figures. *FM*, Ferromagnetic.

conventional spin-valve signals in a FM – graphene structure in magnetic fields, which demonstrates spin could transport in graphene (Fig. 11.2A and B) [27]. The FM electrodes are spin injectors and detectors in this device. Fig. 11.2B shows the temperature dependence of the conventional spin-valve MR. Second, they measured the spin-valve signals in a TI – graphene structure. In contrast, they observed a hysteretic step-like signal. Here, TI and FM are used as a spin injector and a spin detector, respectively. The novel spin-valve signals are dependent on the direction of current(Fig. 11.2C and D), which is due to the spin–momentum locking of the SSs of TIs. For a positive bias current (Fig. 11.2C), the resistance switching is associated with the parallel or antiparallel configuration of the

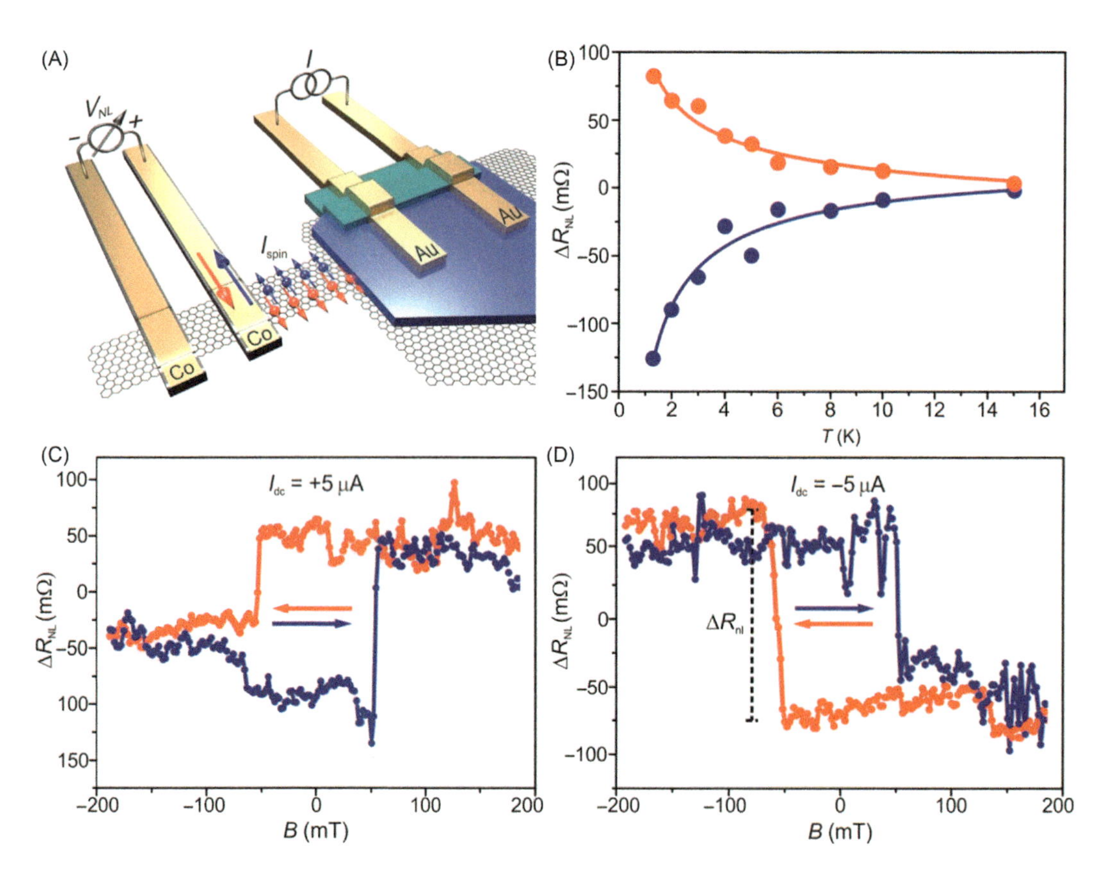

Figure 11.2

TIs as a spin injector [27]. (A) Schematic diagram of a spin-valve device in a TI–graphene structure with Au and Co (FM) electrodes. The TI serves as a spin injector. The graphene serves as a channel for spin transport. The Co electrodes serve as spin detector. (B) The temperature dependence of the nonlocal spin-valve signals in a FM–graphene structure. (C and D) Spin-valve signals in a TI–graphene structure taken at the bias currents of $I_{dc} = +5$ and -5 μA, respectively. *FM*, Ferromagnetic; *TI*, topological insulator.

magnetization directions (FM) and the incoming spins (TI). Upon reversing the bias current (Fig. 11.2D), the step-like signal is also reversed due to the reversal of the incoming spins. Therefore the all-electrical spin injection has been realized based on the spin-polarized current on the surfaces of TIs.

11.1.2 Topological semimetals

Recently, similar to TIs, semimetals with nontrivial topological features stand out, such as Dirac and Weyl semimetals (DSM and WSM) [28]. Topological semimetals (TSMs) are characterized by the topologically protected SSs with metallic bulk states. The SSs in DSM and WSM are called the Fermi arcs with a special unclosed loop. The DSMs (WSM) have the linear bands close to the Dirac (Weyl) points (Fig. 11.3). The effective Hamiltonian for the linear band is given by the gapless Dirac (Weyl) equation. If either the inversion or time-reversal symmetry (TRS) is broken, a Dirac point would separate into a pair of Weyl points with opposite chirality. DSMs (WSM), with a point-like Fermi surface at the Dirac (Weyl) point, are referred as the type-I semimetals (Fig. 11.3A). DSMs (WSM), with the Dirac (Weyl) points at the boundaries between electron and hole pockets, are referred as the type-II semimetals (Fig. 11.3B). Therefore a high carrier density is expected in the type-II semimetals, which induces unusual features in MR.

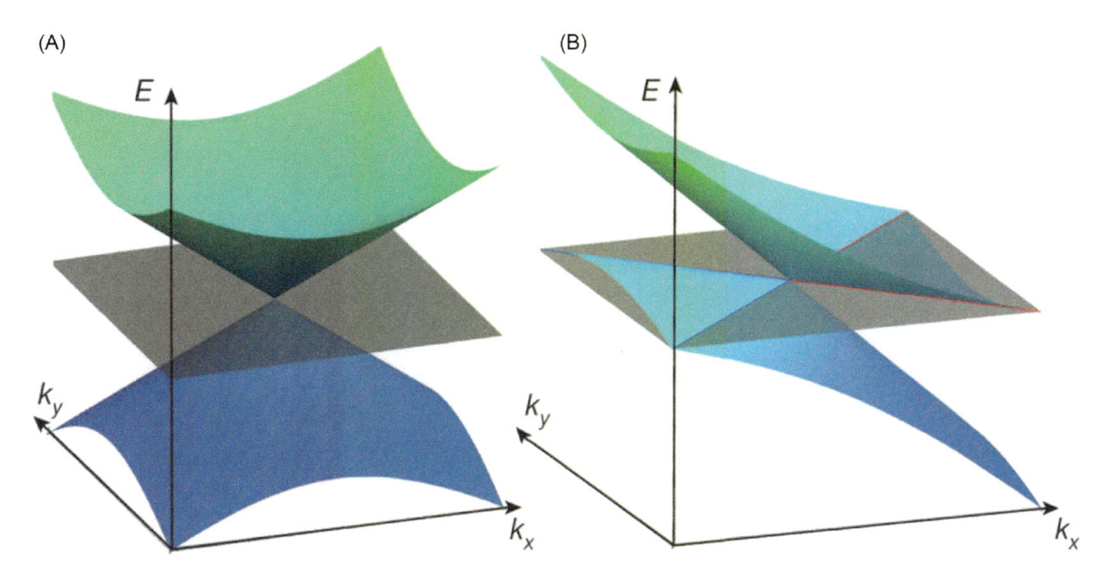

Figure 11.3
The linear band structure near a Dirac (Weyl) point of type-I (A) and type-II (B) semimetals, respectively [28]. In type-II topological semimetals (B) the Dirac (Weyl) points are located at the boundaries between electron and hole pockets.

11.1.2.1 Type-I Dirac semimetals

The type-I DSMs have been realized in the materials, such as $ZrTe_5$ [29–38] or Cd_3As_2 [39–46]. Nontrivial transport has been observed in type-I DSMs [32,33,40–43], which is related to the topologically protected gapless Dirac cones with liner dispersions. Quantum oscillations have been observed in $ZrTe_5$ [33] (Fig. 11.4A) and Cd_3As_2 [40] (Fig. 11.4C), respectively. In particular, a π-Berry phase in the Shubnikov-deHaas (SdH) oscillations gives clear evidence for the nontrivial DSM (Fig. 11.4B) [33]. The quantum oscillations demonstrate the ultrahigh mobility in DSMs.

Besides, negative MR was observed in $ZrTe_5$ [34] and Cd_3As_2 nanostructures [44,45], which is related to the chiral magnetic effect. The chiral magnetic effect could generate an additional current induced by chirality imbalance in presence of a magnetic field.

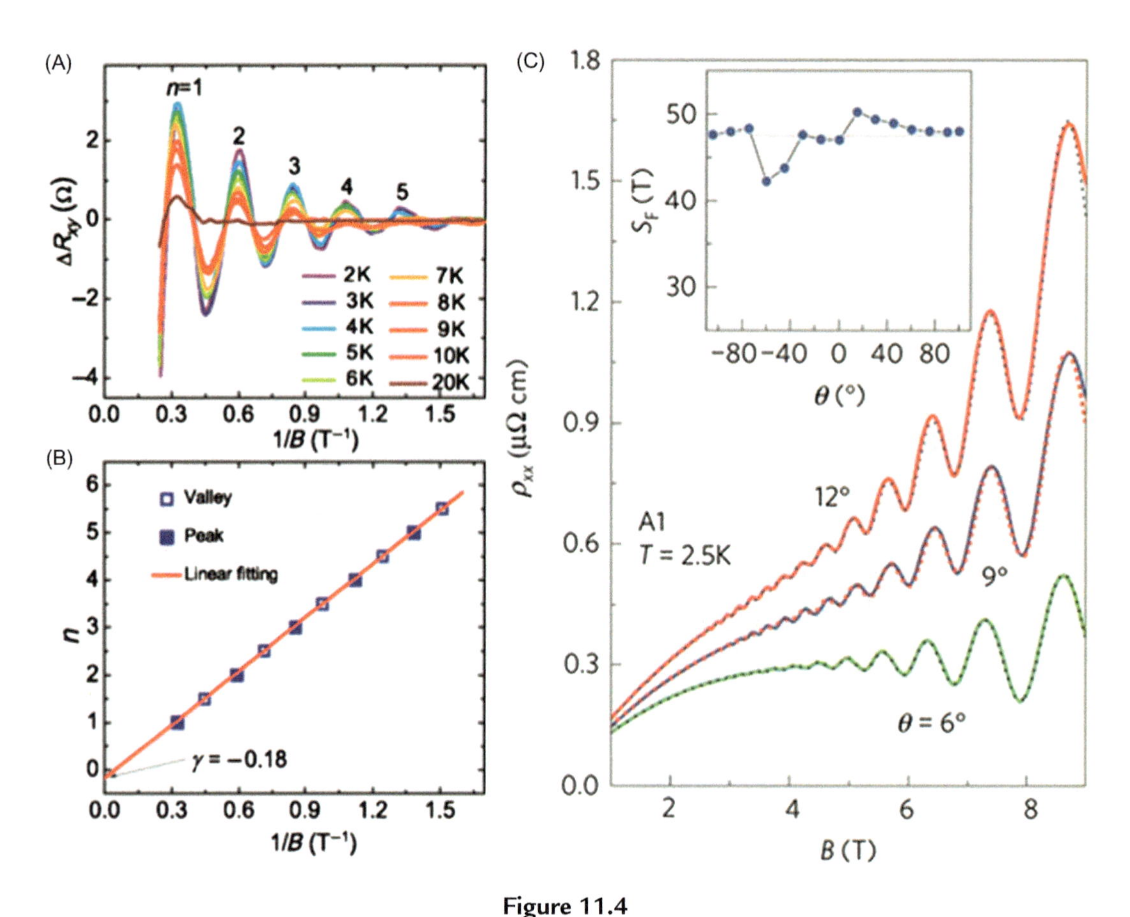

Figure 11.4

Nontrivial transport and quantum oscillations in type-I DSMs [33,40]. (A) The temperature dependence of quantum oscillations of $ZrTe_5$ [33]. (B) Landau fan diagram [33]. (C) The angle dependence of quantum oscillations of Cd_3As_2 at 2.5K [40]. *DSM*, Dirac semimetal.

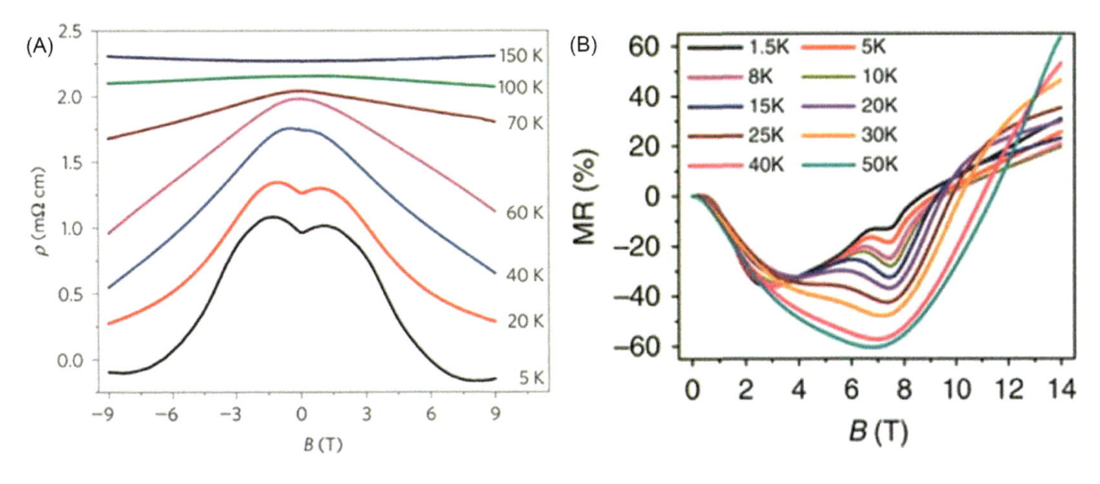

Figure 11.5

Negative MR in $ZrTe_5$ [34] and Cd_3As_2 nanostructures [44]. (A) and (B) The temperature dependence of the negative MR in $ZrTe_5$ and Cd_3As_2 nanostructures, respectively. The negative MR was related to the chiral magnetic effect. The chiral magnetic effect could generate an additional current in presence of a magnetic field. *MR*, Magnetoresistance.

Fig. 11.5A and B shows temperature dependence of the negative MR in $ZrTe_5$ and Cd_3As_2 nanostructures, respectively. The negative MR was only observed when the magnetic field is in parallel with the current. The dominant transport of topological Dirac fermions makes type-I DSMs a promising platform for devices of low dissipation.

However, Cd_3As_2 is not a layered material. Many efforts have been made to develop a facile method for growth of the Cd_3As_2 nanostructures. The Cd_3As_2 nanostructures were successfully synthesized by a chemical vapor deposition method [41–44,46]. Zhang et al. [46] discussed the effects of pressure in the controllable synthesis of Cd_3As_2 nanostructures with different morphologies (Fig. 11.6). Short and long Cd_3As_2 nanowires were grown on the silicon substrates under 100 and 110 Pa, respectively (Fig. 11.6A and B). Thick nanowires, nanobelts, faceted nanoplate, and nano-octahedron were obtained by further increasing the pressure (Fig. 11.6C–G). Clear quantum oscillations have been observed in the as-grown high-quality nanowires, which indicate a very large carrier mobility $\sim 2138\ cm^2/V/\ s$ [43].

11.1.2.2 Type-II Dirac semimetals

Type-II Dirac fermions were characterized with a tilted Dirac cone as the Lorentz invariance is broken in DSMs (e.g., $PdTe_2$ [47–52]). $PdTe_2$ has been found as a superconductor with a $T_c \sim 2K$ [47]. Fei et al. [49] performed detailed transport measurements on the superconductivity property of $PdTe_2$. An anisotropic superconductivity was observed apparently under the different device configurations. The superconductivity under device configurations 1 and 2 are seen in Fig. 11.7A and B,

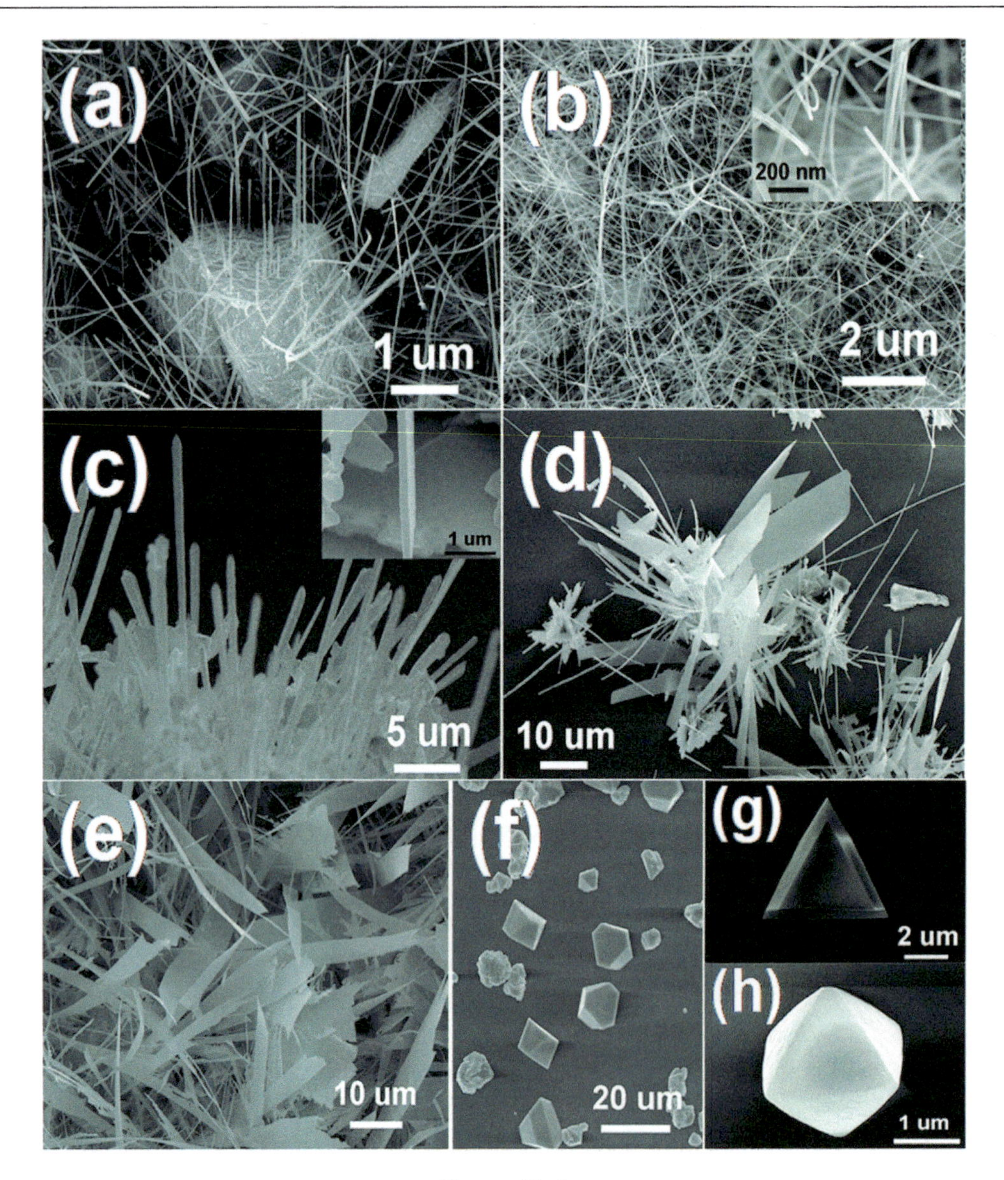

Figure 11.6

SEM images of the Cd_3As_2 nanostructures grown at different pressure [46]. (A–C) The Cd_3As_2 nanowires with different morphologies grown under 100, 110, and 120 Pa, respectively. (D) The coexistence of Cd_3As_2 nanowires and nanobelts grown under 140 Pa. (E) The Cd_3As_2 nanobelts grown under 150 Pa. (F–H) The Cd_3As_2 nanocrystals grown under a pressure over 200 Pa. *SEM*, scanning electron microscope.

respectively. When the current is perpendicular to the field, the superconductivity survives in the field of 2000 Oe (Fig. 11.7A). When the current is parallel to the field, the superconductivity is nearly suppressed as increasing the field to 500 Oe (Fig. 11.7B).

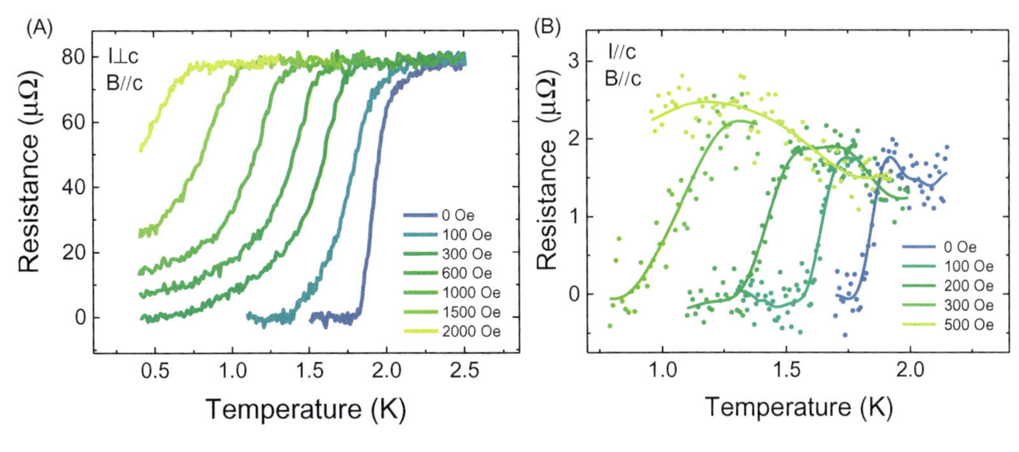

Figure 11.7
Anisotropic superconductivity in $PdTe_2$ [49]. (A) and (B) Temperature dependence of resistance of configurations 1 ($I \perp c$, $B||c$) and 2 ($I||c$, $B||c$), respectively. The superconductivity survives in higher fields in configuration 1, which indicates anisotropic superconductivity in $PdTe_2$.

11.1.2.3 Type-II Weyl semimetals

The type-II WSMs, with the tilted Weyl cones, have been verified in the layered WTe_2 family as the Lorentz invariance is broken [53–69]. Ali et al. reported an extremely large MR in WTe_2 crystals [53]. They observed a large MR ratio of $\sim 10^5\%$ in a magnetic field of 15 T (Fig. 11.8A). In much higher magnetic fields, there is no saturation. In an ultrahigh field (60 T) an extremely large MR ratio of $\sim 10^7\%$ are observed (Fig. 11.8B). The extremely large MR may be explained by an equal number of electron and hole carriers in WTe_2. Besides the extremely large MR, Pan et al. reported the observation of superconductivity in WTe_2 under the different pressure conditions [55]. The resistance changes at low temperature were quite different at various pressures (Fig. 11.8C and D). As the pressure was increased from 2.5 to 16.1 GPa, a maximum superconducting transition temperature (T_c) was observed at $T_c = 3.1$K at a pressure of 16.1 GPa in Fig. 11.8C. When the pressure was further increased to 68.5 GPa, T_c began to decrease monotonically with increasing pressure as shown in Fig. 11.8D.

Gao et al. reported the growth of centimeter-scale WTe_2 ultrathin films on the flexible substrates (the inset of Fig. 11.9A) by pulsed laser deposition technique [69]. (0 0 1) diffraction planes were observed in the XRD patterns, indicating that the WTe_2 film has high quality (Fig. 11.9A). Five phonon modes of WTe_2 were observed in the typical Raman spectra of ultrathin films under different annealing conditions (Fig. 11.9B). The blueshift of the Raman peaks were observed in $WTe_{1.8}$ and $WTe_{1.5}$ samples under poor annealing conditions. Therefore annealing in the appropriate Te vapor pressure can get rid of these structural defects and further improve the film quality.

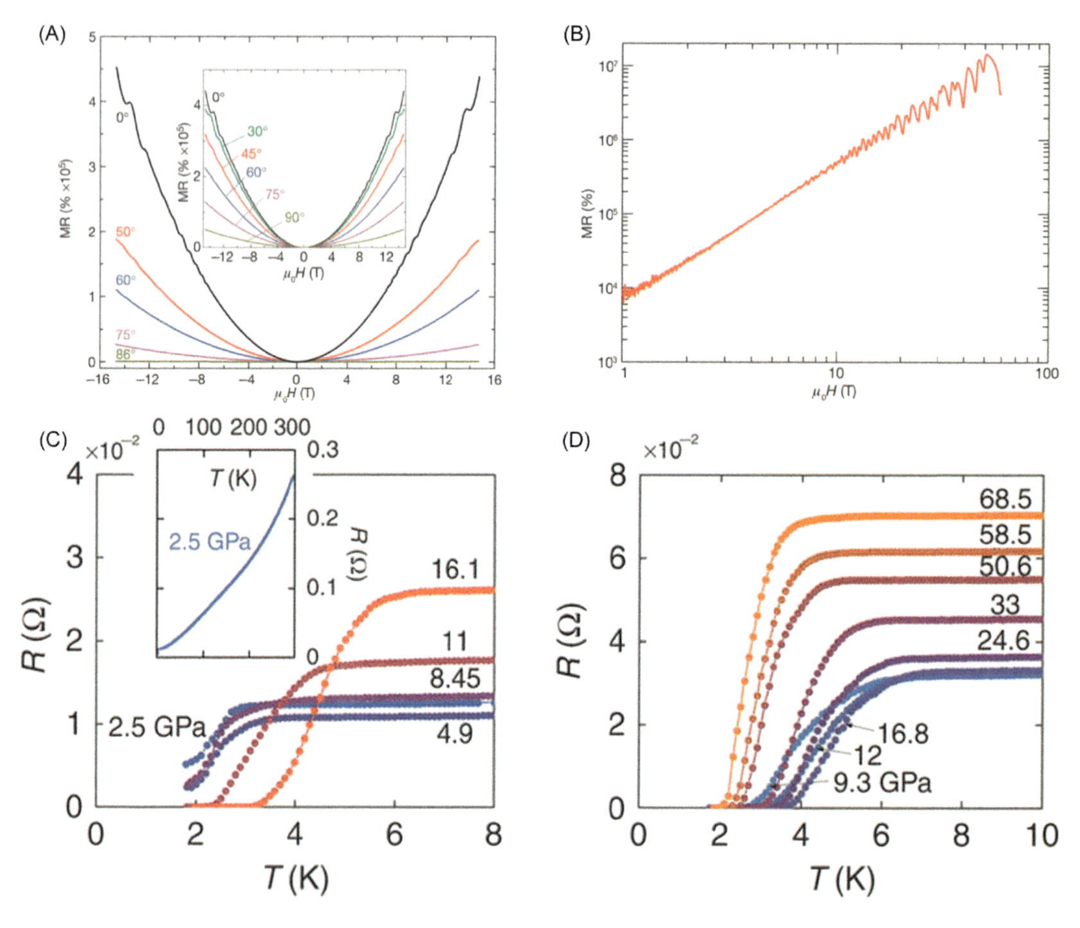

Figure 11.8
Extremely large MR and superconductivity in WTe_2 [53,55]. (A) and (B) MR measured at 15 and 60 T, respectively [53]. An extremely large MR ratio was obtained up to $\sim 10^7\%$ at 60 T. (C) The temperature-dependent resistance under various pressures from 4.9 to 16.1 GPa [55]. The inset shows the temperature-dependent resistance from 1.8K to 300K at 2.5 GPa. (D) Temperature dependence of resistance under various pressures from 9.3 to 68.5 GPa [55]. A maximum superconducting transition temperature (T_c) was observed at $T_c = 3.1$K at a pressure of 16.1 GPa. *MR*, Magnetoresistance.

11.1.2.4 Topological nodal-line semimetals

Topological nodal-line semimetals, a new type of TSMs, have been discovered in crystals with nonsymmorphic symmetry [70,71]. Topological nodal-line semimetals have one-dimensional (1D) Dirac line nodes instead of discrete Dirac points (Fig. 11.10) [70]. Compared to the discrete Dirac points, 1D Dirac nodal line has made a stronger contribution to the charge

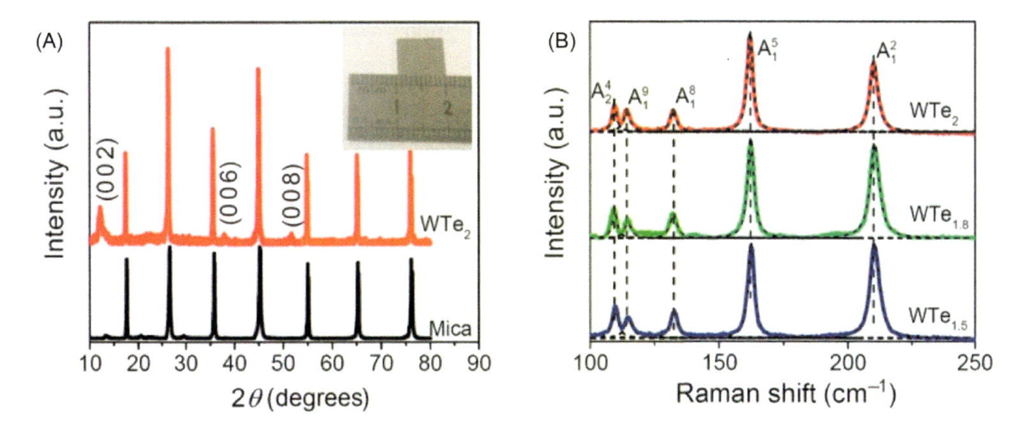

Figure 11.9
Structural characterization of WTe_2 ultrathin films [69]. (A) XRD patterns and (B) Raman spectra of $WTe_{1.5}$, $WTe_{1.8}$, and WTe_2 ultrathin films. Inset shows the photograph of the centimeter-scale WTe_2 sample. *XRD*, X-ray diffraction.

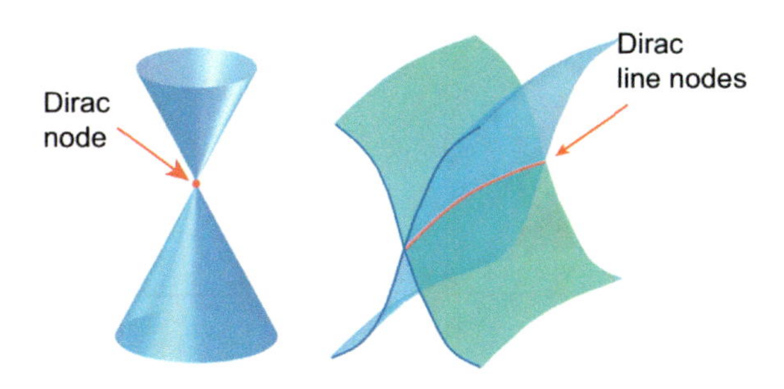

Figure 11.10
Schematic illustration of Dirac point (left panel) and Dirac nodal line (right panel) in the momentum space [70].

carrier density. For example, in the ZrSiS class [71–81], Dirac band crossings take place along a 1D line or loop, contrasted with discrete Dirac (Weyl) points in DSMs (WSM).

Wang et al. observed an anisotropic nonsaturating MR in ZrSiS single crystals [72]. A large MR ratio is observed nearly 170,000% at 53 T at 2K with no saturation (Fig. 11.11A). Interestingly, an anomalous huge anisotropic MR is observed at different angles between the field and crystal cleavage plane (Fig. 11.11B and C). The maximum MR ratio is observed ~200,000% at around 45 degrees, which is more clearly seen in the polar diagram (Fig. 11.11D). The angle-dependent MR shows an interesting butterfly-shaped pattern with a maximum at around 45 degrees (Fig. 11.11D). Besides, they evidenced both the surface and the bulk Dirac bands in ZrSiS single crystals based on the analysis of Berry phase and angle-resolved photoemission spectroscopy (ARPES) spectra.

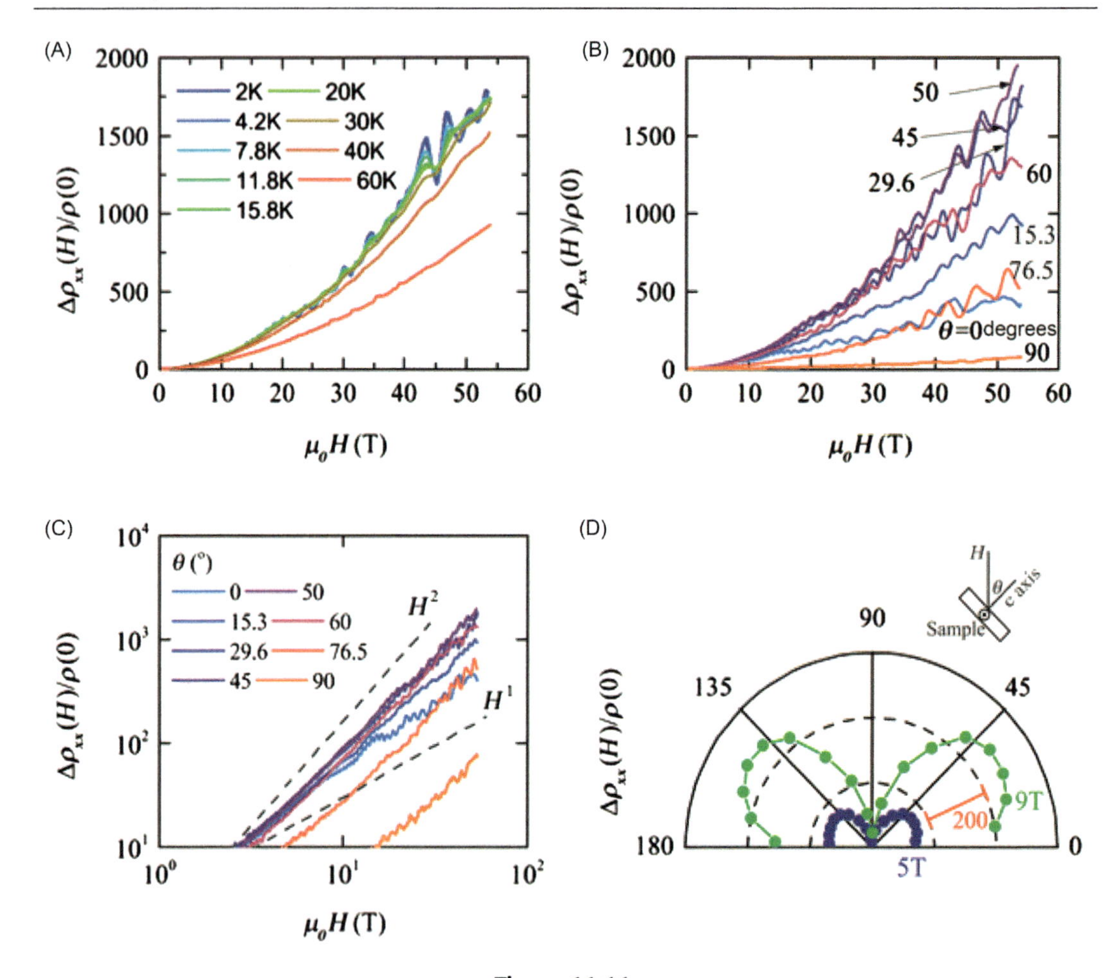

Figure 11.11

Large and butterfly-shaped anisotropic MR of the ZrSiS [72]. (A) The MR curves at different temperatures. (B) The MR curves at different angles. (C) The MR data shown in the log–log frame. (D) Angular dependence of the MR ratio, forming a butterfly shape. *MR*, Magnetoresistance.

Here, we mainly review the latest work of some layered TSMs from the spintronic point of view, which can be fabricated to few layers or even monolayer. Some exotic electron transport properties in TSMs, including ultrahigh charge carrier mobility, chiral magnetic effect, superconductivity, and titanic and butterfly MRs, have been discussed earlier. Due to the strong SOC, Fermi arcs on the surfaces, quantum spin Hall (QSH) effect (QSHE) in monolayer and potential topological superconductivity, TSMs are unique materials that can realize ideally massless and dissipationless states for the next generation of spintronics and quantum computing.

11.2 The strong spin–orbital coupling in topological semimetals

The SOC effect is known as the coupling of the electronic spin and momentum degrees of freedom. It is notable in crystalline lattices with a large atomic number (weight) [11]. When an electron moves across an electric field, it experiences an effective magnetic field $B_{\text{eff}}^{\rightarrow} \sim E^{\rightarrow} \times p^{\rightarrow} /mc^2$ (c is the speed of light). Berry curvature can be viewed as the effective magnetic field associated with the Berry phase [82]. Various properties of Dirac materials are described by the Berry curvature and its associated Berry phase [82]. The effective magnetic field induces a momentum-dependent Zeeman energy, which would result in a spin polarization.

The spin Hall effect is the conversion of an unpolarized charge current into a transverse dissipationless spin current, which is usually observed in materials with strong SOC. Kato et al. detected the spin polarization near the edges of in a GaAs sample based on the Kerr rotation microscopy [83]. The spin polarization is out-of-plane and has opposite sign for the two edges of the sample. Such spin–charge conversion phenomenon is consistent with the predictions of the spin Hall effect, which has direct applications for the injection and detection of a spin current.

The strong SOC was demonstrated in type-II WSMs by using the ARPES technique. Jiang et al. have shown an evidence for the strong SOC of the bulk Fermi pockets in WTe_2 (Fig. 11.12) [54]. The photoemission intensity was inverted between the right-circularly polarized and left-circularly polarized lights (Fig. 11.12A and B), which suggests that the orbital angular momentum varies with Fermi momentum. The band structure was calculated by the density functional theory (DFT) method in Fig. 11.12C. An exotic spin texture was seen on the Fermi surface with $k_z = 0$ (Fig. 11.12D).

The detailed spin texture of WTe_2 has been determined by spin-resolved ARPES [57,63]. Das et al. [57] have performed the spin-resolved measurements at a number of k points in the Brillouin zone (Fig. 11.13A–E). Only large S_y and S_z spin polarizations were measured (Fig. 11.13A–D), which is in agreement with the predictions of only S_y and S_z spin polarizations (Fig. 11.13F). The S_x spin polarization is zero, which means that the electron spin is perpendicular to the tungsten–tungsten zigzag chains.

In ordinary Rashba systems, spin polarization is only in-plane. Here, the spin polarization is observed both in-plane and out-of-plane in WTe_2. Besides, the spin polarization changes sign at positive and negative k values (Fig. 11.13A and C). The amplitude of spin polarization has nonmagnetic origin, which depends primarily on the strength of the SOC in WTe_2. Thus TSMs have been expected to be important for spin Hall devices that can efficiently convert charge current to spin current.

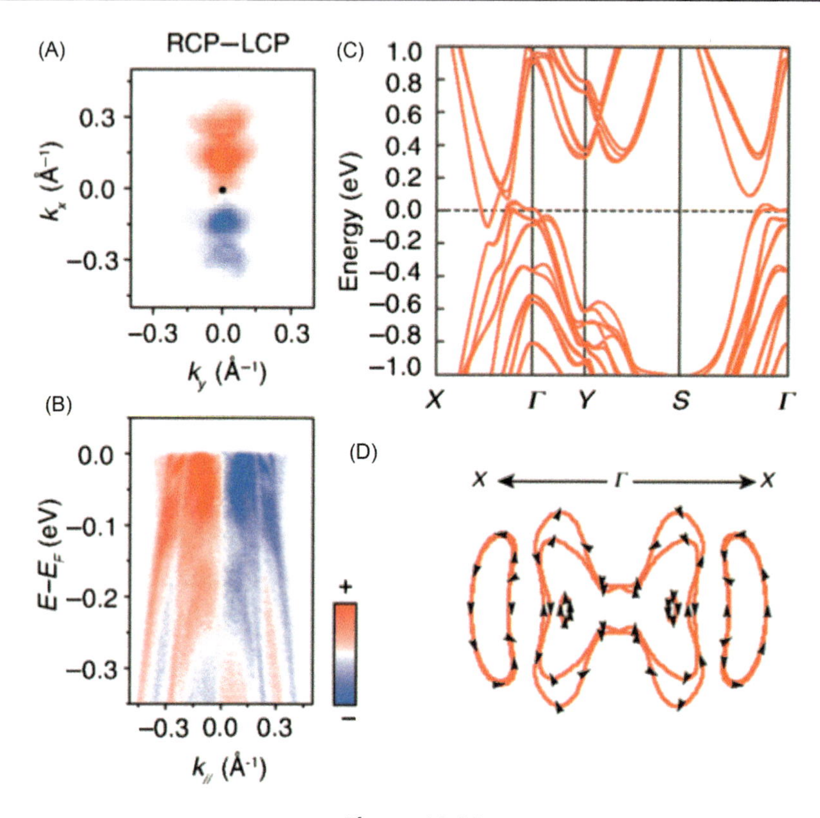

Figure 11.12

The band structures of WTe_2 by ARPES and DFT calculations [54]. (A) The Fermi surface map measured under RCP and LCP light. (B) The corresponding band structure along the Γ−X direction under RCP and LCP light. (C and D) DFT calculations of the electronic structure and spin structure. *ARPES*, Angle-resolved photoemission spectroscopy; *DFT*, density functional theory; *LCP*, left-circularly polarized; *RCP*, right-circularly polarized.

11.3 Fermi arcs

TIs have been viewed as one of the perfect candidates for low-dissipative spintronic devices because of the spin-polarized current at the surface, which is immune to direct backscattering [16−26]. The SSs of TIs commonly exhibit a Dirac cone−type dispersion. The spin and momentum in SSs are locked up and perpendicular to each other [13,84]. Interestingly, similar to the SSs of TIs, spin-polarized SSs (Fermi arcs) also exist on the surface of many WSMs (DSM) (Fig. 11.14) [58−61,85].

Fig. 11.14 shows the difference between the spin texture of the SSs of a 3D TI and a DSM (WSM) [85]. The SSs of a TI have a helical spin texture with a closed loop (Fig. 11.14A). The spin texture of Fermi arcs in a DSM is quite similar to that of SSs in 3D TIs. At the touching point of two arcs, the spin is degenerated (Fig. 11.14B). The Fermi arcs are open

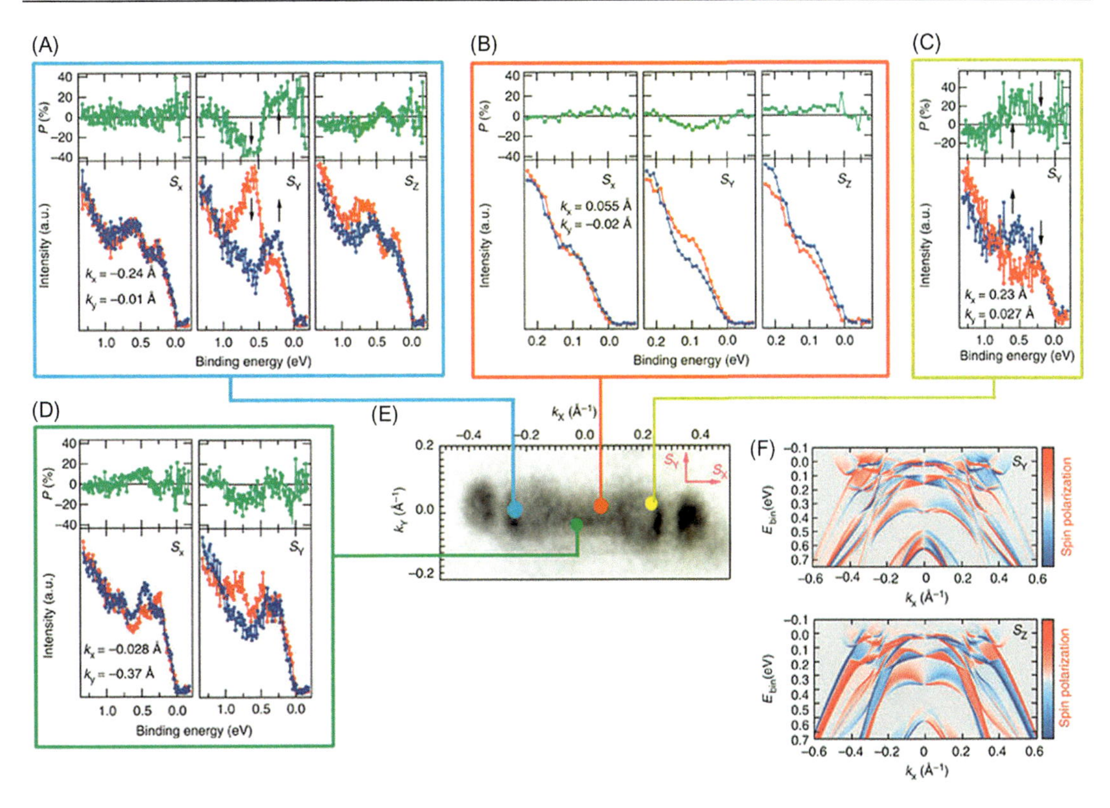

Figure 11.13

Spin polarizations at a number of k points in the Brillouin zone of WTe_2 [57]. (A–D) Measured S_y and S_z spin polarizations at four distinct k points as indicated in (E); (E) Fermi surface with the corresponding spin measurement positions; (F) The band structure by calculation along k_x direction. Only S_y and S_z spin polarizations are observed.

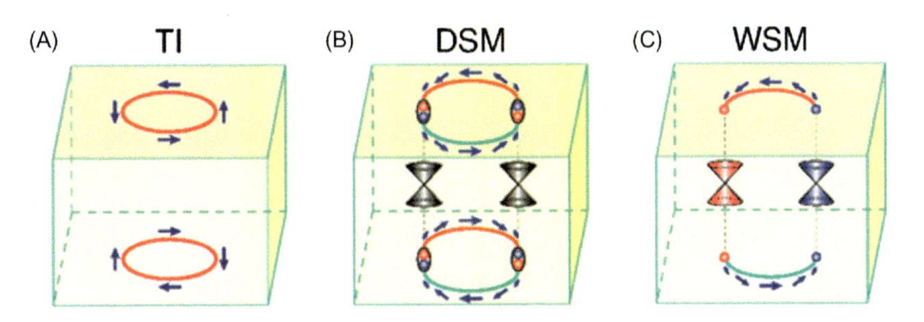

Figure 11.14

The SSs of a TI and a DSM and a WSM [85]. (A) Schematic of the spin–momentum-locked SSs in a 3D TI. (B) Schematic of a DSM with closed Fermi arcs on its surface connecting projections of two bulk Dirac nodes. (C) Schematic of a WSM with open Fermi arcs on its surface connecting projections of two bulk Weyl nodes. *3D*, Three-dimensional; *DSM*, Dirac semimetal; *SSs*, surface states; *TI*, topological insulator; *WSM*, Weyl semimetal.

loops with distinctly different shapes for a WSM (Fig. 11.14C). Each Weyl point acts as a topological charge source. Because the net charge must vanish in the entire Brillouin zone, Weyl points always come in pairs with opposite charge, which lead to various magnetotransport anomalies.

11.3.1 Fermi arcs in WTe_2

The spin-polarized states on the surface of TSMs are called the Fermi arcs. The Fermi arc exhibits an unclosed line that starts from one Weyl point and ends at the other with opposite chirality. Intense efforts have been devoted to study the Fermi arcs of type-II topological WSM WTe_2 by ARPES technique [58–61]. Wang et al. have conducted the ARPES measurements of the band structures of WTe_2 at 100K and 200K, respectively (Fig. 11.15A and B) [58]. Due to

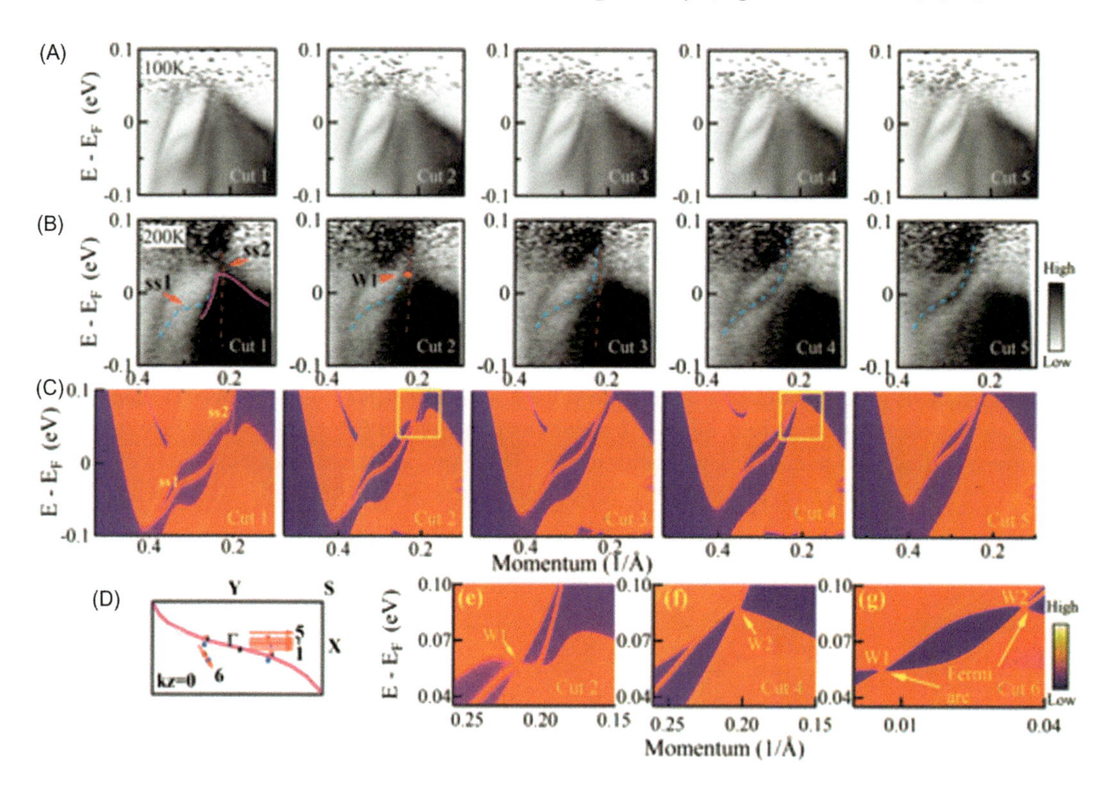

Figure 11.15
Identification of Weyl points and SSs in WTe_2 [58]. (A) and (B) Band structures of WTe_2 measured along different momentum cuts at 100K and 200K, respectively. The location of the momentum Cuts 1–5 are marked in (D). (C) The calculated band structures along the five momentum cuts in (D). (D) A schematic for the distribution of the Weyl points within a bulk Brillouin zone at $k_z = 0$. (E) and (F) Expanded view of the region marked by the yellow rectangle in (C) for Cut 2 and Cut 4, respectively. (G) Calculated band structure for the Cut 6 in (D). *SSs*, Surface states.

the thermal excitations, the SSs and Weyl points were illustrated more clearly along Cut 1 and Cut 2 at 200K (Fig. 11.15B). The calculations of band structures were seen in Fig. 11.15C along the five momentum cuts on the $k_z = 0$ plane (Fig. 11.15D). The two SSs are illustrated in the Cut 1 in Fig. 11.15C. Parts (E) and (F) of Fig. 11.15 are the expanded views of the marked regions along the Cuts 2 and 4 in Fig. 11.15C. The calculations of the band structures along the Cut 6 are seen in the Fig. 11.15G. One Weyl point is located 56 meV above the Fermi level while the other is 89 meV above the Fermi level (Fig. 11.15G).

11.3.2 Fermi arcs in $Mo_xW_{1-x}Te_2$

The short Fermi arcs in WTe_2 cannot be clearly resolved by the ARPES since the Weyl points are above the Fermi level. The Fermi arcs have been found to be elongated by Mo doping in WTe_2, which would make $Mo_xW_{1-x}Te_2$ a promising platform for transport and optics experiments on WSMs [86–88]. Belopolski et al. demonstrated the existence of a group of Weyl points at the Fermi level in $Mo_xW_{1-x}Te_2$ (Fig. 11.16) [88]. They found two kinks at $E_B \sim -0.005$ and ~ -0.05 eV as a smoking-gun signature of a Weyl point (Fig. 11.16A–C). Each kink corresponds to a Weyl point and that the SSs passing through them include a topological Fermi arc (Fig. 11.16D). Besides, the bulk valence and conduction states, and the trivial SSs above the conduction band were all illustrated in Fig. 11.16D.

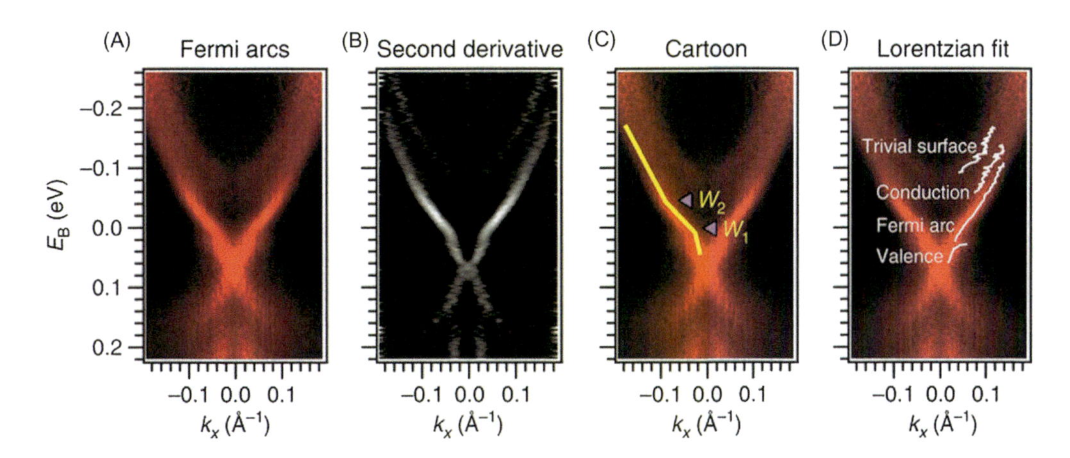

Figure 11.16

Direct experimental observation of Fermi arcs in $Mo_{0.25}W_{0.75}Te_2$ [88]. (A and B) The primal and second-derivative plot of the band structure in $Mo_{0.25}W_{0.75}Te_2$. Two kinks are observed at $E_B \sim -0.005$ and ~ -0.05 eV. (C) Two Weyl points at the locations of the kinks. (D) The bulk conduction and valence bands, topological Fermi arc, and an additional trivial surface state marked by the white lines.

With the support of first-principles calculations, Feng et al. reveal the existence of spin polarization of the Fermi arcs in type-II topological WSM WTe_2 [63]. Based on the spin–momentum locking property of the Fermi arcs, Li et al. observed a hysteretic step-like signal in a WTe_2 – FM structure [89]. The amplitude and direction of the spin-valve signal are directly governed by the charge current effectively (Fig. 11.17A–C), which can be ascribed to the enhanced spin accumulation by the spin–momentum locking effect of the Fermi arcs of WTe_2. The spin accumulation only exists at low temperature (Fig. 11.17D). Thus identified by topological Fermi arcs on the surface, WSMs are interesting for fundamental science and may become the basis for a breakthrough in the spintronic technology.

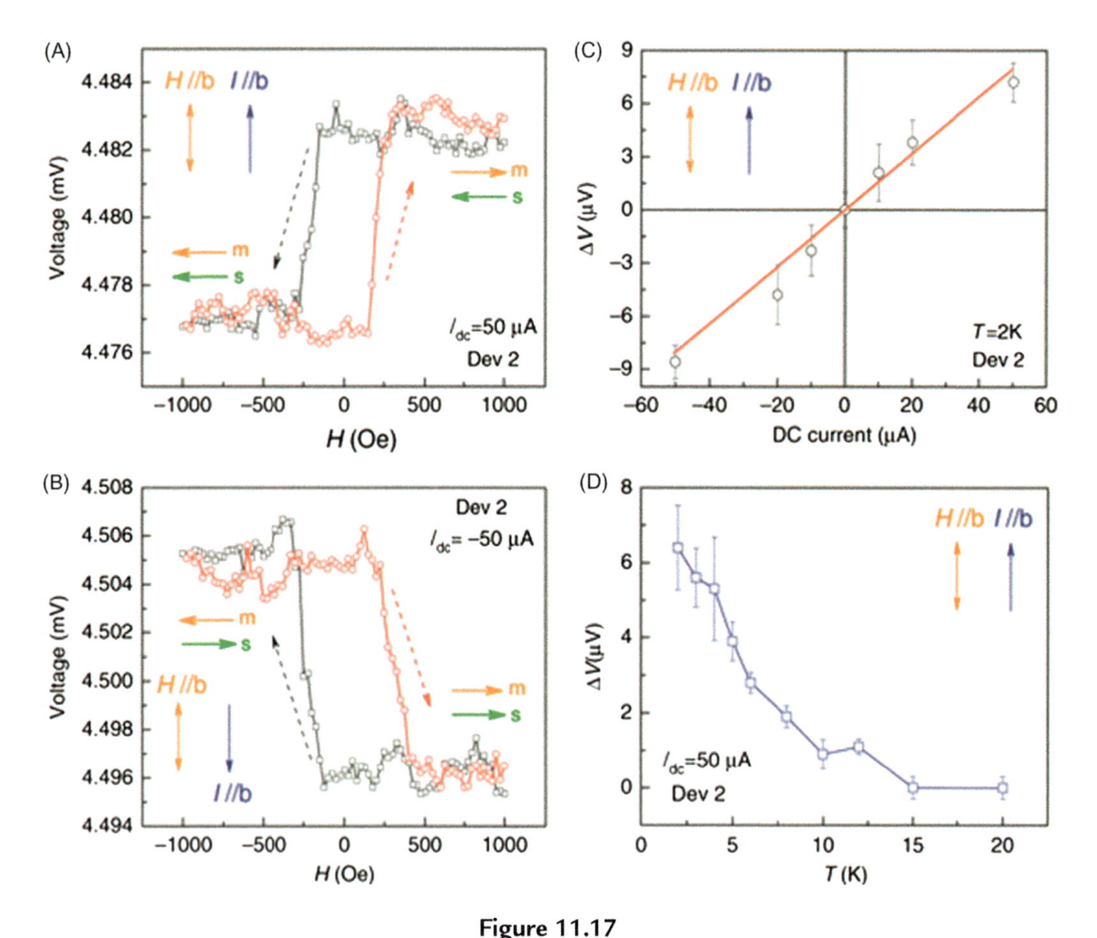

Figure 11.17

WTe_2 as a spin generator [89]. (A and B) Spin-valve signals in a WTe_2–FM structure taken at the bias currents of $I_{dc} = +50$and -50 μA, respectively. (C) The current dependence of the spin signals. (D) The temperature dependence of the spin signals.

11.4 Two-dimensional topological insulators

2D TIs, namely, the QSH insulators, which host helical edge states inside a bulk insulating gap, are regarded as one of the most intriguing materials for dissipationless spintronic devices. QSH systems have an energy bandgap separating the valance and the conduction bands in the bulk. However, they have gapless edge states protected by TRS on the boundary [15]. Particularly in 2D TIs, the low-energy (back)scattering of the edge states is prohibited by the TRS, leading to the dissipationless transport edge channels and the QSHE. However, a major obstacle is the lack of suitable QSH compounds, which can be easily fabricated and have a large bulk gap for experimental studies and possible room-temperature applications.

11.4.1 Two-dimensional topological insulators in quantum well

In 2005 the QSHE was proposed in graphene by Kane and Mele [90]. However, the SOC is too small in graphene, which leads to a negligible bulk gap. The QSHE was experimentally detected in HgTe/CdTe [91,92] and InAs/GaSb [93] quantum wells (QWs). Quantum resistance (e^2/h, where e is the electron charge and h is Planck's constant) was clearly detected under very low temperatures, which proved that electrons do move along a 2D TI's edge without energy dissipation. König et al. reported the observation QSH insulator state in an HgTe QW [92]. In an HgTe QW with a width of 5.5-nm, they measured a resistance of at least several tens of MΩ (i.e., $G_{14,23} = 0.01e^2/h$) when the Fermi level is in the gap. In a 7.3-nm-wide QW, they measured a resistance of ~100KΩ (i.e., $G_{14,23} = 0.3e^2/h$). For much shorter samples with inverted QWs, $G_{14,23}$ actually reaches the predicted value close to $2e^2/h$, which is consistent with the two quantum channels. Besides, the residual resistance of the devices does not depend on the width of the structure, which demonstrates the existence of the QSH insulator state in HgTe QWs.

11.4.2 Two-dimensional topological insulators in WTe_2

However, the bulk gap of HgTe/CdTe is still small ~10 meV as discussed earlier, the dissipationless transport can only occur at extremely low temperature. Thus improving the working temperature of QSH systems has become one of the key issues in the field of 2D TIs. Qian et al. predicted a class of large-gap QSH insulators based on 2D transition metal dichalcogenide monolayers with 1 T′ structure using first-principles calculations [94]. The monolayer of WHM is also predicted to be a 2D TI with achievable bandgap up to 30 meV, with W = (Zr, Hf, or La), H = (Si, Ge, Sn, or Sb), and M = (O, S, Se, and Te) [95].

Several groups have studied electrical transport experiments of WTe_2 monolayer and evidenced the monolayer of WTe_2 is a 2D TI candidate. Fei et al. performed

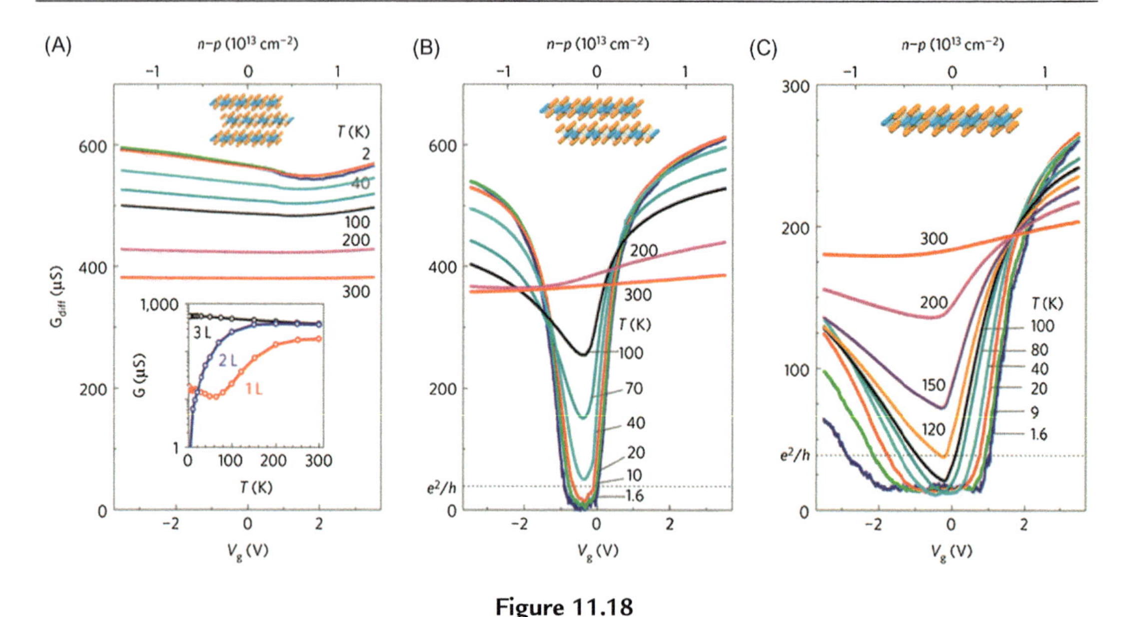

Figure 11.18

Two-terminal characteristics of WTe_2 devices [65]. (A)–(C) Temperature dependence of the resistance on a trilayer, bilayer, and monolayer device, respectively.

low-temperature electrical transport experiment on WTe_2 devices and observed a novel edge conduction in monolayer WTe_2 [65]. Fig. 11.18A–C shows the differential conductance of encapsulated trilayer, bilayer, and monolayer devices, respectively. The trilayer shows a metallic behavior as decreasing the temperature at all V_g (Fig. 11.18A). The conductance saturates at the lowest temperatures. The metallic behavior is consistent with the behavior of bulk WTe_2. The resistance of bilayer shows a strong V_g dependence with a sharp minimum near $V_g = 0$ (Fig. 11.18B). The minimum drops steadily as decreasing the temperature and broadens below 20K. A plateau of conductance is only observed in the monolayer below 100K, which is due to the predominant edge conduction when the bulk becomes insulating (Fig. 11.18C).

Further, Wu et al. have demonstrated the QSHE in monolayer WTe_2 at temperatures up to 100K [96]. The working temperature of QSHE in monolayer WTe_2 is much higher than that in semiconductor QWs. Electronic structure of monolayer WTe_2 was also discovered by the scanning tunneling microscope (STM) [64,67]. Tang et al. have confirmed an edge state in the insulating bulk by STM, which may indicate the existence of a conductive edge state [67].

11.4.3 Two-dimensional topological insulators in $ZrTe_5$

Bulk $ZrTe_5$ has been calculated to be very close to the boundary between weak and strong 3D TIs [29]. $ZrTe_5$ is a good layered compound with very weak interlayer bonding.

The monolayer $ZrTe_5$ has been predicted to be a 2D TI with an energy gap of 0.1 eV [29], which paves a new way for future experimental studies on both the QSHE and topological phase transitions.

The topological properties of monolayer $ZrTe_5$ were confirmed by the observation of a 1D conductive edge state on the surface of bulk $ZrTe_5$. Wu et al. [35] observed a sharp edge on the cleaved surface of bulk $ZrTe_5$ (Fig. 11.19). On the terrace, scanning tunneling spectroscopy shows an energy gap of about 0.1 eV (Fig. 11.19A) that is consistent with DFT calculations (Fig. 11.19B). Within the bulk gap, conductance is zero away from the step edge. But near the edge, conductance is not zero. The nearly constant conductance near the edge explicitly points to the topologically nontrivial nature of the edge states.

Li et al. also confirm that monolayer $ZrTe_5$ is a 2D TI with a large energy gap of 80 meV [36]. They observed the topological edge states at the step edge and an energetic splitting of the topological edge states in the magnetic field. The splitting can be attributed to a strong link between the topological edge states and bulk topology. The relatively large bandgap makes monolayer $ZrTe_5$ a potential candidate for future fundamental studies and device applications. Such a large gap with topological edge states is promising for the QSH device operable at high temperatures. This will stimulate the applications of TSMs as the next generation of microelectronics and spintronics.

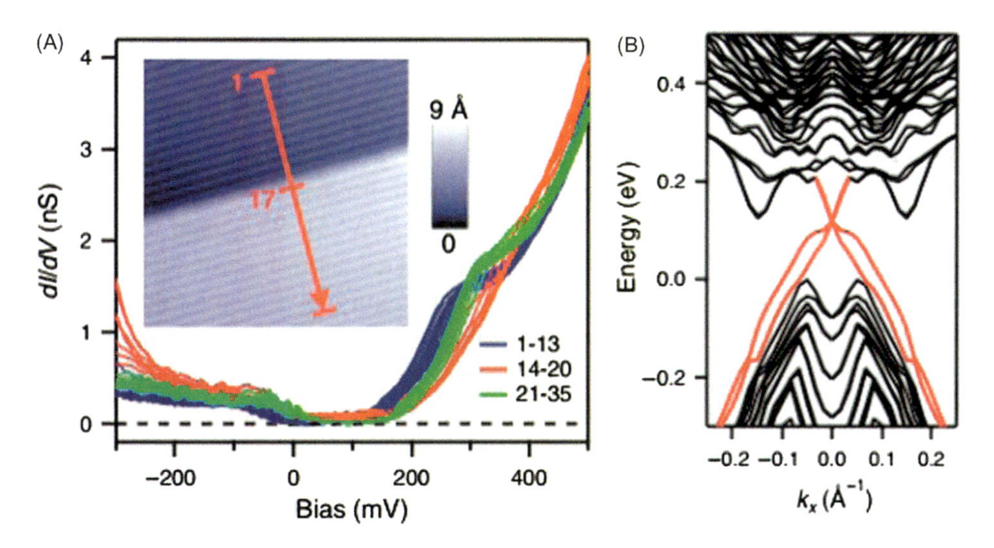

Figure 11.19
1D edge state of a monolayer $ZrTe_5$ [35]. (A) STS along a line perpendicular to a monolayer step edge. Within the bulk gap, conductance is zero away from the step edge. But near the edge, conductance is not zero. The inset shows the image of the sample step. (B) Calculated band structures of the edge states of the monolayer $ZrTe_5$. *1D*, One-dimensional; *STS*, scanning tunneling spectroscopy.

11.5 Majorana fermions

Topological superconductors by chemical doping in TIs [97] or TIs-superconductor structure [98] have attracted particular attention because they have a full pairing gap in the bulk and gapless SSs consisting of Majorana fermions, which have potential applications in spintronics and quantum computing.

Type-II Dirac fermions ($PdTe_2$) [47,49–51], which break Lorentz invariance, have been recently discovered experimentally, which is also a superconductor with a T_c at 1.7K using ARPES and magnetotransport technique. Combined with theoretical calculations, Fei et al. discovered the existence of topologically nontrivial SS with Dirac cone in $PbTe_2$ superconductor (Fig. 11.7) [49]. This establishes that $PdTe_2$ is the first example that has both superconductivity and type-II Dirac fermions, providing a possible platform of research for interactions between superconducting quasiparticles and type-II Dirac fermions. A successful Fermi level tuning has been realized to lower the Fermi level near the Dirac point without losing the superconductivity [99].

$ZrTe_5$ has been discussed as a natural QSH device integrated with multiple edge channels. If the edges are covered by superconducting films, Majorana bound states can be induced through the superconducting proximity effect at the end of the edges, which may lead to potential applications in spintronic devices and quantum computing.

11.6 Summary and outlook

Due to the band inversion, TSMs are characterized by the Fermi arcs with the metallic bulk states. So far, several layered materials have been discovered as TSMs, such as the type-I DSMs $ZrTe_5$, the type-II DSM $PdTe_2$, the type-II WSM WTe_2, and topological nodal-line semimetals ZrSiS. The strong SOC, Fermi arcs, and superconductivity have been demonstrated in these materials. An exotic spin texture was seen on the Fermi surface in the type-II WSM WTe_2. The amplitude of spin polarization depends primarily on the strength of the SOC. The observed spin polarization of bands has nonmagnetic origin. The Fermi arcs on the surface of TSMs are spin-polarized. The spin textures were found to wind against the dispersion of the Fermi arc. 2D TIs, namely, the QSH insulators, are proposed to exist in TSMs. A plateau of edge conductance was observed in the monolayer type-II WSM WTe_2 and the type-I DSMs $ZrTe_5$ as the Fermi level is inside a bulk insulating gap.

However, considerable trivial carrier pockets still coexist with the nontrivial carrier pockets in the Fermi surface. Thus efficient ways to suppress the trivial carrier density are much needed, which aims to locate the linear Dirac (Weyl) bands close to the Fermi energy. High-quality thin films will be favored for building devices.

The exotic properties of TSMs show great potential for applications, such as the high mobility and titanic and negative MRs. WSMs can be employed in high-speed electronics and spintronics. We stress that WSMs may exhibit the strong spin Hall effect, which can be employed in high-speed spintronics owing to the large Berry curvature and SOC. Finally, one vision for TSMs is to utilize their robust Fermi arcs, 2D TI states, and the possible topological superconductivity for quantum computing.

References

[1] S.A. Wolf, et al., Spintronics: a spin-based electronics vision for the future, Science 294 (2001) 1488–1495.
[2] I. Žutić, J. Fabian, S. Das Sarma, Spintronics: fundamentals and applications, Rev. Mod. Phys. 76 (2004) 323–410.
[3] M.N. Baibich, et al., Giant magnetoresistance of (0 0 1)Fe/(0 0 1)Cr magnetic superlattices, Phys. Rev. Lett. 61 (1988) 2472–2475.
[4] G. Binasch, P. Grünberg, F. Saurenbach, W. Zinn, Enhanced magnetoresistance in layered magnetic structures with antiferromagnetic interlayer exchange, Phys. Rev. B 39 (1989) 4828–4830.
[5] S. Datta, B. Das, Electronic analog of the electro-optic modulator, Appl. Phys. Lett. 56 (1990) 665–667.
[6] N. Tombros, C. Jozsa, M. Popinciuc, H.T. Jonkman, B.J. van Wees, Electronic spin transport and spin precession in single graphene layers at room temperature, Nature 448 (2007) 571–574.
[7] W. Han, et al., Spin transport and relaxation in graphene, J. Magn. Magn. Mater. 324 (2012) 369–381.
[8] W. Han, R.K. Kawakami, M. Gmitra, F. Jaroslav, Graphene spintronics, Nat. Nanotechnol. 9 (2014) 794–807.
[9] R. Jansen, Silicon spintronics, Nat. Mater. 11 (2012) 400–408.
[10] G. Schmidt, D. Ferrand, L.W. Molenkamp, A.T. Filip, B.J. van Wees, Fundamental obstacle for electrical spin injection from a ferromagnetic metal into a diffusive semiconductor, Phys. Rev. B 62 (2000) 4790–4793.
[11] A. Manchon, H.C. Koo, J. Nitta, S.M. Frolov, R.A. Duine, New perspectives for Rashba spin-orbit coupling, Nat. Mater. 14 (2015) 871–882.
[12] L. Fu, C.L. Kane, E.J. Mele, Topological insulators in three dimensions, Phys. Rev. Lett. 98 (2007) 106803.
[13] H. Zhang, et al., Topological insulators in Bi_2Se_3, Bi_2Te_3 and Sb_2Te_3 with a single Dirac cone on the surface, Nat. Phys. 5 (2009) 438–442.
[14] M.Z. Hasan, C.L. Kane, Colloquium: topological insulators, Rev. Mod. Phys. 82 (2010) 3045.
[15] X.-L. Qi, S.-C. Zhang, Topological insulators and superconductors, Rev. Mod. Phys. 83 (2011) 1057–1110.
[16] C.H. Li, et al., Electrical detection of charge-current-induced spin polarization due to spin-momentum locking in Bi_2Se_3, Nat. Nanotechnol. 9 (2014) 218–224.
[17] J. Tian, et al., Topological insulator based spin valve devices: evidence for spin polarized transport of spin-momentum-locked topological surface states, Solid State Commun. 191 (2014) 1–5.
[18] Y. Ando, et al., Electrical detection of the spin polarization due to charge flow in the surface state of the topological insulator $Bi_{1.5}Sb_{0.5}Te_{1.7}Se_{1.3}$, Nano Lett. 14 (2014) 6226–6230.
[19] J. Tang, et al., Electrical detection of spin-polarized surface states conduction in $(Bi_{0.53}Sb_{0.47})_2Te_3$ topological insulator, Nano Lett. 14 (2014) 5423–5429.
[20] Y. Wang, et al., Topological surface states originated spin-orbit torques in Bi_2Se_3, Phys. Rev. Lett. 114 (2015) 257202.
[21] A. Dankert, J. Geurs, M.V. Kamalakar, S. Charpentier, S.P. Dash, Room temperature electrical detection of spin polarized currents in topological insulators, Nano Lett. 15 (2015) 7976–7981.

[22] J. Tian, I. Miotkowski, S. Hong, Y.P. Chen, Electrical injection and detection of spin-polarized currents in topological insulator Bi_2Te_2Se, Sci. Rep. 5 (2015) 14293.
[23] E.K. de Vries, et al., Towards the understanding of the origin of charge-current-induced spin voltage signals in the topological insulator Bi_2Se_3, Phys. Rev. B 92 (2015) 201102.
[24] F. Yang, et al., Switching of charge-current-induced spin polarization in the topological insulator $BiSbTeSe_2$, Phys. Rev. B 94 (2016) 075304.
[25] M. Zhang, et al., Unique current-direction-dependent ON–OFF switching in $BiSbTeSe_2$ topological insulator-based spin valve transistors, IEEE Electron Device Lett. 37 (2016) 1231–1233.
[26] J. Tian, S. Hong, I. Miotkowski, S. Datta, Y.P. Chen, Observation of current-induced, long-lived persistent spin polarization in a topological insulator: a rechargeable spin battery, Sci. Adv. 3 (2017) e1602531.
[27] K. Vaklinova, A. Hoyer, M. Burghard, K. Kern, Current-induced spin polarization in topological insulator–graphene heterostructures, Nano Lett. 16 (2016) 2595–2602.
[28] A.A. Soluyanov, et al., Type-II Weyl semimetals, Nature 527 (2015) 495–498.
[29] H. Weng, X. Dai, Z. Fang, Transition-metal pentatelluride $ZrTe_5$ and $HfTe_5$: a paradigm for large-gap quantum spin Hall insulators, Phys. Rev. X 4 (2014) 011002.
[30] L. Zhou, et al., New family of quantum spin Hall insulators in two-dimensional transition-metal halide with large nontrivial band gaps, Nano Lett. 15 (2015) 7867–7872.
[31] G. Manzoni, et al., Evidence for a strong topological insulator phase in $ZrTe_5$, Phys. Rev. Lett. 117 (2016) 237601.
[32] X. Yuan, et al., Observation of quasi-two-dimensional Dirac fermions in $ZrTe_5$, NPG Asia Mater. 8 (2016) e325.
[33] G. Zheng, et al., Transport evidence for the three-dimensional Dirac semimetal phase in $ZrTe_5$, Phys. Rev. B 93 (2016) 115414.
[34] Q. Li, et al., Chiral magnetic effect in $ZrTe_5$, Nat. Phys. 12 (2016) 550–554.
[35] R. Wu, et al., Evidence for topological edge states in a large energy gap near the step edges on the surface of $ZrTe_5$, Phys. Rev. X 6 (2016) 021017.
[36] X.-B. Li, et al., Experimental observation of topological edge states at the surface step edge of the topological insulator $zrte_5$, Phys. Rev. Lett. 116 (2016) 176803.
[37] Y. Zhang, et al., Electronic evidence of temperature-induced Lifshitz transition and topological nature in $ZrTe_5$, Nat. Commun. 8 (2017) 15512.
[38] Z. Fan, Q.-F. Liang, Y. Chen, S.-H. Yao, J. Zhou, Transition between strong and weak topological insulator in $ZrTe_5$ and $HfTe_5$, Sci. Rep. 7 (2017) 45667.
[39] Z.K. Liu, et al., A stable three-dimensional topological Dirac semimetal Cd_3As_2, Nat. Mater. 13 (2014) 677–681.
[40] T. Liang, et al., Ultrahigh mobility and giant magnetoresistance in the Dirac semimetal Cd_3As_2, Nat. Mater. 14 (2015) 280–284.
[41] E. Zhang, et al., Magnetotransport properties of Cd_3As_2 nanostructures, ACS Nano 9 (2015) 8843–8850.
[42] L.-X. Wang, C.-Z. Li, D.-P. Yu, Z.-M. Liao, Aharonov–Bohm oscillations in Dirac semimetal Cd_3As_2 nanowires, Nat. Commun. 7 (2016) 10769.
[43] H. Pan, et al., Quantum oscillation and nontrivial transport in the Dirac semimetal Cd_3As_2 nanodevice, Appl. Phys. Lett. 108 (2016) 183103.
[44] C.-Z. Li, et al., Giant negative magnetoresistance induced by the chiral anomaly in individual Cd_3As_2 nanowires, Nat. Commun. 6 (2015) 10137.
[45] H. Li, et al., Negative magnetoresistance in Dirac semimetal Cd_3As_2, Nat. Commun. 7 (2016) 10301.
[46] K. Zhang, et al., Controllable synthesis and magnetotransport properties of Cd_3As_2 Dirac semimetal nanostructures, RSC Adv. 7 (2017) 17689–17696.
[47] L. Yan, et al., Identification of topological surface state in $PdTe_2$ superconductor by angle-resolved photoemission spectroscopy, Chin. Phys. Lett. 32 (2015) 067303.

[48] Y. Wang, et al., De Hass-van Alphen and magnetoresistance reveal predominantly single-band transport behavior in $PdTe_2$, Sci. Rep. 6 (2016) 31554.
[49] F. Fei, et al., Nontrivial Berry phase and type-II Dirac transport in the layered material $PdTe_2$, Phys. Rev. B 96 (2017) 041201.
[50] H.-J. Noh, et al., Experimental realization of type-II Dirac Fermions in a $PdTe_2$ superconductor, Phys. Rev. Lett. 119 (2017) 016401.
[51] R. Xiao, et al., Manipulation of type-I and type-II Dirac points in $PdTe_2$ superconductor by external pressure, Phys. Rev. B 96 (2017) 075101.
[52] X. Wu, et al., Epitaxial growth and air-stability of monolayer antimonene on $PdTe_2$, Adv. Mater. 29 (2017) 1605407.
[53] M.N. Ali, et al., Large, non-saturating magnetoresistance in WTe_2, Nature 514 (2014) 205–208.
[54] J. Jiang, et al., Signature of strong spin-orbital coupling in the large nonsaturating magnetoresistance material WTe_2, Phys. Rev. Lett. 115 (2015) 166601.
[55] X.-C. Pan, et al., Pressure-driven dome-shaped superconductivity and electronic structural evolution in tungsten ditelluride, Nat. Commun. 6 (2015) 7805.
[56] Y. Zhao, et al., Anisotropic magnetotransport and exotic longitudinal linear magnetoresistance in WTe_2 crystals, Phys. Rev. B 92 (2015) 041104.
[57] P.K. Das, et al., Layer-dependent quantum cooperation of electron and hole states in the anomalous semimetal WTe_2, Nat. Commun. 7 (2016) 10847.
[58] C. Wang, et al., Observation of Fermi arc and its connection with bulk states in the candidate type-II Weyl semimetal WTe_2, Phys. Rev. B 94 (2016) 241119.
[59] Y. Wu, et al., Observation of Fermi arcs in the type-II Weyl semimetal candidate WTe_2, Phys. Rev. B 94 (2016) 121113.
[60] F.Y. Bruno, et al., Observation of large topologically trivial Fermi arcs in the candidate type-II Weyl semimetal WTe_2, Phys. Rev. B 94 (2016) 121112.
[61] J. Sánchez-Barriga, et al., Surface Fermi arc connectivity in the type-II Weyl semimetal candidate WTe_2, Phys. Rev. B 94 (2016) 161401.
[62] Y. Wang, K. Wang, J. Reutt-Robey, J. Paglione, M.S. Fuhrer, Breakdown of compensation and persistence of nonsaturating magnetoresistance in gated WTe_2 thin flakes, Phys. Rev. B 93 (2016) 121108.
[63] B. Feng, et al., Spin texture in type-II Weyl semimetal WTe_2, Phys. Rev. B 94 (2016) 195134.
[64] Z.-Y. Jia, et al., Direct visualization of a two-dimensional topological insulator in the single-layer 1 T′-WTe_2, Phys. Rev. B 96 (2017) 041108.
[65] Z. Fei, et al., Edge conduction in monolayer WTe_2, Nat. Phys. 13 (2017) 677–682.
[66] V. Fatemi, et al., Magnetoresistance and quantum oscillations of an electrostatically tuned semimetal-to-metal transition in ultrathin WTe_2, Phys. Rev. B 95 (2017) 041410.
[67] S. Tang, et al., Quantum spin Hall state in monolayer 1T′-WTe_2, Nat. Phys. 13 (2017) 683–687.
[68] X.-C. Pan, et al., Carrier balance and linear magnetoresistance in type-II Weyl semimetal WTe_2, Front. Phys. 12 (2017) 127203.
[69] M. Gao, et al., Tuning the transport behavior of centimeter-scale WTe_2 ultrathin films fabricated by pulsed laser deposition, Appl. Phys. Lett. 111 (2017) 031906.
[70] C. Chen, et al., Dirac line nodes and effect of spin-orbit coupling in the nonsymmorphic critical semimetals MSiS (M = Hf, Zr), Phys. Rev. B 95 (2017) 125126.
[71] L.M. Schoop, et al., Dirac cone protected by non-symmorphic symmetry and three-dimensional Dirac line node in ZrSiS, Nat. Commun. 7 (2016) 11696.
[72] X. Wang, et al., Evidence of both surface and bulk Dirac bands and anisotropic nonsaturating magnetoresistance in ZrSiS, Adv. Electron. Mater. 2 (2016) 1600228.
[73] M.N. Ali, et al., Butterfly magnetoresistance, quasi-2D Dirac Fermi surface and topological phase transition in ZrSiS, Sci. Adv. 2 (2016) e1601742.
[74] Y.-Y. Lv, et al., Extremely large and significantly anisotropic magnetoresistance in ZrSiS single crystals, Appl. Phys. Lett. 108 (2016) 244101.

[75] M. Neupane, et al., Observation of topological nodal fermion semimetal phase in ZrSiS, Phys. Rev. B 93 (2016) 201104.
[76] R. Sankar, et al., Crystal growth of Dirac semimetal ZrSiS with high magnetoresistance and mobility, Sci. Rep. 7 (2017) 40603.
[77] B. Salmankurt, S. Duman, First-principles study of structural, mechanical, lattice dynamical and thermal properties of nodal-line semimetals ZrXY (X = Si, Ge; Y = S, Se), Philos. Mag. 97 (2017) 175–186.
[78] R. Singha, A.K. Pariari, B. Satpati, P. Mandal, Large nonsaturating magnetoresistance and signature of nondegenerate Dirac nodes in ZrSiS, Proc. Natl. Acad. Sci. U.S.A. 114 (2017) 2468–2473.
[79] J. Hu, et al., Nearly massless Dirac fermions and strong Zeeman splitting in the nodal-line semimetal ZrSiS probed by de Haas–van Alphen quantum oscillations, Phys. Rev. B 96 (2017) 045127.
[80] M.M. Hosen, et al., Tunability of the topological nodal-line semimetal phase in ZrSiX-type materials (X = S, Se, Te), Phys. Rev. B 95 (2017) 161101.
[81] J. Zhang, et al., Transport evidence of 3D topological nodal-line semimetal phase in ZrSiS, Front. Phys. 13 (2018) 137201.
[82] D. Xiao, M.-C. Chang, Q. Niu, Berry phase effects on electronic properties, Rev. Mod. Phys. 82 (2010) 1959–2007.
[83] Y.K. Kato, R.C. Myers, A.C. Gossard, D.D. Awschalom, Observation of the spin Hall effect in semiconductors, Science 306 (2004) 1910.
[84] D. Hsieh, et al., A tunable topological insulator in the spin helical Dirac transport regime, Nature 460 (2009) 1101–1105.
[85] B.Q. Lv, et al., Observation of Fermi-arc spin texture in TaAs, Phys. Rev. Lett. 115 (2015) 217601.
[86] T.-R. Chang, et al., Prediction of an arc-tunable Weyl Fermion metallic state in $Mo_xW_{1-x}Te_2$, Nat. Commun. 7 (2016) 10639.
[87] I. Belopolski, et al., Fermi arc electronic structure and Chern numbers in the type-II Weyl semimetal candidate $Mo_xW_{1-x}Te_2$, Phys. Rev. B 94 (2016) 085127.
[88] I. Belopolski, et al., Discovery of a new type of topological Weyl fermion semimetal state in $Mo_xW_{1-x}Te_2$, Nat. Commun. 7 (2016) 13643.
[89] P. Li, et al., Spin-momentum locking and spin-orbit torques in magnetic nano-heterojunctions composed of Weyl semimetal WTe_2, Nat. Commun. 9 (2018) 3990.
[90] C.L. Kane, E.J. Mele, Quantum spin Hall effect in graphene, Phys. Rev. Lett. 95 (2005) 226801.
[91] B.A. Bernevig, T.L. Hughes, S.-C. Zhang, Quantum spin Hall effect and topological phase transition in HgTe quantum Wells, Science 314 (2006) 1757.
[92] M. König, et al., Quantum spin Hall insulator state in HgTe quantum Wells, Science 318 (2007) 766.
[93] I. Knez, R.-R. Du, G. Sullivan, Evidence for helical Edge modes in inverted InAs/GaSb quantum wells, Phys. Rev. Lett. 107 (2011) 136603.
[94] X. Qian, J. Liu, L. Fu, J. Li, Quantum spin Hall effect in two-dimensional transition metal dichalcogenides, Science 346 (2014) 1344.
[95] Q. Xu, et al., Two-dimensional oxide topological insulator with iron-pnictide superconductor LiFeAs structure, Phys. Rev. B 92 (2015) 205310.
[96] S. Wu, et al., Observation of the quantum spin Hall effect up to 100 kelvin in a monolayer crystal, Science 359 (2018) 76–79.
[97] Y.S. Hor, et al., Superconductivity in $Cu_xBi_2Se_3$ and its implications for pairing in the undoped topological insulator, Phys. Rev. Lett. 104 (2010) 057001.
[98] Q.L. He, et al., Chiral Majorana fermion modes in a quantum anomalous Hall insulator–superconductor structure, Science 357 (2017) 294.
[99] F. Fei, et al., Band structure perfection and superconductivity in type-II Dirac semimetal $Ir_{1-x}Pt_xTe_2$, Adv. Mater. 30 (2018) 1801556.

Index

Note: Page numbers followed by "*f*" and "*t*" refer to figures and tables, respectively.

K

L

M